ILLUSTRATED BIRDS
OF NORTH AMERICA
folio edition

JON L. DUNN AND JONATHAN ALDERFER

 NATIONAL GEOGRAPHIC

WASHINGTON, D.C.

CONTENTS

6 | INTRODUCTION

20 | DUCKS, GEESE, AND SWANS

56 | CURASSOWS AND GUANS

56 | PARTRIDGES, GROUSE, TURKEYS, AND OLD WORLD QUAIL

66 | NEW WORLD QUAIL

70 | LOONS

74 | GREBES

78 | ALBATROSSES

82 | SHEARWATERS AND PETRELS

94 | STORM-PETRELS

98 | FRIGATEBIRDS

98 | TROPICBIRDS

100 | BOOBIES AND GANNETS

102 | PELICANS

104 | DARTERS

104 | CORMORANTS

108 | HERONS, BITTERNS, AND ALLIES

114 | STORKS

114 | FLAMINGOS

116 | IBISES AND SPOONBILLS

118 | NEW WORLD VULTURES

118 | HAWKS, KITES, EAGLES, AND ALLIES

136 | CARACARAS AND FALCONS

146 | LIMPKINS

146 | RAILS, GALLINULES, AND COOTS

152 | CRANES

154 | LAPWINGS AND PLOVERS

160 | JACANAS

160 | OYSTERCATCHERS

160 | STILTS AND AVOCETS

162 | SANDPIPERS, PHALAROPES, AND ALLIES

198 | GULLS, TERNS, AND SKIMMERS

228 | SKUAS AND JAEGERS

232 | AUKS, MURRES, AND PUFFINS

242 | PIGEONS AND DOVES

248 | LORIES, PARAKEETS, MACAWS, AND PARROTS

252 | CUCKOOS, ROADRUNNERS, AND ANIS

256 | OWLS

266 | GOATSUCKERS

270 | SWIFTS

272 | HUMMINGBIRDS

278 | TROGONS

280 | KINGFISHERS

282 | WOODPECKERS AND ALLIES

294 | TYRANT FLYCATCHERS

312 | SHRIKES

314 | VIREOS

320 | CROWS AND JAYS

328 | LARKS

330 | SWALLOWS

334 | BABBLERS

334 | CHICKADEES AND TITMICE

338 | PENDULINE TITS
AND VERDINS

338 | LONG-TAILED TITS
AND BUSHTITS

340 | CREEPERS

340 | NUTHATCHES

342 | WRENS

344 | DIPPERS

346 | KINGLETS

346 | OLD WORLD WARBLERS
AND GNATCATCHERS

350 | OLD WORLD FLYCATCHERS

352 | THRUSHES

362 | MOCKINGBIRDS AND THRASHERS

366 | BULBULS

366 | STARLINGS

368 | ACCENTORS

368 | WAGTAILS AND PIPITS

372 | WAXWINGS

372 | SILKY-FLYCATCHERS

374 | WOOD-WARBLERS

398 | OLIVE WARBLER

400 | TANAGERS

402 | BANANAQUIT

402 | EMBERIZIDS (SPARROWS)

436 | CARDINALS, SALTATORS,
AND ALLIES

442 | BLACKBIRDS

454 | FRINGILLINE AND
CARDUELINE FINCHES
AND ALLIES

464 | OLD WORLD SPARROWS

464 | WEAVERS

464 | ESTRILDID FINCHES

466 | ACCIDENTALS AND
EXTINCT SPECIES

480 | APPENDIX:
AOU AND ABA
CHECKLIST DIFFERENCES;
GREENLAND; BERMUDA

482 | INDEX

500 | ART CREDITS

502 | ACKNOWLEDGMENTS

INTRODUCTION

Summer Tanager:
Now classified with
Cardinals, Saltators, and Allies

The book that you hold in your hands is based on the 5th edition of the *National Geographic Field Guide to the Birds of North America* (2006). Too large to be a field guide, we envision this folio edition residing on bookshelves and coffee tables waiting to be consulted by birders at their leisure. The comprehensive artwork and maps have been enlarged to allow the closest scrutiny of all the details that the artists and cartographers so carefully crafted. We hope this new folio-sized edition brings added insight and enjoyment to your birding pursuits.

The text remains very similar to the 5th edition field guide. Most changes reflect new names, both English and scientific, adopted by the American Ornithologists' Union (AOU) since 2006. Careful readers will notice a revision of the scientific genera in the gulls, redpolls, siskins, and goldfinches. Higher-level taxonomy is also undergoing revision, but this edition continues the arrangement found in the 5th edition field guide. Some of the new findings adopted by the AOU are startling. For example, flamingos are more closely related to grebes than to the much more similar-looking storks and herons. Other findings will change birders' perceptions. North America's tanagers (with the sole exception of Western Spindalis) have been classified with the Cardinalidae family, joining such species as Northern Cardinal, Rose-breasted Grosbeak, and Indigo Bunting. Should we still be calling them tanagers? For the time being, the AOU has made no changes to the familiar English names. A name change that birders will probably embrace is the simplification of the names Nelson's Sharp-tailed Sparrow and Saltmarsh Sharp-tailed Sparrow, now known as Nelson's Sparrow and Saltmarsh Sparrow. In addition, our longtime map consultant Paul Lehman recently updated many of the range maps, and 34 new maps have been added.

Since the publication of the 5th edition in 2006, a small, but impressive number of accidental species have been added to the North American list but do not appear in this book. They include accepted records of Townsend's Shearwater (CA), Tristram's Storm-Petrel (CA), Intermediate Egret (AK), Swallow-tailed Gull (CA), Brown Hawk-Owl (AK), European Turtle-Dove (FL), Loggerhead Kingbird (FL), Sedge Warbler (AK), Pallas's Leaf-Warbler (AK), Song Thrush (PQ), and Yellow-browed Bunting (AK). A handful of other well-documented species await official acceptance, most notably White-crested Elaenia (TX), Crowned Slaty-Flycatcher (LA), Sinaloa Wren (AZ), and Rufous-tailed Robin (AK). A bird guide is always a work in progress.

It has been over 25 years since the publication of the first edition of this guide. In each edition, museum specimens have been an essential tool in preparing detailed and accurate illustrations. Many museums have loaned specimens to the artists (see acknowledgements, page 502), and we are grateful to them.

Saltmarsh Sparrow:
New English name

SPECIES SELECTION

This guide includes all species that have occurred in North America, which is defined here as the land extending northward from the northern border of Mexico, plus adjacent islands and seas within about two hundred miles of the coast, excluding Greenland (see page 480).

For those accidental visitors that are seen in North America only when they wander off course, our standard generally requires that in order to be included in the main text, they have been seen at least three times in the past two decades or five times in the last 100 years. The more accidental species with even fewer records are included in an end section along with the specific details of their occurrence. This section also includes the four species that are known to have gone extinct during the 19th and 20th centuries. In a few cases we have chosen to include those accidentals in the main section (e.g., Parkinson's Petrel, Cape Verde Shearwater, and Asian Brown Flycatcher) because we felt a direct comparison with similar species would be useful. Exotic species from other continents that have been introduced into North America as game, park, or cage birds are seen in the wild frequently, and the list is seemingly endless. We have chosen to show those that have at least some degree of establishment (e.g., Red Bishop). The criterion for acceptance and inclusion differs between the AOU and the ABA. The ABA Checklist Committee is in the process of redefining its policies for treating these "exotics," but in general it requires a significant population that has been present for at least 15 years. Some exotics (e.g., European Starling) have spread continent-wide and are a nuisance, while others have maintained more stable populations or even have declined. One such species, Crested Myna, after thriving for nearly a century, has been extirpated, perhaps as a result of competition with European Starlings.

Accidental:
Asian Brown Flycatcher

For issues of taxonomy and nomenclature—English and scientific names—we strictly adhere to the Committee on Classification and Nomenclature of the American Ornithologists' Union, as determined in the 7th edition of the AOU Check-list (1997) and its six subsequent supplements, published annually in the July issue of the AOU's journal, *The Auk*. Completed and ongoing molecular and traditional studies will refine, and continue to refine, our knowledge of speciation (what is and what is not a species) and higher level systematics (relationships of families to one another and their linear sequence). We follow either the ABA Checklist Committee or the above-mentioned AOU committee on deciding which species to include. Apart from the list of exotics, the differences are few (see page 480).

Introduced:
Orange Bishop

Typically, when a new species is reported from the North American area, it is first reviewed by the respective state or provincial committee. (Nearly all states and provinces now have established committees.) Then it is reviewed by the ABA Checklist Committee. If ABA accepts the species, the AOU committee then reviews it and, if in agreement with the ABA, determines its linear placement within the Check-list. Occasionally, state or provisional committees have accepted records that are subsequently not accepted by both the ABA and the AOU. For instance, the Texas committee has accepted records of Dark-billed Cukoo and Black Catbird; the North Carolina committee has accepted a record of a Swinhoe's Storm-Petrel; and the British Columbia committee has accepted a record of Blue Rock-Thrush. Usually, but not always, these debates concern the origin of the record (identification accepted), often without much evidence pointing one way or the other. Both of the AOU and ABA committees, as well as some state and provincial committees, require verifiable diagnostic evidence (photographic or audio) before adding a new species to a checklist. An appendix (page 480) details the current differences between the AOU and ABA checklists.

Extirpated:
Crested Myna

FAMILIES

Scientists organize animal species into family groups that share certain structural or molecular characteristics. Among the birds, some families encompass more than a hundred members; others have only one. Characteristics within a family are often helpful in identifying birds in the field. Take the members of the Picidae family (page 282), for example. Strong, sharp bills, strong claws, and short legs are among the features that make them easily recognized as woodpeckers. Their family resemblance narrows down the list of possible identifications to a manageable number.

Brief family descriptions can be found at the beginning of each group in this field guide for information applicable to all members of the family. Additionally, a description of a smaller group within a family is provided, such as that given for the sapsuckers (three shown here), which share some unique traits.

Downy Woodpecker
Picoides pubescens

Sapsuckers:
Yellow-bellied, Williamson's, Red-breasted *daggetti*

SCIENTIFIC NAMES

Each species has a two-part scientific name, derived from Greek or Latin, which is given in italics. The first part, always capitalized, indicates the genus, a group of closely related species more narrowly defined than that of the family. For example, nine members of the family Picidae are placed in the genus *Picoides*. Together with the second part of the name (often referred to as the specific epithet), not capitalized, this identifies the species. *Picoides pubescens* is the name of one specific kind of woodpecker, commonly known as Downy Woodpecker. *Picoides dorsalis,* or *P. dorsalis,* is American Three-toed Woodpecker. In scientific nomenclature, no two species share the same two-part scientific name, or binomial.

SUBSPECIES

Since about the latter half of the 19th century, species have been further divided into subspecies (*race* is a synonym) when populations from different geographical regions show recognizable differences. Each subspecies bears a third scientific name, or trinomial. Out of three subspecies (all found in North America) of American Three-toed Woodpecker, *Picoides dorsalis bacatus* (often abbreviated as *P. d. bacatus* or simply *bacatus*) identifies the dark-backed subspecies of Three-toed Woodpecker that inhabits the boreal forest of eastern North America. The Rockies subspecies is the paler-backed race, *dorsalis;* a third and intermediate subspecies, *fasciatus,* is found from Alaska to Oregon.

American Three-toed Woodpecker
Picoides dorsalis fasciatus

If the third part of a scientific name is the same as the second, the subspecies in question is the *nominate* subspecies, the type of the species that was originally described. The subspecies *dorsalis* was named and described in 1858, earlier then *fasciatus* (1870) or *bacatus* (1900). Birders might be surprised to learn that the nominate race of the widespread Red-tailed Hawk, *jamaicensis,* is found only on the island of Jamaica, although one might deduce that from the name. It was described to science in 1788.

American Three-toed Woodpeckers:
bacatus, dorsalis, fasciatus

family	Woodpeckers	Picidae
genus		*Picoides*
species	American Three-toed Woodpecker	*Picoides* **dorsalis**
subspecies or **race**	American Three-toed Woodpecker (eastern subspecies)	*Picoides dorsalis* **bacatus**
nominate race	American Three-toed Woodpecker (Rockies subspecies)	*Picoides dorsalis* **dorsalis**

Scientific names are used in this book for learning and clarity. We have illustrated or described the extremes of variations that look distinctly different from other races of the same species. For example, of the roughly 25 races of Horned Lark in North America, we show the most typical or widespread, the extremes within the species, and an Asian vagrant (*flava*) that occurs casually in Alaska in fall (page 328). For polytypic species where our figures are mainly or entirely of only one subspecies, we italicize that subspecies name under the English name; the other figures will also be of that subspecies, unless indicated otherwise.

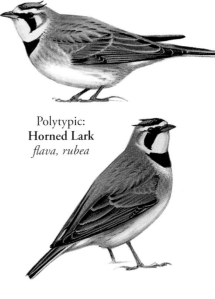

Polytypic:
Horned Lark
flava, rubea

As with species, taxonomists frequently differ as to what constitutes a valid subspecies. One widely used definition is that 75 percent of the individuals from one region differ from all of the individuals from another described subspecies. Some taxonomists would prefer a 90 percent or even 95 percent criterion. Slight differences in size or a very slight and gradual change in coloration, with no sharp breaks, are usually considered clinal differences; and while the differences are important to our understanding the variation within a species as a whole, usually the differing groups are not recognized with separate subspecies names. If the species is considered monotypic (e.g., Cerulean Warbler), with no recognized subspecies, then only the scientific binomial is used (e.g., *Dendroica caerulea*). We have relied on the 5th edition of the AOU Check-list (1957), the last edition in which the AOU treated subspecies. (The next edition of the Check-list will again treat subspecies.) For more recent treatments we have also followed *The Howard and Moore Complete Checklist of the Birds of the World,* 3rd edition (2003).

Since for the most part the artists have used museum specimens for their figures, we have labeled those figures with an italicized subspecies name, if that information is known.

Monotypic:
Cerulean Warbler

HOW TO IDENTIFY BIRDS

Field marks, a bird's physical aspects, are the clues by which birds are identified. Field marks include plumage, or the bird's overall feathering, as well as the shape of the body and its individual parts and any actual markings such as bars, bands, spots, or rings. A field mark

Lesser Scaup

Greater Scaup

can be as obvious as the Northern Cardinal's color or the Killdeer's double breast bands (page 158), or as subtle as the difference in head markings between a Clay-colored and a winter-plumaged Chipping Sparrow or the difference in the head shapes of Greater and Lesser Scaup, a much more reliable field mark than head color. Some field marks are plainly visible only in good light, or from a certain angle, or when the bird is in flight.

The most important thing to do when birding is just to look at the actual bird—there's plenty of time later to consult the field guide. In this guide, the most distinctive features for each species in each plumage are usually listed first in the text account, and boldface terms (e.g., **adult female** or **first-winter male**) in the accounts correspond to the illustrated figures.

Keep in mind that not every bird seen can be identified, even by experts. Remember this the next time you look at a badly worn and bleached immature gull on the beach in July!

PARTS OF A BIRD

Because size, shape, and the configuration of birds' various parts vary from family to family and from species to species, you should become acquainted with all the terms used to describe those parts and their locations—bird topography.

SHAPE

Most birds have a shape that is characteristic of their species or family. Even at great range and in poor light, the short-tailed, pointed-billed, rather chunky body shape of the perched European Starling is distinctive. Shape can change, though, depending on environmental conditions. In cold weather many species fluff their feathers and hunker down. This makes them look bigger, and often shorter, and alters their characteristic posture.

eye line

Chipping Sparrow

eye crescent

Northern Parula

HEAD

A careful look at a bird's head can nearly always provide a correct identification of a species. The boldly marked head of an adult Lark Sparrow offers a nearly complete course in common head markings. Sparrows are a confusing group, and it is the often subtle differences in head markings that can help sort them out. At the top of the head there is often a pale or white *median crown stripe* bordered by dark *lateral crown stripes.* A line running from base of the bill up and over the eye is known as the *eyebrow,* or *supercilium.* Sometimes the area between bill and eye, the *lores,* has its own marking. The front part of the eyebrow, or *supraloral area,* may be of a different color—another useful marking. The eye color itself may be diagnostic in some species. Bird pupils are always black, but the *iris* can be colored, as it is in many subadult and adult gulls. A ring of naked flesh, the *orbital ring,* may surround the eye,

Least Flycatcher

eye ring
scapulars
median coverts
greater coverts
wing bar
flanks
tertials
secondaries
primaries

nape
mantle
scapulars
median coverts
greater coverts
tertials
primary tip projection
tail
tibia
tarsus
undertail coverts

Pacific Golden-Plover

scapulars
greater coverts
median coverts
lesser coverts
marginal coverts
tertials
secondaries
greater primary coverts
primaries

Great Black-backed Gull upper wing

as may a feathered one, usually called an *eye ring*. Least Flycatcher has an unbroken, feathered, white eye ring. In some species the ring is interrupted, forming *eye crescents* instead. The dark stripe extending back from the Lark Sparrow's eye is known as a *postocular stripe*. If it extended through the eye itself, as in Chipping Sparrow, it would be known as an *eye line,* or a *transocular line.* A distinctive brown patch, bordered in black on the Lark Sparrow, marks the *ear coverts,* or *auricular.* Its lower border is a *moustachial stripe.* Below that is a white *submoustachial stripe,* bordered with yet another dark line, the *malar stripe.*

The shape of the bill must not be overlooked. Bills are specialized according to diet and method of feeding. The Lark Sparrow's short, stout bill indicates a seedeater. The Tennessee Warbler's is a fine-pointed insect probe, as is the case with many warblers, but it is also used for sipping nectar. The *gonys,* the junction of the two sides of the lower mandible, forms a prominent ridge in some species, and its shape can be important in identification. For instance, on Western Gull it is more acute, giving the bill a bulbous tip, while on California Gull the angle is much broader and the bill looks almost straight. The top of the bill is known as the *culmen.* In raptors, such as Common Black-Hawk, the patch of bare skin covering the base of the upper mandible is called the *cere.* Some species, such as pelicans or cormorants, have a fold of loose skin called a *gular pouch* hanging from the throat.

gonys

California Gull, Western Gull

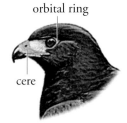

orbital ring
cere

Common Black-Hawk

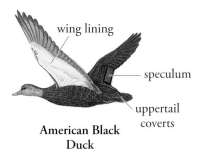

wing lining

speculum

uppertail coverts

American Black Duck

carpal bar

Common Tern,
first summer

axillaries

Black-bellied Plover

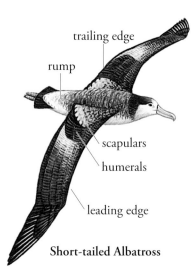

trailing edge

rump

scapulars

humerals

leading edge

Short-tailed Albatross

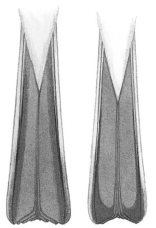

Myiarchus Flycatcher
tails from below:
Brown-crested, Ash-throated

WINGS

Wings are most often observed in the folded position, and while this position provides useful clues, not all of the wing is visible. A bird's wing is composed basically of groups of strong *flight feathers,* covered on the upper and lower side by shorter, protective feathers called *coverts.* Each feather's shaft divides its *inner web* from its *outer web.* The *primaries* comprise the outer 9 or 10 tapering flight feathers; some non-passerines have up to 12. They are followed by the *secondaries.* The three innermost flight feathers are the *tertials.*

Joining the wing and body are rows of short, protective feathers, called the *scapulars.* A series of wing coverts are found on the wing above the secondaries. From *leading edge* to *trailing edge* they are the *marginal, lesser, median,* and *greater* wing coverts. The lesser and marginal coverts have multiple rows of feathers; the others have a single row each. On long-winged species such as albatrosses, another set of feathers, the *humerals,* is well developed. The overall plumage of the back, extending to the scapulars and upperwing coverts, is called the *mantle,* a term used almost exclusively for gulls.

In some species the white tips of the greater and median coverts above the secondaries form *wing bars,* often an important diagnostic feature. In some ducks, such as American Black Duck, the secondaries themselves are uniquely colored, forming a pronounced bar known as a *speculum.* When a bird (e.g, the immature Common Tern) shows a contrasting bar on the forward part of the upper wing, it is referred to as a *carpal bar.*

The spread underwing displays the undersides of the flight feathers and the underwing coverts, and it offers features that prove useful in identifying birds in flight. The coverts of the *wing linings* often contrast with the rest of the underwing, as may the *axillaries,* the bird's "armpit."

A feature to notice in the folded bird wing is how far the primaries extend past the longest tertial. This length—known as the *primary tip projection*—can help distinguish between similar species such as the American and Pacific Golden-Plovers. On such shorebirds, note the prominence of the scapulars and that the secondaries are completely hidden.

TAIL

Lastly, tail shape and pattern can be very important aids to identification. Most bird species have 12 tail feathers, called *rectrices.* A good part of the time we see only the folded tail. From above, usually only the one or two central tail feathers show; from below, only the outermost tail feather on each side is visible. Similar-looking species, such as MacGillivray's and Connecticut Warblers (page 390), may be distinguished from each other from below by the length of tail projecting past the undertail coverts. In order to view the undertail properly, get in front of the bird and look at the tail from below. Subtle differences in undertail pattern, for instance, help distinguish the members of the perplexing flycatcher genus *Myiarchus*.

But sometimes it takes more than the outer tail feathers to discern the pattern of a tail. Then it is best to see the bird in flight, as in the case of the four longspurs (pages 428, 430). In flight, too, tail shape shows well. Thus, we can distinguish the fan-shaped tail of American Crow from the wedge-shaped tail of Common Raven (page 326). The distinctive forked tail of the Barn Swallow (page 332) gets its shape from short central tail feathers and long outer ones.

Scarlet Tanagers:
fall adult ♂, breeding adult ♂

MOLT

Regular renewal of plumage, called *molt,* is essential to a bird's well-being and ability to fly. Yet molt produces variations of plumage within a species and between ages that create additional obstacles to identification. It is not enough for us to know what a bird's plumage looks like in only one season or sex or at one age. Most species undergo a complete molt in late summer or early fall, replacing all their feathers.

Adult male Scarlet Tanagers change their entire body plumage color from summer to fall. Adult Acadian Flycatchers are brightest in spring, becoming worn and faded by late summer. Fresh fall plumage after molt can be brighter than the colors of August. This plumage is usually held through the fall and through most of the winter (and is usually labeled winter plumage).

In some songbirds, such as House Sparrow, the more patterned plumage of spring appears with the gradual wearing away of dull tips on the feathers of the winter plumage, with very little actual molting involved.

Acadian Flycatchers:
spring adult, worn summer adult

Often the molt occurs before the birds migrate, and thus we see the birds in this fresh plumage even if they spend the winter outside North America. In late winter or early spring, many birds undergo a partial molt—usually involving the head, body, and some wing coverts—to the plumage we see in spring and summer. Some species suspend the molt during migration and complete it upon arrival at the wintering or breeding grounds. This applies to many shorebird species. On the whole, fall and winter plumage may vary considerably from spring and summer plumage.

Some species molt over a considerably longer period than others. Birds of prey must rely on their wing feathers for successful hunting and thus molt very gradually so that only a few feathers are missing at any one time.

House Sparrows:
fall ♂, breeding ♂

PLUMAGE VARIATION

In many species, the male and female look quite different, and young birds are unlike either parent. And some species—usually, but not always, of the same genus—occasionally breed with other species, producing *hybrid* offspring that often look intermediate. Different subspecies also interbreed, resulting in *intergrade* individuals or populations.

Some species have two or more *color morphs*—variations, in plumage color occurring regionally or within the same population. Red-tailed Hawk (page 132) comes in multiple morphs, which include *dark, light,* and *rufous* (rust-colored). Common terms, such as *rufous* and *buffy,* alternate with rarer usages, such as *hepatic* (reddish brown), in describing the nuances of shading in a bird's plumage.

Where adult males and females of a species are similar, we show only one. When male (♂) and female (♀) look different, we usually show both. If spring and fall, or *breeding* and *nonbreeding*, plumages differ only slightly, or if only one of these plumages is usually seen in North America, we tend to show only one figure. Juvenile and immature birds are illustrated when they hold a different-looking plumage after they are old enough to be seen away from their more easily recognizable parents.

Painted Redstart:
adult, top, and juvenile

PLUMAGE SEQUENCE

After hatching, not all birds go through the same sequence of plumages. Some young birds, such as Painted Redstart, hold this juvenal plumage for a few weeks and then move right into adult or adultlike plumage. Others, such as Herring Gull (below) and Bald Eagle (page 124), take years to acquire their adult look.

Some nestlings are active immediately after hatching (termed *precocial*) and wear a fluffy *down,* while others are hatched naked, remain in the nest, and are completely helpless (termed *altricial*). The first coat of true feathers, acquired before the bird leaves the nest, is called the *juvenal* plumage, and birds holding it are referred to as *juveniles*. True hawks and

winter adult

juvenile

3rd winter

1st winter

1st winter

2nd winter

Herring Gull:
Plumage sequence

loons, as well as many other waterbirds, hold juvenal plumage well into the winter. In many species, juvenal plumage is replaced in late summer or early fall by a *first-fall* or *first-winter* plumage that more closely resembles the adult. First-fall and any subsequent plumages that do not resemble the adult are termed *immature* plumages. These may continue in a series that includes *first-spring* (when the bird is almost a year old), *first-summer,* and so on, until *adult* plumage is attained. When birds, such as Bald Eagle, take several years to reach adult plumage, we have labeled the interim plumages as *subadult* or with the specific year or season shown. This terminology is often used with gulls.

Some species wear colorful plumage in the breeding season and molt to different colors for fall and winter. In other species, *breeding* plumage looks much like *winter* plumage. Some changes are evident only during the brief period of courtship. In herons, for example, the colors of bill, lores, legs, and feet may change. When these colors are at their height, the birds are said to be in *high breeding* plumage.

After breeding, most ducks molt into an *eclipse* plumage in which males acquire a female-like plumage and females show little change, although some become paler and duller. Eclipse plumage is usually held for several weeks. In eclipse, all the flight feathers are lost simultaneously in some species, rendering them unable to fly until new feathers grow in.

American Golden-Plover:
April male, top, and
breeding male

MEASUREMENTS

Knowing the size of a bird is another key. When identifying a kinglet, for example, it helps to know that the bird in question is tiny (only about four inches long). Relative size is also important. In mixed flocks, Greater and Lesser Yellowlegs (page 162) are easily distinguished from each other by size alone.

The measurements in this book come from a large number of published works. Technical books give a variety of precise measurements, such as the *wing chord* (the measurement of a folded wing). Body length and wingspread measurements are by their nature more imprecise and may vary slightly from one source to another. For birders in the field, however, we felt that these were the most helpful aids. Thus, average length (L) from tip of bill to tip of tail for each species is given, with figures rounded off. Where size varies greatly within a species, either because of sex or geographical variation among subspecies, a range of smallest to largest is provided. And for large birds, often seen in flight, we give the wingspan (WS), measured from wing tip to wing tip.

VOICE

A bird's songs and calls reveal not only its presence, but also, in many cases, its identity. Some species—particularly nocturnal or secretive birds such as owls, nightjars, and rails— are more often heard than seen. A few species are most reliably identified by voice even when they are seen well. Willow and Alder Flycatchers, for example, are nearly identical in appearance, but they have different songs as well as call notes: Willow gives a liquid *wit* call, while Alder delivers a loud *pip*.

When birds assemble or travel in flocks, they often keep in touch with a *contact* or *flight call* that may be markedly different from their other calls. Flight calls are especially impor-

Willow Flycatcher (left) song:
a sneezy *fitz-bew*
Alder Flycatcher (right) song:
a falling, wheezy *weeb-ew*

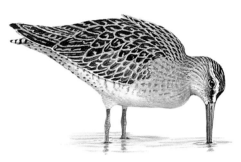

Short-billed Dowitcher

Abundant:
European Starling

Common:
Indigo Bunting

Rare: **Ruff**

tant for identifying individuals (or flocks) overhead. Flying Blackpoll Warblers and other warblers, for example, announce their identity with buzzy *zeet* notes.

Birds sensing danger give an *alarm call,* which may be the same as or a variant of their basic call, though delivered in a more urgent tone. Dowitchers execute a series of call notes that increases in rapidity according to the birds' degree of alarm. Bushtits give an excited twittering when raptors are about.

While bird species have a number of vocalizations that are used for different purposes, some are multipurpose. The primary song is sung mainly by males on or near the breeding grounds, although some species sing year-round; and its purpose is to announce territorial claims to attract females. Females of some species, such as Northern Cardinal, sometimes sing also. Many of our songbirds (where song is learned) sing two or more different types of songs (e.g., Black-throated Green Warbler), and these are often used for different purposes.

Learning songs and calls makes birding a more rewarding experience. Distinctive songs and calls are described in our text. Transcription into words, such as *cheerily cheer-up cheerio* for American Robin's song, helps to express tone and pattern but is very subjective. You may find it best to transcribe sounds into your own words as you hear them; one birder's *chip* is another's *tsip* or *chik* or even *peek.*

BEHAVIOR

Behavioral traits also provide many clues to species identity. Look, for example, at a bird's flight. Is it direct or undulating? Does the bird beat its wings rapidly or slowly? Turkey Vultures soar with their wings held slightly raised in a shallow V, known as a *dihedral,* while Black Vultures soar on level wings.

Observe feeding. Does the bird forage on the ground or in the treetops? Among shorebirds, Solitary Sandpiper and the two species of Yellowlegs peck at the water to get food, while the dowitchers drill the mud like the needle in a sewing machine.

Look for personality clues. Is the bird usually visible and approachable, or is it shy and difficult to locate? How does the bird move? Mourning and MacGillivray's Warblers hop; Connecticut Warblers walk deliberately. Black-and-white Warblers climb up and down and around tree trunks; the superficially similar Blackpoll Warbler does not. How does the bird use its tail? Among the hard-to-identify flycatchers of the genus *Empidonax,* nearly all species flick their tails up—except for Gray Flycatcher, which drops its tail down.

ABUNDANCE AND HABITAT

Abundance must be considered in relation to habitat. Habitat information included in the text will help you find a particular species within its range. Under the heading RANGE in our species accounts, we include supplemental information about habitat, abundance (such as casual or accidental), and seasonal status that cannot be shown in a single map. Some species are highly local, found only in a very specialized habitat. Bank Swallows, for example, are common in summer only near the steep sandy or gravelly banks they require for nesting. Common Loons are fairly common to common on large bodies of water; but away from large lakes, as in parts of the Southwest, they are very rare.

Seasons of the year, of course, also affect numbers. In spring, Semipalmated Sandpiper is abundant on its northern breeding grounds; during migration it is common in the Midwest and on the East Coast but rare in the West; in the winter a few are still found, in North America, in southernmost Florida only.

The following categories of abundance are used in this guide:

Abundant means a species is present in great numbers in an area. European Starling is abundant in varied habitats throughout most of North America.

Common means that a species is very likely to be seen in a given area, but in fewer numbers than an abundant one. Indigo Bunting is common over much of eastern North America.

Fairly common is rather numerous, but there are fewer still than those designated as common.

Casual: **Redwing**

Uncommon species can have a large range and still be seen infrequently. Northern Goshawk, for example, is uncommon over much of its resident range.

Rare are species that occur in very low numbers, but annually in North America. These include visitors and rare breeding residents. For instance Ruff, an Old World visitor, is rare over much of North America.

Very rare species are annual in North America, but only one or a few will occur in a season.

Casual is used for species that do not occur annually in North America, but over decades there is a pattern of their occurrence, as with Redwing in Newfoundland.

Accidental:
Eurasian Hoopoe

Accidental refers to species that have been seen only once or a few times in an area that is far out of their normal range. In fact, it may be decades, or even centuries, before another one is seen there again. Eurasian Hoopoe has been recorded only once in North America (from coastal southwestern Alaska in fall).

Other references may use different abundance terms, or they may give the same terms a slightly different meaning, but it is worth noting that both the AOU and ABA committees have adopted the same general definitions for the terms *rare, casual,* and *accidental.*

Some other terms that are used herein include the following:

Vagrant is a bird that is found well off its usual migration route.

Irruptive species, such as Snowy Owl or Bohemian Waxwing, are erratic in their movements over much of their winter range. One year they may be numerous in a given region and the next year, or even the next decade, totally absent.

Irruptive:
Snowy Owl

RANGE MAPS

Maps are provided for all species with two general exceptions. One is for certain introduced exotic species with very limited ranges that instead are described in the text. The other is those species that do not breed in North America and are of only rare, casual, or accidental occurrence.

On each map, range boundaries are drawn where the species ceases to be regularly seen. Keep in mind that nearly every species will be rare at the edges of its range. The sample map (next page) explains the colors and symbols used on our maps; the map on the inside back cover gives an overview of North America.

Recent Arrival:
Shiny Cowbird

DUCKS, GEESE, AND SWANS (FAMILY ANATIDAE)

Worldwide family. Web-footed, gregarious birds, ranging from small ducks to large swans.
Largely aquatic, but geese, swans, and some "puddle ducks" also graze on land. **Species:** *158 World, 62 N.A.*

GREATER WHITE-FRONTED GOOSE
Anser albifrons | L 28" (71 cm)

Adult has distinctive white band at base of bill. Medium-size, grayish brown goose, with irregular black barring on underparts; orange feet and legs. Bill pink or orangish with white tip. In flight, note grayish blue wing coverts and white, U-shaped rump band. Most **immatures** acquire white feathering around bill and white bill tip during first winter; acquire black belly bars by second fall; distinguished from bean-geese by bill color; from Pink-footed Goose by bill and leg color. Color and size vary in adults: tundra-breeding birds are small and pale with heavy barring; taiga-breeding race (*elgasi*, central Alaska) is larger, with bigger bill, darker neck, and less barring; Greenland race (*flavirostris*), rare in the Northeast, is darker, with the heaviest barring and an orange bill. Call is a high-pitched, laughing *kah-lah-aluck*.
RANGE: Large numbers winter on the Great Plains (fewer east to Mississippi Valley) and in Pacific states. Rare on western Great Plains and in the East.

TAIGA BEAN-GOOSE
Anser fabalis | L 30–35" (76–90 cm)

Similar to immature Greater White-fronted Goose, but larger with blackish base to long, wedge-shaped bill with variable amounts of yellow-orange before tip. **Calls** are loud and deep, very different from Greater White-fronted.
RANGE: Eurasian, taiga-breeding species. Accidental on the Pribilofs and Aleutians, Alaska; a possible record from western Washington.

TUNDRA BEAN-GOOSE
Anser serrirostris | L 28–33" (71–84 cm)

Similar to Taiga Bean-Goose, but slightly smaller and with shorter, thicker-based bill that is extensively black at base with a yellow-orange ring near the tip. **Calls** are said to be higher pitched than Taiga.
RANGE: Eurasian, tundra-breeding species. Casual to the western Aleutians; also recorded from the Pribilofs and St. Lawrence Island; one record from the southern Yukon. The two bean-geese have been recently split and the majority of previous Alaska records of bean-geese are best left as unidentified.

PINK-FOOTED GOOSE
Anser brachyrhynchus | L 26" (66 cm)

Similar to immature Greater White-fronted Goose, but with pink legs and variable dark base and tip to stubbier pinkish bill; neck shorter than Greater White-fronted. Juvenile is browner, looks more scaly than barred; legs duller. In flight, upperwings and back pale bluish gray; darker head and neck contrast with paler body; tail base grayer and paler than other geese on this page.
RANGE: Old World species; closest breeding grounds are in eastern Greenland. Casual, now nearly annual, to East Coast region with a scattering of records between eastern Pennsylvania and Newfoundland; most sightings thought to involve wild birds.

GREATER
WHITE-FRONTED
GOOSE

adult

adult

immature

taiga adult
elgasi

tundra
adult

Greenland adult
flavirostris

adult

TUNDRA
BEAN-GOOSE
serrirostris

TAIGA
BEAN-GOOSE
middendorffii

adult

adult

adult

adults

PINK-FOOTED
GOOSE

SNOW GOOSE
Chen caerulescens | L 26–33" (66–84 cm)

Two color morphs. Adults distinguished from smaller Ross's Goose by larger, pinkish bill with black "grinning patch." Flies with slower wingbeat than Ross's; rusty stains often visible on face in summer. **White-morph juvenile** is grayish above, with dark bill. **Blue-morph adult** has mostly white head and neck, brown back, variable amount of white on underparts. **Blue-morph juvenile** has dark head and neck, overall slaty body coloring; distinguished from Greater White-fronted Goose (preceding page) by dark legs and bill and lack of white on face. Intermediates between white and blue morphs have mainly white underparts and whitish wing coverts.

RANGE: Abundant. Occasionally hybridizes with Ross's Goose. A larger subspecies, the "Greater Snow Goose" (not shown), breeds around Baffin Bay, winters only along mid-Atlantic coast; blue morph almost unknown. Smaller race, the "Lesser Snow Goose" (morphs shown here), is rare in winter throughout interior U.S. and in southern Canada outside mapped range. Blue morph is abundant in central North America; rare west of the Great Plains; uncommon in East.

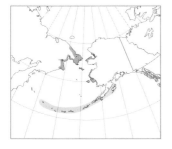

ROSS'S GOOSE
Chen rossii | L 23" (58 cm)

Stubby, triangular bill, mostly deep pinkish red. Neck shorter, head rounder than Snow Goose, white head generally lacks rusty stains; **calls** higher pitched. Ross's has two color morphs. **White-morph juvenile** may have very pale gray wash on head, back, and flanks, but far less than juvenile white-morph Snow Goose. Extremely rare blue morph is darker than **blue-morph** Snow Goose; face and belly are white.

RANGE: High-Arctic breeder; winters mainly west of Mississippi Valley. Seen during migration in grasslands and grainfields in eastern Great Plains and Mississippi Valley; rare but regular farther east to western Vermont. Rare winter visitor to mid-Atlantic states; rare visitor also to much of West outside mapped range. Usually seen with Snow Geese.

EMPEROR GOOSE
Chen canagica | L 26" (66 cm)

Fairly stocky, small goose with short, thick neck. Head and back of neck white; chin and throat black; face often stained rusty in summer. Bill pinkish; lower mandible is sometimes black. Black-and-white edging to silvery gray plumage creates a scaled effect below; upperparts appear barred. **Juvenile** has dark head and bill. During first fall, acquires white flecking on head; resembles adult by first winter.

RANGE: Emperor Goose breeds in tidewater marsh and tundra; winters on seashores, reefs. Casual south on Pacific coast to central California and inland to Sacramento Valley.

BARNACLE GOOSE
Branta leucopsis | L 27" (69 cm)

Note distinctive head pattern and stubby bill. Bluish gray upperparts, barred with black, and white, U-shaped rump band. Silver-gray wing linings show in flight.

RANGE: More of a land goose than the Brant, feeding in fields near the ocean. Fairly common in captivity. Numerous sightings from eastern North America are of uncertain origin; perhaps most are escapes, especially away from Atlantic seaboard. Part of world population breeds in northeastern Greenland; accidental vagrant in Maritime Provinces.

adults

SNOW GOOSE
caerulescens

blue-morph
adults

white-morph
juvenile

blue-morph
juvenile

white-morph adult

ROSS'S
GOOSE

adult

white-morph
juvenile

white-
morph
adult

blue-morph
adult

EMPEROR
GOOSE

juvenile

adult

BARNACLE
GOOSE

CACKLING GOOSE
Branta hutchinsii | L 23–33" (58–84 cm)

Formerly considered part of the Canada Goose complex, this species is smaller with a rounder head and a stubbier bill. Its shorter neck is most obvious in flight. Four subspecies vary from smallest and darkest *minima,* breeding in southwest Alaska, to largest *taverneri,* breeding in western and northern Alaska; both winter in the Pacific states. Aleutian breeding *leucopareia,* with a prominent white neck ring (sometimes present on other subspecies), winters in California's Central Valley. Intermediately sized and palest *hutchinsii,* breeding in Canada's Arctic archipelago and wintering from southern Great Plains to western Gulf Coast region, is generally otherwise rare to casual in eastern North America. Size of *parvipes* Canada and *taverneri* Cackling overlap; *taverneri* averages darker with stubbier bill, but identification criterion is still not adequately worked out. **Calls** are much higher pitched than Canada Goose.

CANADA GOOSE
Branta canadensis | L 30–43" (76–109 cm)

Our most common and familiar goose. Black head and neck marked with distinctive white "chin strap," stretching from ear to ear. In flight, shows large dark wings; white undertail coverts; white, U-shaped rump band; and long neck. The approximately seven named subspecies vary in overall color, generally darker in West; ranges from pale *canadensis* of eastern seaboard to dark *occidentalis,* which breeds on eastern Gulf of Alaska coast. Size decreases northward: The smallest subspecies, *parvipes* ("**Lesser Canada Goose**") breeds from central Alaska to north-central Canada and winters in central portions of U.S. with some in the Pacific states. **Call** is a deep, musical *honk-a-lonk.*
RANGE: Flocks usually migrate in V-formations, stopping to feed in wetlands, grasslands, or agricultural areas. Breeding programs have now produced expanding populations that are resident south of mapped range and along Pacific and Atlantic coasts; explosive increase has prompted control measures in **East.**

BRANT
Branta bernicla | L 25" (64 cm)

A small, dark, stocky sea goose. Note whitish patch on side of neck. White uppertail coverts almost conceal black tail. White undertail coverts conspicuous in flight. Wings comparatively long and pointed, wingbeat rather rapid. **Immature** birds show bold white edging to wing coverts and secondaries, and fainter neck patches than **adults.** Juveniles usually lack neck patches entirely. In more easterly subspecies *hrota,* "**American Brant**," pale belly contrasts with black chest, and neck patches do not meet in front. Western *nigricans,* "**Black Brant**," has dark belly, and neck patches meet in front. **Call** is a low, rolling, slightly upslurred *raunk-raunk.*
RANGE: Primarily a sea goose; flocks fly low in ragged formation and feed on aquatic plants of shallow bays and estuaries. Locally common. Western *nigricans* is regular at Salton Sea in spring and sometimes summers. Brant is rare inland, but some *hrota* are sighted during migrations through the Great Lakes region, particularly in fall. Both races are casual during migration and winter on opposite coasts. Eastern *hrota* also casual south of mapped range.

hutchinsii

CACKLING
GOOSE

"Aleutian"
leucopareia

"Cackling"
minima

"Taverner's"
taverneri

"Richardson's"
hutchinsii

"Lesser"
parvipes

CANADA
GOOSE

"Dusky"
occidentalis

"Atlantic"
canadensis

adult

adults

adult

BRANT
"American Brant"
hrota

"Black Brant" adult
nigricans

immature

SWANS

Large, long-necked birds; North American species are white overall as adults; browner immatures are difficult to identify.

TUNDRA SWAN
Cygnus columbianus | L 52" (132 cm)

In **adult**, black facial skin tapers to a point in front of eye and cuts straight across forehead; most birds have a yellow spot of variable size in front of eye. Head is rounded, bill slightly concave. In Eurasian race, "**Bewick's Swan**," seen very rarely in the Pacific states, facial skin and base of bill are yellow, but usually only behind the nostril; compare with Whooper Swan. **Juvenile** Tundras molt earlier than immature Trumpeter and Whooper Swans; appear much whiter by late winter. Immature "Bewick's" has whitish bill patch. Tundra's **call** is a noisy, high-pitched whooping or yodeling.
RANGE: Nests on tundra or sheltered marshes; winters in flocks on shallow ponds, lakes, estuaries. Rare to uncommon in winter over parts of interior U.S.; casual south to the Gulf Coast and north to the Maritimes.

TRUMPETER SWAN
Cygnus buccinator | L 60" (152 cm)

Adult's black facial skin tapers to broad point at the eye, dips down in a V on forehead. Forehead slopes evenly to straight bill. **Juveniles** retain gray-brown plumage through first spring. Common **call** is a single or double *honk* like an old car horn.
RANGE: Locally fairly common in its breeding areas. Reintroduced into parts of former range and introduced elsewhere. Rare in winter south to California and in western interior south of mapped range.

WHOOPER SWAN
Cygnus cygnus | L 60" (152 cm)

Large yellow patch on lores and bill usually extends in a point to the nostrils; compare with "Bewick's Swan." Forehead slopes evenly to the straight bill. Common **call** is a buglelike double note. **Juvenile** retains dusky plumage through first winter; by first fall, bill attains whitish patches in same shape as adult's bill patch.
RANGE: Eurasian species closely related to Trumpeter Swan. Regular winter visitor to outer and central Aleutians; has bred on Attu Island. Casual elsewhere in northwestern North America (south to northern California). Records from eastern North America are all treated as escapes.

MUTE SWAN
Cygnus olor | L 60" (152 cm)

Prominent black knob at base of orange bill. **Juvenile** may be white or brownish; bill gray with black base. Darker juvenile begins to molt to white plumage by midwinter; bill becomes pinkish. Mute Swan often holds its long neck in an S-shaped curve, with bill pointed down. Often swims with wings arched over back. Gives a variety of hisses and snorts, but generally silent.
RANGE: An Old World species, introduced in the U.S. Seen in parks. East Coast populations are increasing. Mute Swan is now established on the southeastern portion of Vancouver Island, British Columbia, and in parts of the Midwest. It is being systematically removed from the Great Lakes and other areas.

TUNDRA SWAN
columbianus

juvenile

"Bewick's Swan"
adult
bewickii

adult

TRUMPETER
SWAN

juvenile

adult

WHOOPER
SWAN

adult

juvenile

MUTE SWAN

adult

juvenile

WHISTLING-DUCKS

Named for their whistling calls, these gooselike ducks have long legs and long necks. Wing-beats are slower than ducks, faster than geese.

BLACK-BELLIED WHISTLING-DUCK
Dendrocygna autumnalis | L 21" (53 cm)

Gray face with white eye ring, red bill. Legs red or pink; belly, rump, and tail black. White wing patch shows as broad white stripe in flight. **Juvenile** is paler, with gray bill. **Call** is a high-pitched, four-note whistle.

RANGE: Inhabits wetlands; nests in trees, nest boxes. Casual west to southeastern California; recent sightings in the East, north to Canada, may include wild birds.

FULVOUS WHISTLING-DUCK
Dendrocygna bicolor | L 20" (51 cm)

Rich tawny color overall; back darker, edged with tawny. Dark stripe along hindneck is continuous in female, usually broken in male. Bill and legs dark. Whitish rump band conspicuous in flight. **Call** is a squealing *pe-chee*.

RANGE: Forages in rice fields, marshes, shallow waters; often dives to feed. More active at night than day. Irregular summer wanderer north to dashed line on map; casual farther north. Nearly extirpated in the West.

PERCHING DUCKS

These surface-feeding, woodland ducks are equipped with sharp claws for perching in trees. They nest in tree cavities or nest boxes.

WOOD DUCK
Aix sponsa | L 18½" (47 cm)

Male's glossy, colorful plumage and sleek crest are distinctive. Head pattern and bill colors are retained in drab eclipse plumage. **Female** identified by short crest and large, white, tear-drop-shaped eye patch; compare with female Mandarin Duck (page 50). **Juvenile** resembles female but is spotted below. In all plumages, flight profile (page 52) is distinctive: large head with bill angled downward; long, squared-off tail. Male gives soft, upslurred whistle when swimming; female's squealing flight **call** is a rising *oo-eek*.

RANGE: Fairly common in open woodlands near water. Rare during winter throughout most of breeding range.

MUSCOVY DUCK
Cairina moschata | L 26–33" (66–84 cm)

Large, blackish duck with green and purple gloss above with white patches on upper- and underwing. **Male** has blackish to dark reddish knob at base of bill, bare facial skin (usually brighter red in **domestic male**, whose color varies; can be all-white). Female is smaller, duller; lacks knob and bare facial skin. **Juvenile** even duller; slowly acquires wing patches in first winter. Wild Muscovies are shy, usually silent; seen mostly in slow, gooselike flight at dawn and dusk.

RANGE: Tropical species; tame escaped birds found in parks across North America (now established in Florida). A nest box program in northeastern Mexico helped spread of wild Muscovies to Rio Grande area, where they are now present near Falcon Dam, Texas.

adults

FULVOUS
WHISTLING-DUCK

BLACK-BELLIED
WHISTLING-DUCK
autumnalis

juvenile

WOOD DUCK

♀

juvenile ♂

♂

juvenile ♀

adult ♂

adult ♂

MUSCOVY
DUCK

domestic
variety ♂

DABBLING DUCKS

Surface-feeding members of the genus *Anas*: the familiar "puddle ducks" of freshwater shallows and, chiefly in winter, salt marshes. Dabblers feed by tipping, tail up, to reach aquatic plants, seeds, and snails. They require no running start to take off but spring directly into flight. Most species show a distinguishing swatch of bright color, the speculum, on the secondaries. Many are known to hybridize.

MALLARD
Anas platyrhynchos | L 23" (58 cm)

Male readily identified by metallic green head and neck, yellow bill, narrow white collar, chestnut breast. Black central tail feathers curl up. Both sexes have white tail, white underwings, bright blue speculum with both sides bordered in white (see also page 53). **Female's** mottled plumage resembles other *Anas* species; look for orange bill marked with black. **Juvenile** and **eclipse male** resemble female but bill is dull olive.
Range: Abundant and widespread. Mallards in central Mexico, formerly considered a separate species, "Mexican Duck," are darker, lack distinctive male plumage; **intergrades** occur in southwestern U.S.

MOTTLED DUCK
Anas fulvigula | L 22" (56 cm)

Both sexes closely resemble American Black Duck but body is slightly paler, throat and face are buffy and unstreaked; speculum bluish. Differs from female Mallard by darker plumage and absence of white in tail or black on bill. Western Gulf Coast race, *maculosa,* is darker than nominate Florida race.
Range: Mottled Duck has been introduced to coastal South Carolina. Common year-round in coastal marshes. Very rare north to Kentucky and Kansas. Begins pairing in Jan. or Feb., earlier than the migratory American Black Ducks and Mallards. Some authorities consider the Mottled Duck to be a subspecies of the Mallard.

AMERICAN BLACK DUCK
Anas rubripes | L 23" (58 cm)

Blackish brown, paler on face and foreneck. In flight (page 52), white wing linings contrast more sharply with otherwise dark plumage than in similar female Mallard. Violet speculum is bordered in black, may show a thin white trailing edge. **Male's** bill is yellow; **female's** dull green, may be flecked with black.
Range: Nesting pairs favor woodland lakes and streams, freshwater or tidal marshes. Small introduced populations from western British Columbia and Washington appear to be extirpated. In many parts of range, especially deforested areas, Mallards are replacing American Black Ducks; the two species **hybridize** frequently.

EASTERN SPOT-BILLED DUCK
Anas zonorhyncha | L 22" (56 cm)

Pale tertials and sharply defined yellow tip of black bill, visible at a great distance, set this species apart from similar American Black Duck and female Mallard. This species was recently split from Spot-billed Duck (*A. poecilorhyncha*) and lacks the red spots at base of bill for which that duck is named.
Range: Asian species, casual vagrant to Aleutians and Kodiak Island.

MALLARD

eclipse ♂

juvenile

♂

♀

"MEXICAN DUCK"–MALLARD
INTERGRADE

MOTTLED
DUCK
maculosa

♂

♂

AMERICAN BLACK
DUCK x MALLARD
HYBRID

♂

♀

EASTERN
SPOT-BILLED DUCK

♂

AMERICAN
BLACK DUCK

GADWALL
Anas strepera | L 20" (51 cm)

Male is mostly gray, with white belly, black tail coverts, pale chestnut on wings. Female's mottled brown plumage resembles **female** Mallard (preceding page), but belly is white, forehead steeper, upper mandible gray with orange sides. Both sexes have white inner secondaries that may show as small patch on swimming bird and identify the species in flight (page 53).
RANGE: Common and widespread across Northern Hemisphere.

FALCATED DUCK
Anas falcata | L 19" (48 cm)

Named for **male**'s long, falcated (sickle-shaped) tertials that overhang tail. Both sexes are chunky, with large head. **Female**'s all-dark bill distinguishes her from female wigeons (next page) and Gadwall; note slight bump on back of head. In flight (page 53), both sexes show a broad, dark speculum bordered in white.
RANGE: East Asian species, rare visitor to the western Aleutians; casual to Pribilofs and from West Coast region.

GREEN-WINGED TEAL
Anas crecca | L 14½" (37 cm)

Our smallest dabbler. **Male**'s chestnut head has dark green ear patch outlined in white. Female distinguished from other **female** teals (see also page 36) by smaller bill and by largely white undertail coverts that contrast with mottled flanks. A fast-flying, agile duck. In flight (page 52), shows green speculum bordered in buff on leading edge, white on trailing edge. In the subspecies seen in most of North America, *carolinensis*, male has vertical white bar on side. Eurasian race, *crecca*, is fairly common on the Aleutians and Pribilofs; rare to very rare elsewhere on West and East Coasts; was formerly considered a separate species, the Common Teal. It lacks the vertical bar, but has white stripe on scapulars.

BAIKAL TEAL
Anas formosa | L 17" (43 cm)

Adult male's intricately patterned head is distinctive. Long, dropping dark scapulars are edged in rufous and white. Gray sides are set off front and rear by vertical white bars. **Female** similar to smaller female Green-winged Teal, but tail appears a bit longer; note distinct face pattern with a well-defined white spot at base of bill and white throat that angles up to rear of eye. Distinct eyebrow is bordered by darker crown. "Female" birds with **bridle** marking may be immature males. In flight (page 52), Baikal Teal's underwing is like Green-winged, but has more extensive and darker leading edge. Green speculum has an indistinct cinnamon-buff inner border.
RANGE: Asian species. After several decades of decline, Asian populations appear to have rebounded recently with a corresponding increase of records from western Alaska, and two others on West Coast (Washington and California). Records away from Pacific region may be of escapes.

GADWALL

♀

♂

FALCATED
DUCK

♂

♀

GREEN-WINGED
TEAL

♂ *crecca*

♀ *carolinensis*

♂ *carolinensis*

BAIKAL
TEAL

bridled

♂

♀

AMERICAN WIGEON
Anas americana | L 19" (48 cm)

Male's white forehead and cap are conspicuous in mixed flocks foraging in fields, marshes, and shallow waters; in flight (page 52), identified by mainly white wing linings and, in **adult males**, by large white patches on upperwing. Wing patches are grayish on **adult female** and immatures. Female lacks white on head, closely resembles gray-morph female Eurasian Wigeon; distinguishing field marks in flight are female American's white wing linings and contrast between gray throat and brown breast; Eurasian female's throat and breast are of uniform color.

RANGE: Common. A recently established breeder on the East Coast. Fairly common during winter north to Lake Erie, rare on northern Great Lakes; rare to the Aleutians.

EURASIAN WIGEON
Anas penelope | L 20" (51 cm)

Dark rufous head and gray back and sides make males conspicuous in flocks of American Wigeons; dusky wing linings are distinctive in all plumages (page 52). Hybridizes regularly with American Wigeon. **Adult male** has reddish brown head and neck with creamy crown; large white patches on upperwings. Many fall males retain some brown eclipse feathers but show distinctive reddish head. **Immature male** begins to acquire adult head and breast color but retains some brown juvenal plumage, particularly on forewing, similar to American Wigeon. **Gray-morph female** more closely resembles female American. **Rufous morph** has a more reddish head.

RANGE: Eurasian Wigeon is a rare but regular winter visitor along both coasts, more common in the West; rare in interior of North America. Regular migrant and winter visitor on western and central Aleutians and Bering Sea islands.

NORTHERN PINTAIL
Anas acuta | ♂ L 26" (66 cm) ♀ L 20" (51 cm)

Male's chocolate brown head tops long, slender white neck, the white extending in a thin line onto head. Black central tail feathers extend far beyond rest of long, wedge-shaped tail. **Female** is mottled brown, paler on head and neck; bill uniformly grayish. In both sexes, flight profile (page 52) shows long neck; slender body; long, pointed wings; dark speculum bordered in white on trailing edge. In flight, female's mottled brown wing linings contrast with white belly; long, wedge-shaped tail lacks male's extended feathers.

RANGE: A common, widespread duck, found in marshes and open areas with ponds, lakes; in winter often feeds in grainfields. Much more common in West than in East. Rare in winter north to southern Alaska and Great Lakes region.

WHITE-CHEEKED PINTAIL
Anas bahamensis | L 17" (43 cm)

White cheeks and throat contrast with dark forehead and cap; blue bill has a red spot near base. Long, pointed tail is buffy; tawny or reddish underparts are heavily spotted. Female is paler than **male**; tail slightly shorter. In flight, both sexes show green speculum broadly bordered on each side with buff.

RANGE: Casual vagrant from the West Indies to southern Florida. Sightings even from Florida are of uncertain origin; those away from Florida are most likely birds escaped from captivity.

AMERICAN
WIGEON

eclipse
adult ♂

adult ♂

♀

rufous-morph ♀

gray-morph ♀

EURASIAN
WIGEON

adult ♂

immature ♂

NORTHERN
PINTAIL

♀

♂

WHITE-CHEEKED
PINTAIL
bahamensis

♂

NORTHERN SHOVELER
Anas clypeata | L 19" (48 cm)

Large, spatulate bill, longer than head, identifies both sexes. **Male** distinguished by green head, white breast, brown sides; in early **fall** has a white crescent on each side of face, like Blue-winged Teal. **Female**'s grayish bill is tinged with orange on cutting edges and lower mandible. In flight (page 53), both sexes show blue forewing patch.
RANGE: Common to abundant in the West; increasing in the East. Found in marshes, ponds, and bays.

BLUE-WINGED TEAL
Anas discors | L 15½" (39 cm)

Violet-gray head with white crescent on each side identifies **male**. **Female** distinguished from smaller female Green-winged Teal (page 32) by larger bill, more heavily spotted under-tail coverts, yellowish legs. Compare also with female Cinnamon Teal; note Blue-winged's grayer plumage, smaller bill, and bolder facial markings, including whiter lore spot and more prominent, broken eye ring. Male in eclipse plumage resembles female. In flight (page 53), wing patterns of both sexes match those of Cinnamon.
RANGE: Blue-winged Teal is fairly common in marshes and on ponds and lakes in open country. Uncommon on the West Coast.

CINNAMON TEAL
Anas cyanoptera | L 16" (41 cm)

Cinnamon head, neck, and underparts identify **male**. **Female** closely resembles female Blue-winged Teal but plumage is a richer brown; lore spot, eye line, and broken eye ring less distinct; bill longer and more spatulate. Compare also with Green-winged Teal (page 32). Young birds and males in eclipse plumage resemble female. Males more than a couple months old have red-orange eyes; Blue-winged's eyes are dark. Wing pattern (page 53) is almost identical to Blue-winged.
RANGE: Common in marshes, ponds, and lakes. Regular from eastern Great Plains south to eastern Texas. Casual to eastern Midwest and farther east. Some sightings may be escaped birds. Cinnamon Teal is known to interbreed with Blue-winged Teal.

GARGANEY
Anas querquedula | L 15½" (39 cm)

Prominent whitish edge to tertials is an important field mark in swimming birds; head shape is less rounded than in Blue-winged and Cinnamon Teal. **Male**'s bold white eyebrows separate dark crown, red-brown face; in flight (page 53), shows gray-blue forewing and green speculum bordered fore and aft with white. Wing pattern is retained when male acquires femalelike supplemental plumage, held into winter. **Female** has strong facial pattern: dark crown, pale eyebrow, dark eye line, white lore spot bordered by a second dark line; note also dark bill and legs, dark undertail coverts. Larger and paler overall than female Green-winged Teal (page 32). Female in flight shows gray-brown forewing, dark green speculum bordered in white. Note pale gray inner webs of primaries, visible in flight from above.
RANGE: Old World species. Regular migrant on western Aleutians; very rare on Pribilofs and in Pacific states; casual elsewhere in North America.

NORTHERN
SHOVELER

fall ♂

♀

♂

BLUE-WINGED
TEAL

♀

♂

CINNAMON
TEAL

♀

♂

GARGANEY

♀

fall ♂

♂

POCHARDS

Diving ducks of the genus *Aythya* have legs set far back and far apart, which makes walking awkward. Heavy bodies require a running start on water for takeoff. Various species hybridize. Always carefully check potential vagrants to make sure they are not hybrids.

CANVASBACK
Aythya valisineria | L 21" (53 cm)

Forehead slopes to long, black bill. **Male**'s head and neck are chestnut, back and sides whitish. **Female** and eclipse male have pale brown head and neck, pale brownish gray back and sides. In flight (page 54), whitish belly contrasts with dark breast, dark undertail coverts. Wings lack contrasting pale stripe of smaller Common Pochard and Redhead.
RANGE: Locally common in marshes, lakes; feeds in large flocks; has decreased significantly but decline has stabilized. Migrating flocks fly in irregular V-formations or in lines.

COMMON POCHARD
Aythya ferina | L 18" (46 cm)

Resembles Canvasback in plumage and head shape. Bill similar to Redhead but dark at base and tip, gray in center. Gray on upperparts immediately distinguishes **female** from female Redhead and Ring-necked Duck (next page). In flight (page 54), wings show gray stripe along trailing edge.
RANGE: Eurasian species, rare migrant to Pribilofs and to western and central Aleutians; accidental to south coastal Alaska and southern California.

REDHEAD
Aythya americana | L 19" (48 cm)

Rounded head and shorter, tricolored bill separate this species from Canvasback. Bill is mostly pale blue (male) or slate (female), with narrow white ring bordering black tip. **Male**'s back and sides are smoky gray. **Female** and eclipse male are tawny brown, with slightly darker crown, pale patch bordering black bill tip; compare female scaup (next page). Redheads in flight (page 54) show gray stripe on trailing edge of wings.
RANGE: Locally common in marshes, ponds, and lakes; is declining in the East.

CANVASBACK

COMMON
POCHARD

REDHEAD

RING-NECKED DUCK
Aythya collaris | L 17" (43 cm)

Peaked head; bold white ring near tip of bill. **Male** has second white ring at base of bill; white crescent separates black breast from gray sides. Cinnamon collar is often hard to see in the field. **Female** has dark crown, white eye ring; may have a pale line extending back from eye; face is mainly gray. In flight (page 54), all plumages show a gray stripe on secondaries. RANGE: Fairly common in freshwater marshes and on woodland ponds, lakes; during winter, found also in southern coastal marshes. Range is variable; may breed south or winter north of mapped range.

TUFTED DUCK
Aythya fuligula | L 17" (43 cm)

Head is rounded; crest distinct in **male**, smaller in **female** and immatures; may be absent in eclipse male. Gleaming white sides further distinguish male from male Ring-necked Duck. **First-winter male** has gray sides but lacks the white crescent conspicuous in male Ring-necked. Female is blackish brown above; lacks white eye ring and white bill ring of female Ring-necked. Bills of both male and female Tufted have a wide black tip. Some females also have a small white area at base of bill. In flight (page 54), all plumages show a broad white stripe on secondaries and extending onto primaries.
RANGE: Found on ponds, rivers, bays, often with Ring-necked Ducks and especially scaup. Old World species; regular visitor to western Alaska. Rare winter visitor along East Coast as far south as Maryland, and on West Coast to southern California; casual elsewhere in West; very rare in Great Lakes region.

GREATER SCAUP
Aythya marila | L 18" (46 cm)

Larger size and smoothly rounded head help distinguish this species from Lesser Scaup. In close view, note Greater Scaup's slightly larger bill with wider black tip. **Male** averages paler on back and flanks; in good light, head may show a green gloss. In both species, **female** has bold white patch at base of bill. Some female Greater Scaup, especially in spring and summer, have a paler head with a distinct whitish ear patch. In flight (page 54), Greater Scaup typically shows a bold white stripe on secondaries and well out onto primaries, unlike Lesser Scaup.
RANGE: Locally common; found on large, open lakes and bays. Migrates and winters in small or large flocks, often with Lesser Scaup. Rare to uncommon winter visitor throughout the Gulf Coast states.

LESSER SCAUP
Aythya affinis | L 16½" (42 cm)

Smaller size and peaked crown help distinguish from Greater Scaup. In close view, note Lesser Scaup's slightly smaller bill with smaller black tip. In good light, **male**'s head may show a purple gloss, sometimes mixed with green. **Female** is brown overall, with bold white patch at base of bill. In some females, especially in spring and summer, head is paler, with whitish ear patch less distinct than in female Greater Scaup. In flight (page 54), Lesser Scaup shows bold white stripe on secondaries only.
RANGE: Common; breeds in marshes, small lakes, and ponds. In winter, found in large flocks on sheltered bays, inlets, and lakes.

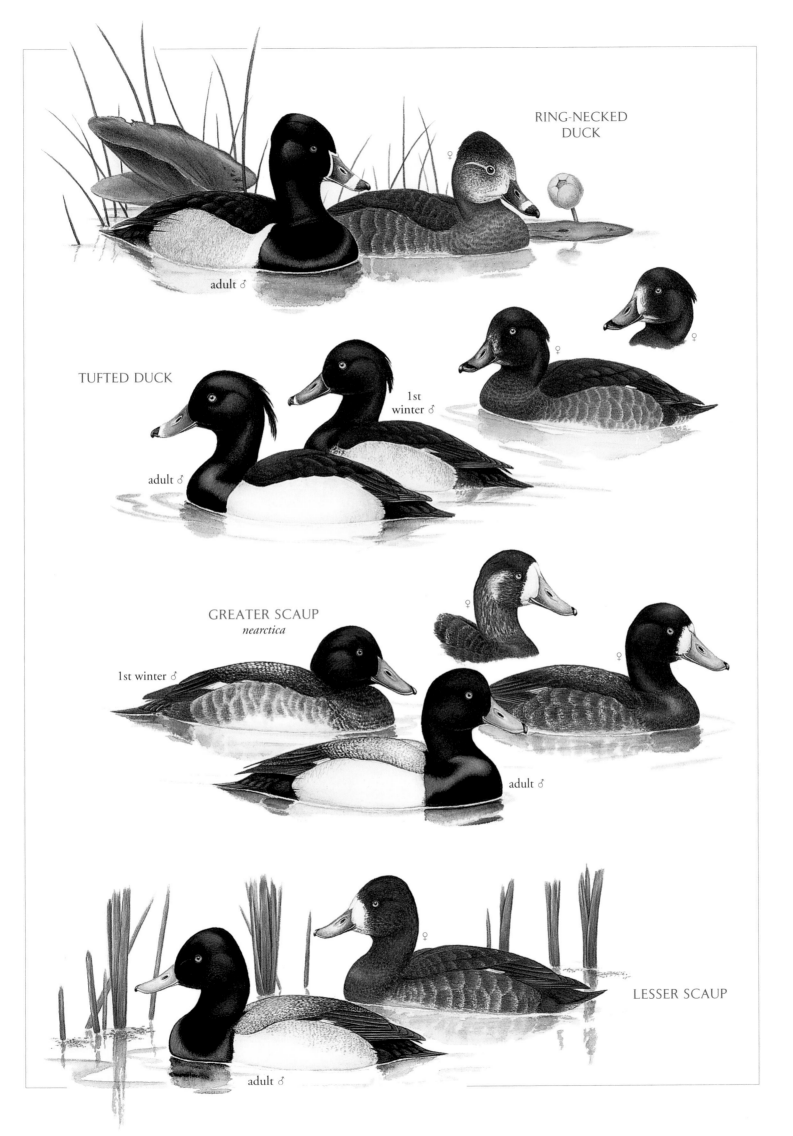

RING-NECKED
DUCK

adult ♂

♀

TUFTED DUCK

adult ♂

1st
winter ♂

♀

♀

GREATER SCAUP
nearctica

1st winter ♂

♀

♀

adult ♂

♀

LESSER SCAUP

adult ♂

EIDERS

These large, bulky diving sea ducks have dense down feathers that help insulate them from the cold northern waters. Females pluck their own down to line nests.

COMMON EIDER
Somateria mollissima | L 24" (61 cm)

Female distinguished from female King Eider by larger size, sloping forehead, and evenly barred sides and scapulars. Feathering extends along sides of bill to or beyond nostril, with minimal feathering on top of bill. Females range in overall color from rust to gray. Eastern *dresseri* is reddish brown. Western *v-nigrum* is duller brown. **Male's** head pattern is distinctive. Most western and a few eastern males show a thin black V on throat; *v-nigrum* male has orange-yellow bill. **Eclipse** and **first-winter males** are dark; first-winter has white on breast; full adult plumage is attained by fourth winter. In flight (page 54), adult male shows solid white back and wing coverts.

RANGE: Locally abundant on shallow bays, rocky shores; casual on Great Lakes. Rare in winter to North Carolina; casual on East Coast to Florida; accidental on West Coast.

KING EIDER
Somateria spectabilis | L 22" (56 cm)

Female distinguished from female Common Eider by smaller size, more rounded head, and crescent or V-shaped markings on sides and scapulars. Feathering extends only slightly along sides of bill but extensively down the top, making bill look stubby. **Male's** head pattern is distinctive. In flight (page 55), shows partly black back, black wings with white patches. **First-winter male** has brown head, pinkish or buffy bill, buffy eye line; lacks white wing patches; full adult plumage attained by third winter.

RANGE: Common on tundra and coastal waters in northern part of range; generally very rare on Great Lakes except on Lake Ontario, where rare and increasing. Rare in winter on East Coast to Virginia; casual to Florida and on West Coast.

SPECTACLED EIDER
Somateria fischeri | T | L 21" (53 cm)

Male has green head with white, black-bordered eye patches and orange bill. In flight (page 55), black breast separates adult male from other eiders, smaller size from Common Eider. Drab **female** has fainter spectacle pattern; bill is gray-blue; feathering extends far down upper mandible.

RANGE: Uncommon and declining; found on coastal tundra near lakes and ponds. Flocks winter in openings in ice pack on Bering Sea. Casual on Aleutians.

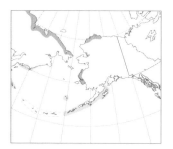

STELLER'S EIDER
Polysticta stelleri | T | L 17" (43 cm)

Greenish head tufts, black eye patch, chin, and collar identify **male. Female** is dark cinnamon brown with pale eye ring, unfeathered dark bill. In flight (page 55), adults, immature males, and some immature females show blue speculum bordered fore and aft in white.

RANGE: Found along rocky coasts; nests on inland grassy areas or tundra. Winters casually south to northern California coast. Numbers reduced over last two decades.

COMMON
EIDER

♀ *v-nigrum*

♀ *dresseri*

eclipse adult ♂
v-nigrum

1st winter ♂
dresseri

adult ♂
dresseri

adult ♂
v-nigrum

1st winter ♂

adult ♂

KING EIDER

♀

SPECTACLED
EIDER

adult ♂

♀

STELLER'S
EIDER

1st
winter ♂

♀

adult ♂

SEA DUCKS

Stocky, short-necked diving ducks, most species breed in the far north and migrate in large, compact flocks to and from their coastal wintering grounds.

BLACK SCOTER
Melanitta nigra | L 19" (48 cm)

Male is black, with orange-yellow knob at base of dark bill. **Female**'s dark crown and nape contrast with pale face and throat; feathering does not extend onto bill. In both, forehead is rounded. In flight (page 55), adult male's blackish wing linings contrast with paler flight feathers. Juveniles resemble females but are whitish on belly; **first-winter male** has some yellow at base of bill by winter.

RANGE: Nests on tundra. Small numbers seen in fall on Great Lakes (common on Lake Ontario, where some winter); rare elsewhere in eastern interior; casual in western interior.

WHITE-WINGED SCOTER
Melanitta fusca | L 21" (53 cm)

White secondaries, conspicuous in flight (page 54), may show as a small white patch on swimming bird. Forehead slightly rounded. Feathering extends almost to nostrils on top and sides of bill. **Female** and juveniles lack contrasting dark crown and paler face of other scoters; white facial patches are distinct on juveniles, often indistinct on adult female. Juveniles and immatures are whitish below. **Adult male** has black knob at base of colorful bill, crescent-shaped white patch below white eye, brownish flanks. Black-flanked adult male Asian *stejnegeri,* casual to western Alaska, has more obvious nasal hook and different bill, color pattern.

RANGE: Fairly common on inland lakes in breeding season, coastal areas in winter. Uncommon inland migrant. Large numbers winter on Lake Ontario since introduction of zebra mussels. Rarest scoter in the South.

SURF SCOTER
Melanitta perspicillata | L 20" (51 cm)

Male's black plumage sets off colorful bill, white eye, white patch on forehead and nape; forehead is sloping, not rounded. **Female** is brown, with dark crown; usually has two white patches on each side of face; feathering extends down top of bill only. Adult female and **first-winter male** may have whitish nape patch. All juveniles have whitish belly, usually white face patches. In flight (page 54), more uniform color of underwings helps distinguish Surf from Black Scoter; also orangish, not dark, legs and feet.

RANGE: Common; nests on tundra and in wooded areas near water. Rare inland migrant. A few winter on Great Lakes; most in coastal waters.

HARLEQUIN DUCK
Histrionicus histrionicus | L 16½" (42 cm)

Small duck, with rounded head, stubby bill. **Male**'s colorful plumage appears dark at a distance. **Female** has three white spots on each side of head. Juvenile resembles adult female. Flight is rapid, low. Compare female in flight (page 55) with female Bufflehead. Male's **call** is a high-pitched nasal squeaking.

RANGE: Locally common on rocky coasts; moves inland along swift streams for nesting. Rare on Great Lakes in migration and winter.

BLACK SCOTER
americana

adult ♂

adult ♀

1st winter ♂

1st winter ♂

1st winter ♀

adult ♂

WHITE-WINGED
SCOTER
deglandi

adult ♂
stejnegeri

adult ♀

adult ♂

adult ♀

1st winter ♀

SURF SCOTER

1st winter ♂

♀

HARLEQUIN
DUCK

1st winter ♂

adult ♂

LONG-TAILED DUCK

Clangula hyemalis | ♂ L 22" (56 cm) ♀ L 16" (41 cm)

Formerly Oldsquaw. **Male**'s long tail is conspicuous in flight, may be submerged in swimming bird. Male in winter and spring is largely white; breast and back dark brown, scapulars pearl gray; stubby bill shows pink band. By late spring, male becomes mostly dark, with pale facial patch, bicolored scapulars; in later, supplemental molt, acquires paler crown and shorter, buff-edged scapulars. Molt into full eclipse plumage continues until early fall. **Female** lacks long tail; bill is dark; plumage whiter in winter, darker in summer. First-fall birds (see page 55) are even darker. Long-taileds are identifiable at some distance by their swift, careening flight and loud, yodeling, three-part **calls**, heard all year. Both sexes show uniformly dark underwings.

RANGE: Away from Great Lakes, rare in the interior and south to Gulf Coast.

BARROW'S GOLDENEYE

Bucephala islandica | L 18" (46 cm)

Male has white crescent on each side of face; white patches on scapulars show on swimming bird as a row of spots; dark color of back extends forward in a bar partially separating white breast from white sides. **Female** and male in eclipse plumage closely resemble Common Goldeneye. Puffy, oval-shaped head, steep forehead, and stubby, triangular bill help identify Barrow's. Adult female's head is slightly darker than female Common; bill mostly yellow, except in young females, which may have only a yellow band near tip of bill. In all plumages, white wing patches visible in flight (page 55) differ subtly between the two species. **Hybrids** between the two goldeneye species are regularly noted.

RANGE: Both goldeneyes summer on open lakes and small ponds; winter in sheltered coastal areas, inland lakes, and rivers. Overall, Barrow's is much less common; rare to casual outside mapped winter range.

COMMON GOLDENEYE

Bucephala clangula | L 18½" (47 cm)

Male has round white spot on each side of face; scapulars are mostly white. **Female** and eclipse male closely resemble Barrow's Goldeneye. Head of Common is more triangular; forehead more sloped; bill longer. Female's head is slightly paler than female Barrow's; bill generally all-dark or with yellow near tip only; rarely completely dull yellow. In all plumages, there are subtle differences between the two species, white wing patches visible in flight (page 55).

RANGE: Fairly Common. Nests in tree cavities near water, as do Buffleheads and most Barrow's Goldeneyes.

BUFFLEHEAD

Bucephala albeola | L 13½" (34 cm)

A small duck with a large, puffy head, steep forehead, short bill. **Male** is glossy black above, white below, with large white patch on head. **Female** is duller, with small, elongated white patch on each side of head. **First-winter male** and male in eclipse resemble female. In flight (page 55), males show white patch across entire wing; female has white patch only on inner secondaries. Flight is fast and direct with rapid wingbeats.

RANGE: Generally common, Buffleheads nest in woodlands near small lakes and ponds. During migration and winter, found also on sheltered bays, rivers, and lakes.

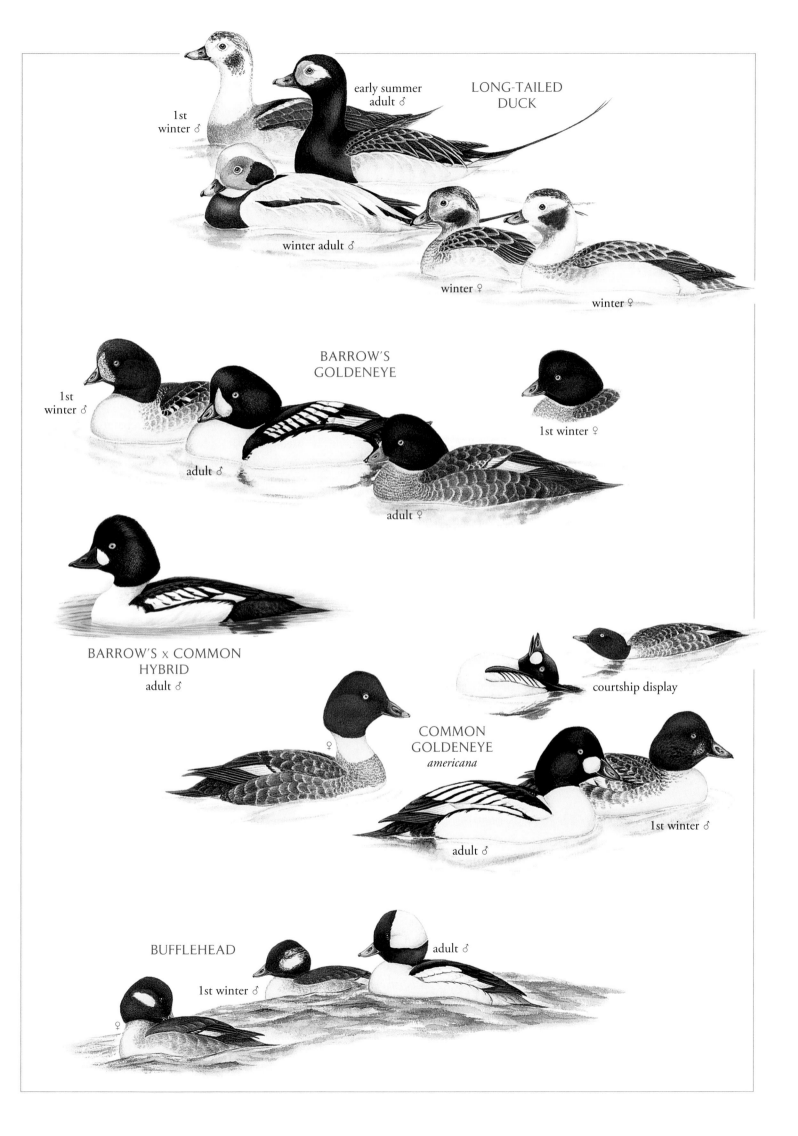

1st winter ♂

early summer adult ♂

LONG-TAILED DUCK

winter adult ♂

winter ♀

winter ♀

1st winter ♂

BARROW'S GOLDENEYE

adult ♂

adult ♀

1st winter ♀

BARROW'S x COMMON HYBRID
adult ♂

courtship display

♀

COMMON GOLDENEYE
americana

adult ♂

1st winter ♂

BUFFLEHEAD

1st winter ♂

adult ♂

♀

MERGANSERS

Long, thin, serrated bills help these divers catch fish, crustaceans, and aquatic insects. Mergansers in flight show pointed wings.

COMMON MERGANSER
Mergus merganser | L 25" (64 cm)

Large duck with long, slim neck and thick-based, hooked, red bill. White breast and sides, often tinged with pink, and lack of crest distinguish **male** from Red-breasted Merganser. **Female**'s bright chestnut, crested head and neck contrast sharply with white chin, white breast. Adult male in flight (page 54) shows white patch on upper surface of entire inner wing, partially crossed by a single black bar. Eclipse male resembles female but retains wing pattern. Female's white inner secondaries and greater coverts are partially crossed by a black bar. Old World "**Goosander**" (nominate *merganser*), recorded from western Aleutians, lacks dark bar on wing. Note different bill shape. As in all species on this page, young male resembles adult female.
RANGE: Common Mergansers nest in woodlands near lakes and rivers; in winter, sometimes also found on brackish water. Casual to Gulf Coast.

RED-BREASTED MERGANSER
Mergus serrator | L 23" (58 cm)

Shaggy double crest, white collar, and streaked breast distinguish **male** from male Common Merganser. **Female**'s head and neck are paler than female Common; chin and foreneck whitish. Adult male in flight (page 54) shows white patch on upper surface of inner wing, partly crossed by two black bars. Eclipse male resembles female but retains male wing pattern. Female's white inner secondaries and greater coverts are crossed by a single black bar. Smaller size and thinner bill help distinguish Red-breasted Merganser in mixed flocks.
RANGE: Nests in woodlands near fresh water or in sheltered coastal areas; prefers brackish or salt water in winter. Abundant migrant on Great Lakes, where moderate numbers winter; elsewhere, fairly common to common migrant in interior.

HOODED MERGANSER
Lophodytes cucullatus | L 18" (46 cm)

Puffy, rounded crest; thin bill. **Male**'s bill is dark; white head patches are fan shaped and conspicuous when crest is raised. Compare with male Bufflehead (preceding page). **Female** brownish overall; upper mandible dark, lower yellowish. Rapid wingbeats in flight (page 55); both sexes show black-and-white inner secondaries. In flight, crest is flattened and male's head patch shows only as a white line.
RANGE: Uncommon in West; common over much of East. In breeding season, found on woodland ponds, rivers, and backwaters. Winters chiefly on fresh water.

SMEW
Mergellus albellus | L 16" (41 cm)

Bill is dark and relatively short. In **female**, white throat and lower face contrast sharply with reddish head and nape. **Adult male** is white with black markings; black-and-white wings are conspicuous in flight (page 55).
RANGE: Eurasian species, rare visitor to Aleutians; casual on Pribilofs; accidental in Pacific states and in East.

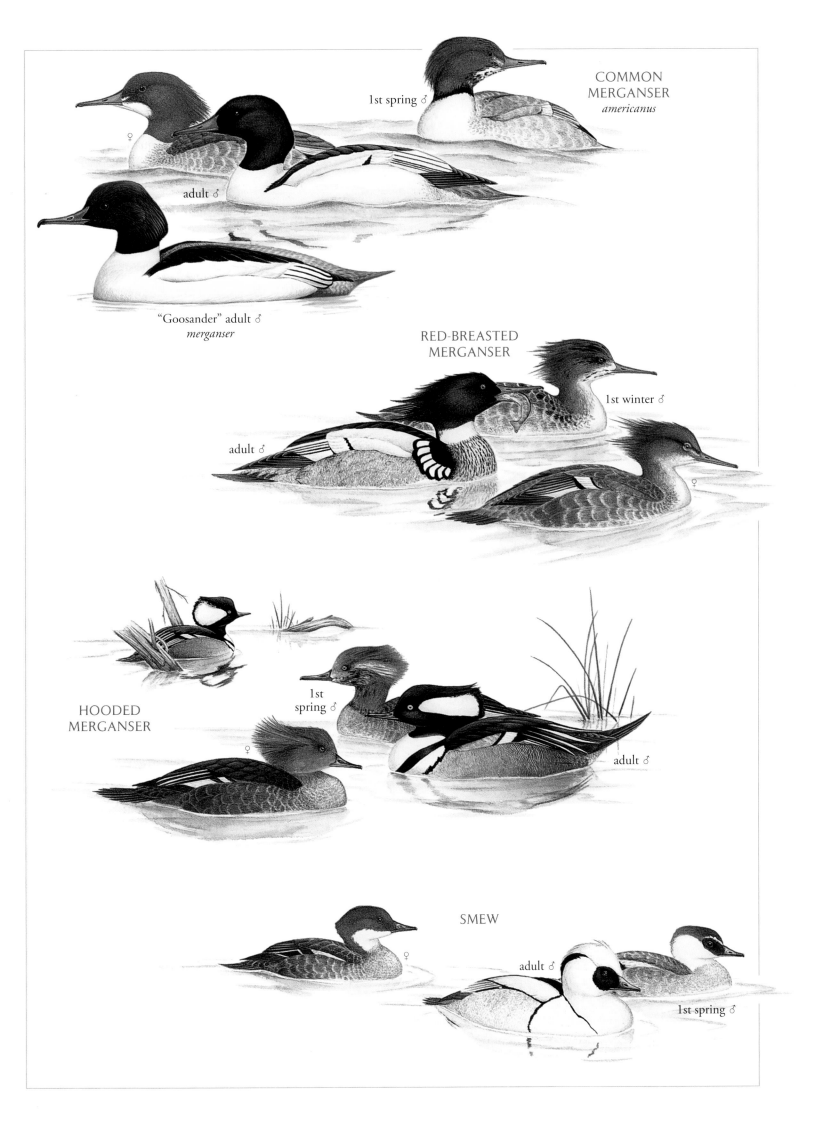

COMMON
MERGANSER
americanus

1st spring ♂

♀

adult ♂

"Goosander" adult ♂
merganser

RED-BREASTED
MERGANSER

adult ♂

1st winter ♂

♀

HOODED
MERGANSER

1st
spring ♂

♀

adult ♂

SMEW

♀

adult ♂

1st spring ♂

STIFF-TAILED DUCKS

Long, stiff tail feathers serve as a rudder for these diving ducks. In both species, male's bill is blue in breeding season.

RUDDY DUCK *Oxyura jamaicensis* | L 15" (38 cm)

Chunky, with large head, broad bill, long tail, often cocked up. **Male**'s white cheeks are conspicuous both in **breeding** plumage and in dull **winter** plumage. In **female**, single dark line crosses cheek. Young resemble female through first winter.
RANGE: Common; nests in dense vegetation of freshwater wetlands. During migration and winter, found on lakes, bays, and salt marshes.

MASKED DUCK *Nomonyx dominicus* | L 13½" (34 cm)

Shy; found on densely vegetated ponds. **Male**'s black face on reddish brown head is distinctive. In **female**, **winter male**, and **juvenile**, two dark stripes cross face and barred back. White wing patches show in flight (page 53).
RANGE: Tropical species, rare and irregular visitor to southern and southeastern Texas; casual in Louisiana and Florida. Accidental in East, north to Wisconsin and New England.

EXOTIC WATERFOWL

Many waterfowl species are brought into North America from other continents for zoos and private collections. Escapes are frequent. The species shown here are among those seen most frequently.

RUDDY SHELDUCK *Tadorna ferruginea* | L 26" (66 cm)
Afro-Eurasian species often kept in captivity. A record of a flock of six on 23 July 2000 from Southampton Island, Nunavut, Canada, may have been vagrants from the Old World.

COMMON SHELDUCK *Tadorna tadorna* | L 25" (64 cm)
Eurasian species. Female smaller, lacks knob on bill.

EGYPTIAN GOOSE *Alopochen aegyptiacus* | L 27" (68 cm)
African species. Widespread escape. Note white wing patches.

SWAN GOOSE *Anser cygnoides* | L 45" (114 cm)
Asian species, domesticated type often called Chinese Goose. Wild race is slim, has long, swanlike bill lacking knob at base.

MANDARIN DUCK *Aix galericulata* | L 16" (41 cm)
Asian species. Compare female to female Wood Duck (page 28).

BAR-HEADED GOOSE *Anser indicus* | L 30" (76 cm)
Asian species. Fairly common in zoos and private collections.

GRAYLAG GOOSE *Anser anser* | L 34" (86 cm)
Eurasian species, progenitor of most domestic geese. A recent record of a wild bird off Newfoundland (page 466). Compare carefully to Greater White-fronted Goose (page 20).

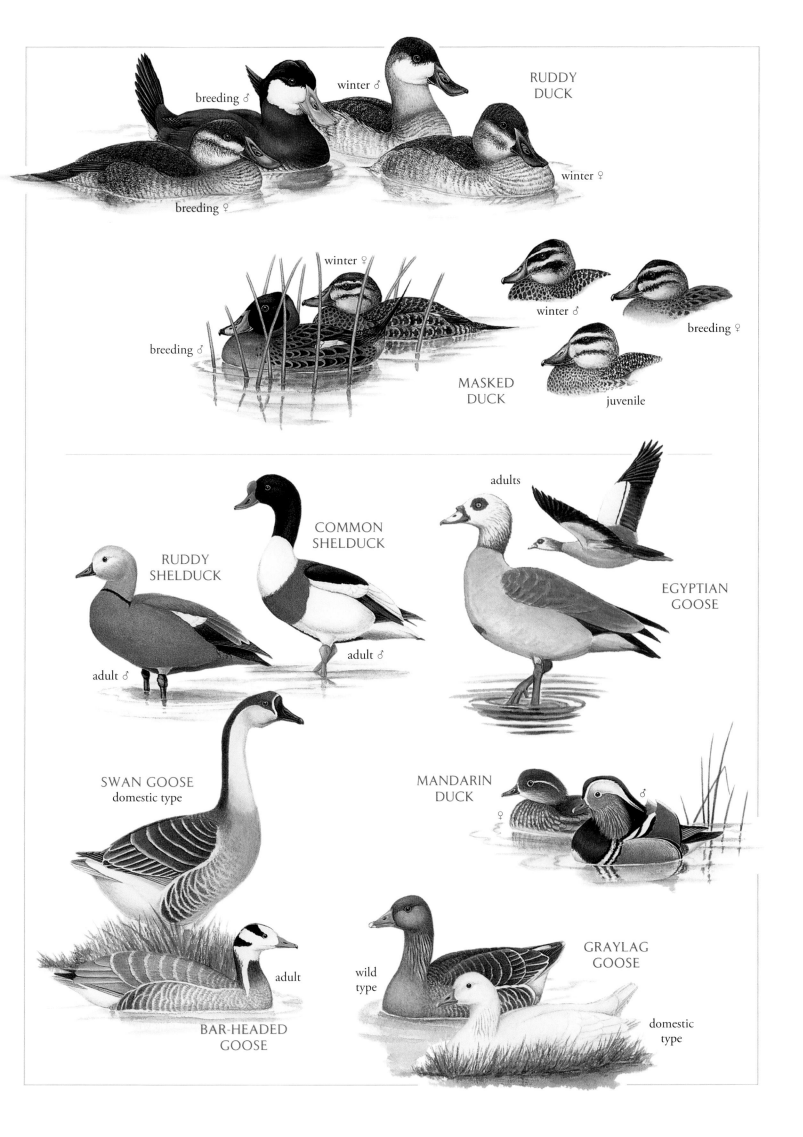

RUDDY DUCK

breeding ♂

winter ♂

breeding ♀

winter ♀

MASKED DUCK

winter ♀

breeding ♂

winter ♂

breeding ♀

juvenile

RUDDY SHELDUCK

adult ♂

COMMON SHELDUCK

adult ♂

adults

EGYPTIAN GOOSE

SWAN GOOSE
domestic type

MANDARIN DUCK

♀

♂

BAR-HEADED GOOSE

adult

GRAYLAG GOOSE

wild type

domestic type

NORTHERN
PINTAIL

AMERICAN
BLACK DUCK

EURASIAN
WIGEON

adult ♂

gray-morph ♀

AMERICAN
WIGEON

adult ♂

WOOD
DUCK

BAIKAL
TEAL

RUDDY
DUCK

breeding ♂

GREEN-WINGED
TEAL
carolinensis

DUCKS IN FLIGHT

MALLARD

GADWALL

NORTHERN
SHOVELER

FALCATED
DUCK

CINNAMON
TEAL

BLUE-WINGED
TEAL

GARGANEY

fall ♂

breeding ♂   MASKED
DUCK

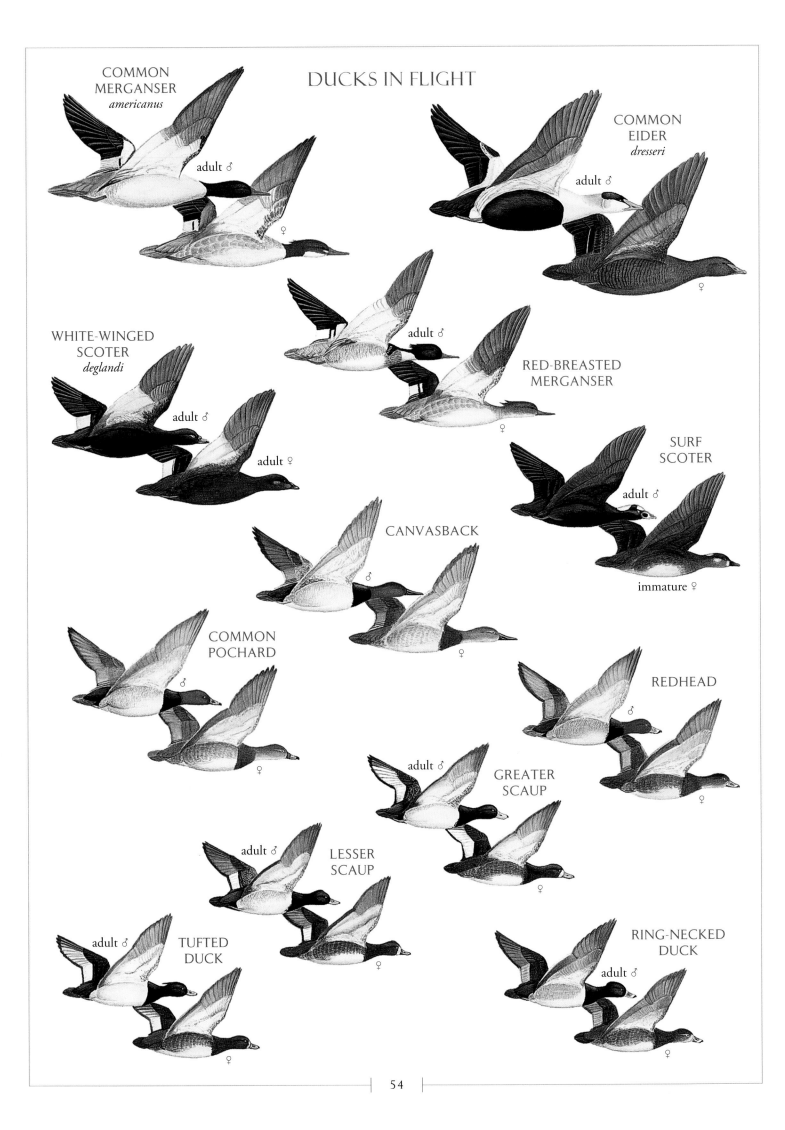

DUCKS IN FLIGHT

COMMON
MERGANSER
americanus

adult ♂

COMMON
EIDER
dresseri

adult ♂

♀

WHITE-WINGED
SCOTER
deglandi

adult ♂

adult ♂

adult ♀

RED-BREASTED
MERGANSER

♀

SURF
SCOTER

adult ♂

immature ♀

CANVASBACK

♂

♀

COMMON
POCHARD

♂

♀

REDHEAD

♂

♀

GREATER
SCAUP

adult ♂

♀

LESSER
SCAUP

adult ♂

♀

TUFTED
DUCK

adult ♂

♀

RING-NECKED
DUCK

adult ♂

♀

DUCKS IN FLIGHT

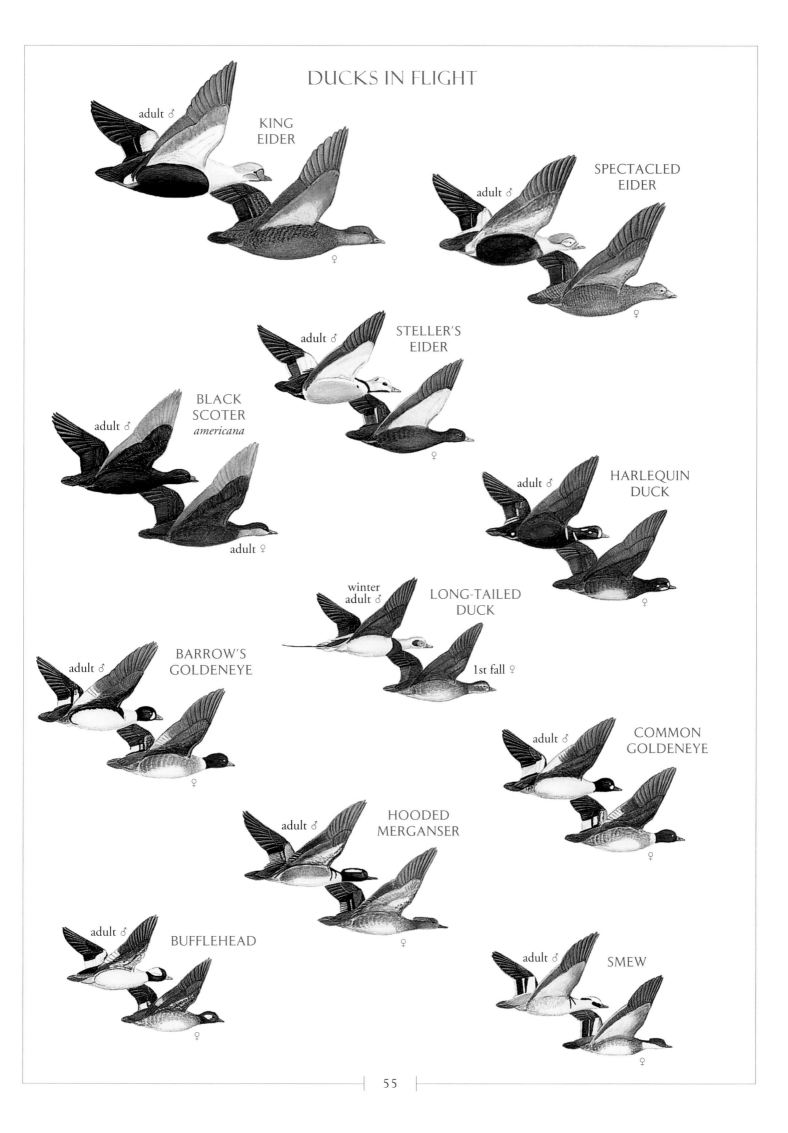

KING
EIDER

adult ♂

♀

SPECTACLED
EIDER

adult ♂

♀

STELLER'S
EIDER

adult ♂

♀

BLACK
SCOTER
americana

adult ♂

adult ♀

HARLEQUIN
DUCK

adult ♂

♀

LONG-TAILED
DUCK

winter
adult ♂

1st fall ♀

BARROW'S
GOLDENEYE

adult ♂

♀

COMMON
GOLDENEYE

adult ♂

♀

HOODED
MERGANSER

adult ♂

♀

BUFFLEHEAD

adult ♂

♀

SMEW

adult ♂

♀

CURASSOWS AND GUANS (Family Cracidae)

These tropical-forest birds have short, rounded wings and long tails. Generally secretive but highly vocal. One species of this family is found in the United States. **Species: *50 World, 1 N.A.***

PLAIN CHACHALACA
Ortalis vetula | L 22" (56 cm)

Gray to brownish olive above, with small head, slight crest; long and rounded, lustrous, dark green tail tipped with white. Patch of bare skin on throat, usually grayish, is pinkish red in **breeding male**. Male's **call** is a deep, ringing *cha-cha-lac,* often given in a loud chorus with other birds; female's voice is higher pitched.

Range: Inhabits tall chaparral thickets along the Rio Grande; feeds in trees, chiefly on leaves and buds; often best seen at feeding stations, which many habituate. Introduced to Georgia's Sapelo Island.

PARTRIDGES, GROUSE, TURKEYS, AND OLD WORLD QUAIL (Family Phasianidae)

Ground dwellers with feathered nostrils, short, strong bills, and short, rounded wings. Flight is brief but strong. Males perform elaborate courting displays. In some species, birds gather at the same strutting grounds, known as leks, every year. **Species: *180 World, 17 N.A.***

CHUKAR
Alectoris chukar | L 14" (36 cm)

Old World species, introduced in North America as a game bird in the 1930s. Gray-brown above; flanks boldly barred black and white; buffy face and throat outlined in black; breast gray; belly buff; outer tail feathers chestnut (best seen in flight just prior to landing). Bill and legs are red. Sexes are similar, but males are slightly larger and have small leg spurs. **Juvenile** is smaller and mottled; lacks bold black markings of **adults**. **Calls** include a series of loud, rapid *chuck chuck chuck* notes and a shrill *whitoo* alarm note.

Range: Chukars have become established in rocky, arid, mountainous areas of the West. Game farm Chukars or hybrids with Rock Partridge (*A. graeca*) are released for hunting in the East. In fall and winter, Chukars feed in coveys of 5 to 40 birds.

GRAY PARTRIDGE
Perdix perdix | L 12½" (32 cm)

Grayish brown bird with rusty face and throat, paler in **female**. **Male** has dark chestnut patch on belly; patch is smaller or absent in females. Flanks are barred with reddish brown; outer tail feathers rusty. **Calls** include a hoarse *kee-uck,* likened to a rusty gate.

Range: Widely introduced from Europe in early 1900s. Has declined over parts of North American range. Inhabits open farmlands, grassy fields. In fall, forms coveys of 12 to 15 birds.

breeding ♂

PLAIN
CHACHALACA

GRAY
PARTRIDGE

CHUKAR

adults

juvenile

♀

♂

RING-NECKED PHEASANT

Phasianus colchicus | ♂ L 33" (84 cm) ♀ L 21" (53 cm)

Introduced from Asia, this large, flashy bird has a long, pointed tail and short, rounded wings. **Male** is iridescent bronze overall, mottled with brown, black, and green; head varies from dark, glossy green to purplish, with fleshy red eye patches and iridescent ear tufts. Often shows a broad white neck ring. **Female** is buffy overall, much smaller and duller than male. Distinguished from female Sharp-tailed Grouse (page 64) by larger size, longer tail, lack of barring below, and white in tail. Male's territorial **call** is a loud, penetrating *kok-cack*. Both sexes give hoarse, croaking alarm notes. When flushed, rise almost vertically with a loud whirring of wings.

RANGE: Locally common; declining in parts of the East. Found in open country, farmlands, brushy areas, and woodland edges. A group of subspecies with white wing coverts (not shown) has become established in parts of the West. The Japanese subspecies "**Green Pheasant**," *versicolor*, introduced in tidewater Virginia and southern Delaware, is apparently gone.

WILD TURKEY

Meleagris gallopavo | ♂ L 46" (117 cm) ♀ L 37" (94 cm)

Largest game bird in North America; slightly smaller, more slender than the domesticated bird. **Male** has dark, iridescent body, flight feathers barred with white, red wattles, blackish breast tuft, spurred legs; bare-skinned head is blue and pink. Tail, uppertail coverts, and lower rump feathers are tipped with chestnut on eastern birds, buffy white on western birds. **Female** and immature are smaller and duller than male, often lack breast tuft. Of the races seen in North America, *silvestris* predominates in the East, *merriami* in the West. Birds from Kansas to Mexico (*intermedia*) are intermediate. Birds from Peninsular Florida (*osceola*) are like *silvestris,* but smaller. Birds of the open forest and forest openings, Wild Turkeys forage mostly on the ground for seeds, nuts, acorns, and insects. At night they roost in trees. In spring, male's gobbling **call** may be heard a mile away.

RANGE: Restocked in much of its former range and introduced in other areas.

HIMALAYAN SNOWCOCK

Tetraogallus himalayensis | L 28" (71 cm)

Gray-brown overall, with tan streaking above. Whitish face and throat, outlined with chestnut stripes; undertail coverts white. Note white in wing in flight. **Male** almost identical to female, except female slightly smaller, lacks spurs, forehead buff and area around eye grayer. Inhabits mountainous terrain; flies downhill in the morning, then walks back up, feeding. **Calls** include various clucks and cackles while feeding; one advertising call suggestive of Long-billed Curlew (male's call rises, female's descends).

RANGE: Large Asian bird, successfully established (introduced 1963) only at high elevations in the Ruby Mountains of northeastern Nevada.

RING-NECKED
PHEASANT

"Green Pheasant"

eastern
silvestris

WILD
TURKEY

HIMALAYAN
SNOWCOCK

western
merriami
displaying ♂

RUFFED GROUSE
Bonasa umbellus | L 17" (43 cm)

Small crest; black ruff on sides of neck, usually inconspicuous; banded tail with wide dark band near tip, incomplete in **female**. The two color **morphs**, **red** and **gray**, are most apparent by tail color. Red morphs predominate in the humid Pacific Northwest and Appalachian region; gray morphs in the North and West outside the Pacific Northwest. Ten subspecies. In spring, the **male** displays by raising ruff and crest, fanning tail, and beating wings to make a hollow, accelerating, drumming noise.
RANGE: Uncommon to fairly common in deciduous and mixed woodlands.

SPRUCE GROUSE
Falcipennis canadensis | L 16" (41 cm)

Male has dark throat and breast, edged with white; red eye combs. Over most of range, both sexes have black tail with chestnut tip. Birds of the northern Rockies and Cascades, "**Franklin's Grouse**," *franklinii*, have white spots on uppertail coverts; **male**'s tail is all-dark. In all subspecies, **females** have two color **morphs**, **red** and **gray**; resemble female Sooty and Dusky Grouse but are smaller and have black barring and white spots below. Juveniles resemble red-morph female. Female's high-pitched **call** is thought to be territorial. In courtship display, male spreads his tail, erects the red eye combs, and rapidly beats wings. In territorial flight display, male flutters upward on shallow wingstrokes; "Franklin's Grouse" ends this performance by beating wings together, making a clapping sound.
RANGE: Spruce Grouse inhabit open coniferous forests with dense undergrowth. Frequents roadsides, especially in fall.

SOOTY GROUSE
Dendragapus fuliginosus | L 20" (51 cm)

Formerly (with Dusky Grouse) known as Blue Grouse. Larger than Spruce Grouse. **Male**'s sooty gray plumage sets off yellow-orange eye comb. On neck, white-based feathers cover an inflatable bare yellow sac. Female is mottled brown above, with plain gray belly. On both sexes, the 18 tail feathers are round and tipped with a gray terminal band. Chicks are yellowish. Male display **call** is series of loud low hoots audible at considerable distance, usually delivered from perch in tree; neck sacs are inflated to amplify sound.
RANGE: Inhabits coniferous forest but will forage at meadow edges. Believed extirpated from mountains of southern California.

DUSKY GROUSE
Dendragapus obscurus | L 20" (51 cm)

All plumages similar to Sooty Grouse, but paler overall; closed tail squarer (less graduated), the 20 tail feathers are more square tipped. Northern subspecies (*richardsonii* and *pallidus*) lack or virtually lack the gray terminal band. **Male** neck sac is purplish and smoother, with broader white-feathered border than in Sooty. Male's display, usually from ground, often involves low fluttering or making short circular flights, then strutting with tail fanned, body tipped forward, head drawn in, wings dragging. Male's display **call**, usually given from ground, is softer and lower pitched than Sooty and audible only at close range. Chicks are grayish.
RANGE: Often prefers more open forest than Sooty; sometimes found in sagebrush. Range almost entirely separate from Sooty; hybrids recorded from interior of British Columbia.

gray-morph
displaying ♂

RUFFED
GROUSE

red-morph ♂

red-morph ♀

SPRUCE
GROUSE

red-morph ♀

gray-morph ♀

"Franklin's
Grouse"
franklinii

♂

displaying ♂

DUSKY
GROUSE

SOOTY GROUSE

northern Rockies
richardsonii
displaying ♂

southern
Rockies ♀
obscurus

coastal
fuliginosus
displaying ♂

WHITE-TAILED PTARMIGAN
Lagopus leucura | L 12½" (32 cm)

As with all ptarmigans, legs and feet are feathered and plumage is molted three times a year, matching seasonal changes in habitat. Distinguished from other ptarmigans in all seasons by white tail. **Winter** bird is white except for small dark bill and eyes and red eye combs. In **summer**, body is mottled blackish or brown with white belly, wings, and tail. Spring and **fall molts** give a patchy appearance. **Calls** include a henlike clucking and soft, low hoots.
RANGE: Locally common on rocky alpine slopes, high meadows. Small numbers have been successfully introduced in the central Sierra Nevada, Wallowa, and Uinta Mountains and on Pike's Peak. Reintroduced into northern New Mexico. Moves to lower elevations during severe weather.

ROCK PTARMIGAN
Lagopus muta | L 14" (36 cm)

Mottled **summer** plumage is black, dark brown, or grayish brown; **male** generally lacks the reddish tones of male Willow Ptarmigan. There are many recognized subspecies, with color variations according to geography. In **winter** plumage, **male** has a black line from bill through eye, lacking in male Willow. Acquires breeding plumage later in spring than does Willow. In both sexes, bill and overall size are slightly smaller than in Willow. **Females** are otherwise difficult to distinguish from Willows. Plumage is patchy white during spring and fall molts. Both species retain white wings and black tail year-round. **Calls** include low growls and croaks and noisy cackles.
RANGE: Rock Ptarmigan is common on high, rocky slopes and tundra. In breeding season, generally prefers higher and more barren habitat than does Willow Ptarmigan. Accidental in northern Minnesota and Queen Charlotte Islands, British Columbia, in spring.

WILLOW PTARMIGAN
Lagopus lagopus | L 15" (38 cm)

Largest ptarmigan. Mottled **summer** plumage of **male** is generally redder than in Rock Ptarmigan. White **winter** plumage lacks the black eye line of male Rock Ptarmigan; bill and overall size are slightly larger in Willow Ptarmigan. **Female** is otherwise difficult to distinguish from Rock Ptarmigan. Both species retain white wings and black tail year-round. Plumage is patchy white during **spring** and fall **molts**. **Calls** include low growls and croaks, noisy cackles. In courtship and territorial displays, male utters a raucous *go-back go-back go-backa go-backa go-backa*. Ptarmigans' red eye combs can be concealed or raised during courtship and aggression.
RANGE: Willow Ptarmigan is common on tundra, especially in thickets of willow and alder. In breeding season, generally prefers wetter, brushier habitat than Rock Ptarmigan. Casual in spring and winter to northern tier of U.S. states.

WHITE-TAILED
PTARMIGAN

molting fall ♂

winter

summer ♀

summer ♂

winter ♀

winter ♂

ROCK
PTARMIGAN

summer ♀

summer ♂

fall ♂

molting
spring ♂

winter

summer ♀

summer ♂

WILLOW
PTARMIGAN

summer ♂

GREATER PRAIRIE-CHICKEN

Tympanuchus cupido | L 17" (43 cm)

Heavily barred with dark brown, cinnamon, and pale buff above and below. Short, rounded tail is all-dark in **male**, barred in **female**. Male has fleshy yellow-orange eye combs. Both sexes have elongated dark neck feathers, longer in males and erected during courtship to display inflated golden neck sacs. Courting males make a deep *oo-loo-woo* sound known as "booming," like blowing over top of an empty bottle.

RANGE: Uncommon, local, and declining. Found in areas of natural tallgrass prairie interspersed with cropland. A smaller, darker race, endangered "Attwater's Prairie-Chicken," *attwateri* (**E**) of southeastern Texas, is nearly extinct. The "Heath Hen" (nominate *cupido*), formerly resident along the Atlantic seaboard from Massachusetts to Virginia, is now extinct —last record on Martha's Vineyard in 1932.

LESSER PRAIRIE-CHICKEN

Tympanuchus pallidicinctus | L 16" (41 cm)

Resembles Greater Prairie-Chicken, but slightly smaller, paler, less heavily barred below. Male's courtship notes are higher pitched than Greater. Courting male displays dull orange-red neck sacs and erects dark neck tufts.

RANGE: Uncommon, local, and declining; found in sagebrush and shortgrass prairie country, especially where shinnery oak grows.

SHARP-TAILED GROUSE

Tympanuchus phasianellus | L 17" (43 cm)

Similar to prairie-chickens, but underparts are scaled and spotted; tail is mostly white and pointed; yellowish eye combs are less prominent. Compare with female Ring-necked Pheasant (page 58). Birds are darkest in Alaska and northern Canada (standing figure), palest in the Plains (flying figure). **Male**'s purplish neck sacs are inflated during courtship display. His courting notes include cackling and a single, low *coo-oo* **call** accompanied by the rattling of wing quills.

RANGE: Inhabits grasslands, sagebrush, woodland edges, and river canyons. Fairly common over range; rare in western U.S. Where ranges overlap, can hybridize with Greater Prairie-Chicken and Dusky Grouse.

GUNNISON SAGE-GROUSE

Centrocercus minimus | ♂ L 22" (56 cm) ♀ L 18" (46 cm)

Distinguished from Greater Sage-Grouse by smaller size and more strongly white-banded tail. Longer, denser filoplumes are erected to form a distinct, recurved crest on **displaying male**.

RANGE: Small, declining population in south-central Colorado and southeastern Utah is geographically isolated from Greater Sage-Grouse.

GREATER SAGE-GROUSE

Centrocercus urophasianus | ♂ L 28" (71 cm) ♀ L 22" (56 cm)

Blackish belly, long pointed tail feathers, and large size are distinctive. **Male** is larger than **female** and has yellow eye combs, black throat and bib, and large white ruff on breast. In flight, dark belly, absence of white outer tail feathers, and larger size distinguish it from Sharp-tailed Grouse. **Displaying male** fans tail and rapidly inflates and deflates air sacs, emitting a loud, bubbling popping.

RANGE: Fairly common but local; found in sagebrush areas of foothills and plains.

displaying ♂

displaying ♂

LESSER
PRAIRIE-CHICKEN

GREATER
PRAIRIE-CHICKEN

♀

♀

SHARP-TAILED
GROUSE

displaying ♂

GUNNISON
SAGE-GROUSE

♀

GREATER
SAGE-GROUSE

displaying ♂

♀

♂

♀

displaying males
on lek

NEW WORLD QUAIL (Family Odontophoridae)

Scientific evidence has recently placed the New World Quail in their own family. All have chunky bodies and crests or head plumes. In North America, most live in the West. **Species:** *32 World, 6 N.A.*

GAMBEL'S QUAIL
Callipepla gambelii | L 11" (28 cm)

Grayish above, with prominent teardrop-shaped plume or double plume. Chestnut sides and crown, and lack of scaling on underparts, distinguish Gambel's from California Quail. **Male** has dark forehead, black throat, black patch on belly. Smaller **juvenile** is tan and gray with pale mottling and streaking. Shows less scaling and streaking than darker California juvenile; nape and throat are grayer. **Calls** include varied grunts and cackles and a plaintive *qua-el;* loud, querulous *chi-ca-go-go* call is similar to California Quail but higher pitched and usually has four notes.

RANGE: Common in desert scrublands and thickets, usually near permanent water source. Gregarious; in fall and winter, forms large coveys. Sometimes **hybridizes** with Scaled Quail (next page) and California Quail where ranges overlap. Introduced populations exist in Idaho and on California's San Clemente Island.

CALIFORNIA QUAIL
Callipepla californica | L 10" (25 cm)

Gray and brown above, with prominent teardrop-shaped plume or double plume. Scaled underparts and brown sides and crown separate California from Gambel's Quail. Body color varies from grayish, seen over most of range, to brown in coastal mountains of California; extremes are shown here in **females**. **Male** has pale forehead, black throat, and chestnut patch on belly. **Juvenile** is smaller; resembles Gambel's juvenile, but is darker, with traces of scaling on underparts. **Calls** include varied grunts and cackles; loud, emphatic *chi-ca-go* call is similar to Gambel's Quail but lower pitched and usually has three notes rather than four.

RANGE: Common in open woodlands, brushy foothills, stream valleys, suburbs, usually near permanent water source. Gregarious; in fall and winter, assembles in large coveys. Populations in northeastern portion of range and Utah are probably introduced.

MOUNTAIN QUAIL
Oreortyx pictus | L 11" (28 cm)

Gray and brown above, with two long, thin head plumes that often appear to be one plume. Gray breast; chestnut sides boldly barred with white; chestnut throat outlined in white. **Male** and **female** are alike; female has shorter head plumes. Amount of brown and gray in upperparts varies in different races; birds of humid coastal Northwest are browner than three gray interior subspecies. Smaller **juvenile** has grayer underparts and longer head plumes than Gambel's or California Quail juveniles. Male's mating **call**, a loud, clear, descending *quee-ark;* both sexes give whistled notes.

RANGE: Uncommon to fairly common, but declining in parts of range; in chaparral, brushy ravines, mountain slopes, at altitudes up to 10,000 feet. Nonmigratory but descends to lower altitudes in winter. Gregarious, forming small coveys in fall and winter. Secretive; best seen in late summer in family groups along roadsides.

GAMBEL'S
QUAIL

juvenile

♂

♀

SCALED x GAMBEL'S
HYBRID

♀ *californica*

CALIFORNIA
QUAIL

coastal ♀
brunescens

♂

♂ *californica*

coastal
juvenile
brunescens

coastal ♂
palmeri

MOUNTAIN
QUAIL

interior ♀

juvenile

NORTHERN BOBWHITE

Colinus virginianus | L 9¾" (25 cm)

Mottled, reddish brown quail with short gray tail. Flanks are striped with reddish brown. Throat and eye stripe are white in **male**, buffy in **female**. **Juvenile** is smaller and duller. Male's **call** is a rising, whistled *bob-white*, heard chiefly in late spring and summer; whistled *hoy* call is heard year-round.

RANGE: Uncommon to common in brushlands and open woodlands; feeds and roosts in coveys except during nesting season. The population in the Northwest is introduced. At northern edge of range, numbers have greatly declined over the last couple of decades. "**Masked Bobwhite**," *ridgwayi* (**E**), which was formerly found from southeastern Arizona to central Sonora, Mexico, was eliminated from the U.S. part of its range by the early 1900s. Reintroduced in 1970 from Mexico (Sonora) to Altar Valley in southeastern Arizona, it remains endangered. Male has black throat and cinnamon underparts.

MONTEZUMA QUAIL

Cyrtonyx montezumae | L 8¾" (22 cm)

Plump, short-tailed, round-winged quail. **Male** has distinctive facial pattern and rounded pale brown crest on back of head. Back and wings mottled black, brown, and tan; breast dark chestnut; sides and flanks dark gray with white spots. **Female** is mottled pinkish brown below with less-distinct head markings. **Juvenile** is smaller, paler, with dark spotting on underparts. **Call** given by male in breeding season is a loud, quavering, descending whistle.

RANGE: Uncommon, secretive, and local in grassy undergrowth of open juniper-oak or pine-oak woodlands on semiarid mountain slopes. Recently rediscovered in Chisos Mountains of southwest Texas.

SCALED QUAIL

Callipepla squamata | L 10" (25 cm)

Grayish quail with conspicuous white-tipped crest. Bluish gray breast and mantle feathers have dark edges, creating a shingled or scaly effect. Female's crest is buffy and smaller. **Males** in southernmost Texas (*castanogastris*) tend to show a dark chestnut patch on belly, unlike the common subspecies, *pallida*, found over much of the U.S. range. **Juvenile** resembles adult but is more mottled above, with less conspicuous scaling. During breeding season, both sexes give a location **call** when separated, a low, nasal *chip-churr,* accented on the second syllable.

RANGE: Fairly common; found on barren mesas and plateaus, semidesert scrublands, and grasslands with mixed scrub; often frequents roadsides. In fall, forms large coveys.

NORTHERN
BOBWHITE

"Masked Bobwhite" ♂
ridgwayi

♀

♂

juvenile

MONTEZUMA
QUAIL

juvenile

♀

♂

SCALED
QUAIL

juvenile

♂ *pallida*

south Texas ♂
castanogastris

LOONS (Family Gaviidae)

In all species, juvenal-like plumage is held through the first summer. **Species:** *5 World, 5 N.A.*

RED-THROATED LOON
Gavia stellata | L 25" (64 cm)

Thin bill often appears slightly upturned; tends to hold head tilted up. **Breeding adult** has gray head with brick red throat patch that appears dark in flight; dark brown upperparts with no contrasting white patches on scapulars as in all other loons in breeding plumage. Winter adult has sharply defined white on face and extensive white spotting on back. **Juvenile**'s head is grayish brown; throat may have dull red markings. In all plumages, white on flanks extends upward a bit on sides of rump, which may cause confusion with Arctic Loon. In flight, shows smaller head and feet than Common and Yellow-billed Loon (next page); wingbeat is quicker; often flies with drooping neck, unlike other loons. Flight **call**, heard on breeding range, is a rapid, gooselike *kak-kak-kak*.
Range: Migrates coastally; also overland in the East, where most numerous on northern and eastern Great Lakes. Always casual inland in western North America, and in the eastern interior during winter.

PACIFIC LOON
Gavia pacifica | L 26" (66 cm)

In all plumages, has dark flanks, with no white extending upward on sides of rump. Bill is slim and straight; head smoothly rounded and held level. **Breeding adult**'s head and nape are pale gray; white stripes on sides of neck show only moderate contrast; throat's iridescent purple patch, sometimes washed with green, usually appears black unless seen clearly on swimming bird. **Juvenile**'s crown and nape are slightly paler than back, unlike Common Loon (next page); in juveniles and **winter adults**, dark cap extends to eye. Winter adults and most juveniles have a thin, brown "chin strap," though it may be faint in juveniles. In flight, resembles Common, but head and feet are smaller.
Range: A coastal and offshore migrant; unlike other loons, often migrates in small to moderate-size flocks closer to the water's surface than migrating Common Loons. Rare inland throughout the West; very rare in Midwest; casual on East Coast.

ARCTIC LOON
Gavia arctica | L 28" (73 cm)

Larger than Pacific Loon, with less-rounded head; best distinguished in all plumages from Pacific by more extensive white on flanks, coming up over sides of rump. Visibility of white area depends on how buoyantly the bird is swimming. When diving, often only a small white rump patch is evident. At rest, Arctic Loon shows much more white; note Pacific can also show some white. Nape in **breeding adult** is darker, and black-and-white stripes are bolder, than in Pacific; white stripes on face connect more to sides of neck; greenish on throat very hard to see.
Range: Old World species. To date, only the larger Siberian race, *viridigularis*, has been recorded in North America. Breeds in northwestern Alaska. Seen in migration in coastal western Alaska, especially at St. Lawrence Island. Casual elsewhere on West Coast.

winter
adult

1st
spring

RED-THROATED
LOON

breeding
adult

winter
adult

juvenile

Red-throated

breeding adults
(for comparison)

PACIFIC LOON

Pacific

Arctic

breeding
adult

winter
adult

winter
adult

juveniles

winter
adult

breeding
adults

ARCTIC LOON
viridigularis

juvenile

winter
adult

winter
adult

COMMON LOON
Gavia immer | L 32" (81 cm)

Large, thick-billed loon with slightly curved culmen. Bill is black in **breeding** plumage, blue-gray in **winter adults** and **juveniles**, but the culmen remains dark. In winter plumage, crown and nape are darker than back; dark on nape extends around sides of neck, but note the white indentation above this. In winter adults the white extends up and around the eye; the face pattern is more blended in juveniles. Forehead is steep, crown is peaked at front. Holds head level. Juvenile Common and Yellow-billed Loons have whitish scalloping on their scapulars, distinguishing them from the plainer-backed winter adults. Full juvenal plumage is kept through most of the winter, with a partial molt in spring. Most winter adults retain at least a few spotted coverts, often visible on swimming birds. In flight, large head and feet help distinguish Common from Arctic, Pacific, and Red-throated Loons (preceding page). Loud yodeling **calls** delivered on water and in flight are heard all year, but most often on breeding grounds.
RANGE: Fairly common; nests on large lakes. Migrates overland as well as coastally. Under most conditions, Common and Yellow-billed fly quite high above the water when migrating, while the other loon species fly lower. Note their slower wingbeats and the paddle-shaped feet that are usually visible beyond the tail. Winters mainly in coastal waters or on large, ice-free inland bodies of water.

YELLOW-BILLED LOON
Gavia adamsii | L 34" (86 cm)

Breeding adult has straw-yellow bill, usually longer than in Common Loon; culmen is straight, giving bill a slightly uptilted look; head often tilted back, which enhances this effect. Crown is peaked at front and rear, giving a subtle double-bump effect. Bill is duskier at the base in **winter adults** and **juveniles**, but always shows strong yellow cast toward the tip. Note also pale face and distinct dark mark behind eye; eye is smaller, and back and crown are paler and browner, than in Common. **Calls** are similar to Common.
RANGE: Yellow-billed Loon breeds on tundra lakes and rivers. Migrates coastally; rare south of Canada on West Coast, where it is recorded annually south to northern California, casually to southern California. Very rare inland in West; casual east to Great Lakes region and south to Texas and Georgia.

winter
adult

breeding
adult

COMMON
LOON

winter
adult

juvenile

breeding
adult

swimming juveniles
(for comparison)

Common   Yellow-billed   Arctic   Pacific   Red-throated

winter
adult

breeding
adults

YELLOW-BILLED
LOON

winter
adult

juvenile

GREBES (Family Podicipedidae)

A worldwide family of aquatic diving birds. Lobed toes make them strong swimmers. Grebes are infrequently seen on land or in flight. **Species:** *22 World, 7 N.A.*

LEAST GREBE
Tachybaptus dominicus | L 9¾" (25 cm)

A small, short-necked grebe with golden yellow eyes, a slim, dark bill, and purplish gray face and foreneck. **Breeding adult** has blackish crown, hindneck, throat, and back. **Winter** birds have white throat, paler bill, less black on crown. In flight, shows large white wing patch.
RANGE: Rather uncommon and local; may hide in vegetation near shores of ponds, sloughs, ditches. May nest at any season on any quiet, inland water. Casual straggler to southern Arizona, southeast California, south Florida, and upper Texas coast.

PIED-BILLED GREBE
Podilymbus podiceps | L 13½" (34 cm)

Breeding adult is brown overall, with black ring around stout, whitish bill; black chin and throat; pale belly. **Winter** birds lose bill ring; chin is white, throat tinged with pale rufous. **Juveniles** resemble winter adult but throat is much redder, eye ring absent, head and neck streaked with brown and white. First-winter birds lack streaking; throat is duller. A short-necked, big-headed, stocky grebe. In flight, shows almost no white on wing.
RANGE: Nests around marshy ponds and sloughs; sometimes hides from intruders by sinking until only its head shows. Common but not gregarious. Winters on fresh or salt water. Casual to Alaska.

HORNED GREBE
Podiceps auritus | L 13½" (34 cm)

Breeding adult has chestnut foreneck, golden "horns." In **winter** plumage, white cheeks and throat contrast with dark crown and nape; some are dusky on lower foreneck. Black on nape narrows to a thin stripe. All birds show a pale spot in front of eye. In flight (next page), white secondaries show as patch on trailing edge of wing. Bill is short and straight, thicker than Eared Grebe's; neck is thicker too, crown flatter. Smaller size and shorter, dark bill most readily separate winter Horned from Red-necked Grebe (next page).
RANGE: Breeds on lakes and ponds. Winters mostly on salt water but also on ice-free lakes of eastern North America; a few winter inland in West. Casual in Newfoundland.

EARED GREBE
Podiceps nigricollis | L 12½" (32 cm)

Breeding adult has blackish neck, golden "ears" fanning out behind eye. In **winter** plumage, throat is variably dusky; cheek dark; whitish on chin extends up as a crescent behind eye; compare with Horned Grebe. Note also Eared Grebe's longer, thinner bill; thinner neck; more peaked crown. Lacks pale spot in front of eye. Generally rides higher in the water than Horned Grebe, exposing fluffy white undertail coverts. In flight, white secondaries show as white patch on trailing edge of wing.
RANGE: Usually nests in large colonies on freshwater lakes. Rare in eastern North America.

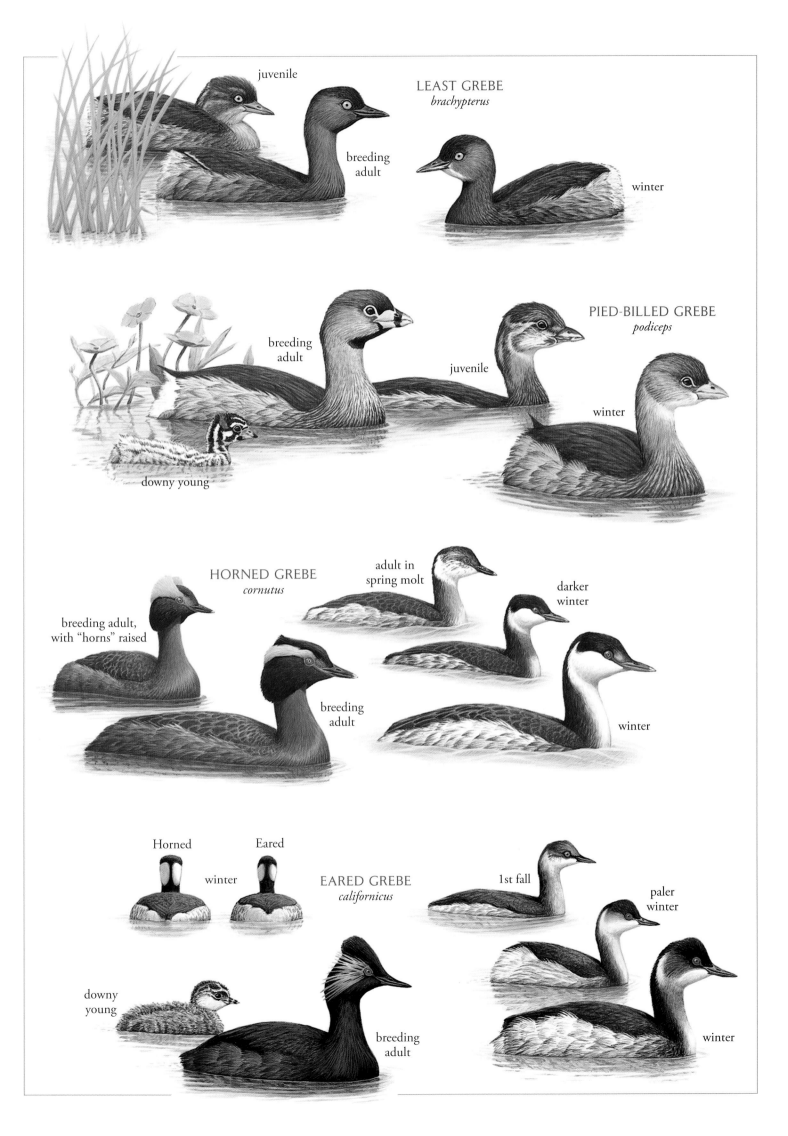

juvenile

LEAST GREBE
brachypterus

breeding
adult

winter

breeding
adult

PIED-BILLED GREBE
podiceps

juvenile

downy young

winter

HORNED GREBE
cornutus

adult in
spring molt

darker
winter

breeding adult,
with "horns" raised

breeding
adult

winter

Horned   Eared

winter

EARED GREBE
californicus

1st fall

paler
winter

downy
young

breeding
adult

winter

RED-NECKED GREBE

Podiceps grisegena | L 20" (51 cm)

Large grebe with heavy, tapered, yellowish bill almost as long as the head. **Breeding adult**'s whitish throat and cheeks contrast with reddish foreneck. In **winter** plumage, throat is dusky, white of chin extends onto rear of face in a crescent. **First-winter** bird has rounder head, paler eye; lacks strong facial crescent. **Juvenile** has striped head. In flight, Red-necked Grebe shows a white leading and trailing edge on inner wing; thick neck is often held slouched down. **Calls**, usually heard only on breeding grounds, include a *crick-crick* note and drawn out braying calls. Generally solitary.

RANGE: Breeds on shallow lakes; winters mostly along coasts. Rare in interior south of northern tier of states; occasionally, moderate numbers winter in mid-Atlantic region, especially in years when Great Lakes freeze. Casual south to Southwest and Gulf Coast states.

CLARK'S GREBE

Aechmophorus clarkii | L 25" (64 cm)

Resembles Western Grebe but bill is orange; back and flanks are paler; black cap does not extend to eye in **breeding** plumage; **downy young** are paler. In **winter adult**, lore region acquires more dark color, pattern looks more like Western; best distinction then is bill color. In flight, Clark Grebe's white wing stripe is more extensive than on Western. **Call** is a single, two-syllabled, upslurred *kree-eek* note.

RANGE: Limits of range in both species are not well known; Clark's occupies same general area and habitat as Western but is much less common in northern and eastern part of range. Accidental to eastern North America. Formerly considered one species with Western.

WESTERN GREBE

Aechmophorus occidentalis | L 25" (64 cm)

Large, gregarious grebe, strikingly black and white, with a long, thin neck and long bill. Resembles Clark's Grebe but bill is yellow-green; black cap extends to include eyes; back and flanks are darker; **downy young** are darker. In **winter adult**, lore region acquires more whitish color, and pattern can closely resemble winter Clark's. In flight, Western Grebe's white wing stripe is less extensive than Clark's. **Call** is a loud, two-note *crick-kreek*.

RANGE: Nests in reeds along broad, freshwater lakes. Winters on seacoasts and sheltered bays and large inland bodies of water. Occupies same general range and habitat as Clark's but greatly predominates in northern and eastern part of range. Casual during migration and winter to eastern North America. Formerly considered one species with Clark's; hybrids are regularly noted.

Western

winter birds
(for comparison)

Horned

Red-necked

RED-NECKED
GREBE
holboellii

1st winter

juvenile

winter
adult

breeding
adult

CLARK'S GREBE
transitionalis

winter
adult

breeding adult with
downy young

WESTERN GREBE
occidentalis

winter
adult

courtship
display

breeding adult
with downy young

ALBATROSSES (Family Diomedeidae)

Gliding on extremely long narrow wings, these largest of seabirds spend most of their lives at sea, alighting on the water when becalmed or when feeding on squid, fish, and refuse. Pelagic; most species nest in colonies on oceanic islands; pairs mate for life. A number of species, especially those in Southern Hemisphere, are threatened by long-line fishing. **Species:** *13 World, 8 N.A.*

SHORT-TAILED ALBATROSS
Phoebastria albatrus | **E** | L 36" (91 cm) WS 85–91" (215–230 cm)

Large size, but best field mark is the long and massive pink (initially dark on very young juveniles) bill with pale bluish tip. Dark humerals and pale feet distinctive in postjuvenal plumages. **Adult** is mostly white, with golden wash on head. **Juvenile** is blackish brown, except for traces of white below and behind eye and on chin. Older juvenile has more white around bill; compare with Black-footed Albatross. **Subadult** shows white forehead and face, dark cap; acquires white patches on scapulars and inner secondary coverts; with age becomes progressively white, but retains dark hindneck. Full adult plumage takes more than a decade to acquire, but can breed in subadult plumages.

RANGE: Common until end of 19th century, then decimated and was on verge of extinction by 1930s. Very small numbers of breeders reappeared on Torishima Island, beginning in 1951. Now protected and population slowly recovering—global population believed to number about 2,000. Presently breeds on Torishima (most) and Minami-kojima Islands off southern Japan. Still very rare to rare, but sightings are increasing in North Pacific off North America from Aleutians to central California. The great majority of these sightings are of juveniles and subadults.

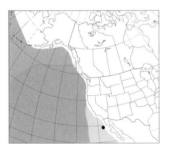

LAYSAN ALBATROSS
Phoebastria immutabilis | L 32" (81 cm) WS 77–80" (195–203 cm)

Back, upperwing blackish brown, except for white flash in primaries; underwings with black margins and variable internal markings. Note blurry dark mark surrounding eye; pinkish bill. Adults and juveniles similar. Occasionally hybridizes with Black-footed Albatross.

RANGE: Most numerous spring through summer off Alaska; rare to uncommon off West Coast from late fall through spring; casual inland in winter and spring; most records from southeastern California in spring. Breeds mainly on Hawaiian Islands; small colonies recently established in the Revillagigedo Archipelago and Isla Guadalupe, Mexico.

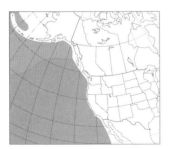

BLACK-FOOTED ALBATROSS
Phoebastria nigripes | L 32" (81 cm) WS 80" (203 cm)

Mostly dark in all plumages. White area around bill is more extensive on **old** birds; reduced or absent on immatures. Grayish black bill, sometimes paler with yellow or pink tones. Most birds of all ages have dark undertail coverts. Some **adults** have white undertail coverts, and white may extend onto belly; these birds can be confused with subadult Short-tailed Albatross, but lack white upperwing patches and have thinner, shorter, darker bills.

RANGE: Seen year-round off West Coast; most common in spring and summer. Breeds mainly on Hawaiian Islands; overall population declining.

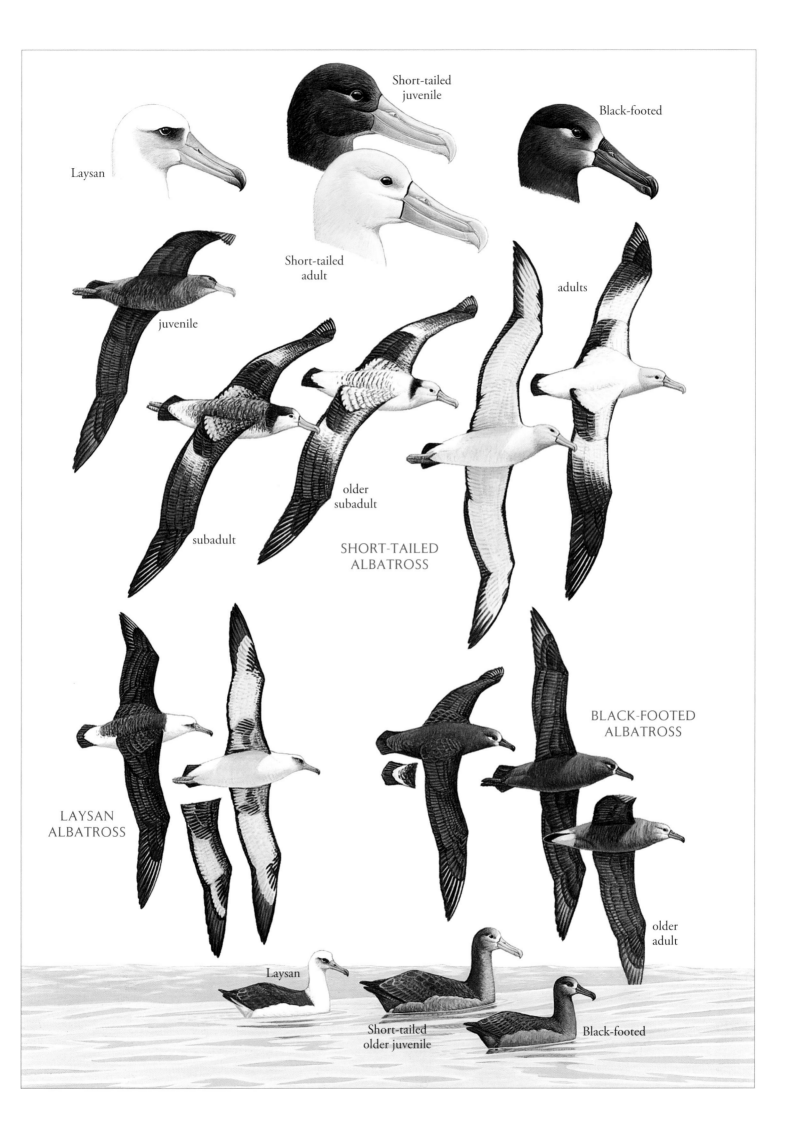

Laysan

Short-tailed
juvenile

Black-footed

Short-tailed
adult

juvenile

adults

subadult

older
subadult

SHORT-TAILED
ALBATROSS

BLACK-FOOTED
ALBATROSS

LAYSAN
ALBATROSS

older
adult

Laysan

Short-tailed
older juvenile

Black-footed

SHY ALBATROSS

Thalassarche cauta | L 35–39" (90–99 cm) WS 87–101" (220–256 cm)

Adults of *cauta/steadi* are white-headed, unlike all ages of *salvini,* which have a grayish washed head; younger *cauta/steadi* head also grayish. Distinctive in all ages is underwing pattern: primaries more extensively dark in *salvini* (and *eremita*); larger than Laysan, which has smaller, thinner, mostly pinkish bill. Note Shy's paler back, longer and grayer tail, more extensive white on the rump, and more languid flight style; all ages show extensively white underwings and characteristic dark "thumb mark" at beginning of leading edge. Yellow-tipped bill in *cauta* adults (*steadi* very similar); grayer and darker-tipped bill in all younger birds. Bill is dusky with a dull yellow ridge and dark lower tip in adult *salvini.*

RANGE: Casual off Pacific coast from Washington to northern California; once off Kasatochi Island, Aleutians. Only specimen (nominate *cauta*) was off Washington in 1951; nine other records since 1996. Taxonomic opinions differ, but presently divided into four subspecies breeding in the following locations: *cauta* breeds on islands off Tasmania; nearly identical *steadi* breeds on Auckland, Antipodes, and Chatham Islands, New Zealand; *salvini* breeds on Snares and Bounty Islands off New Zealand, and on Iles Crozet in southwest Indian Ocean; and unrecorded *eremita* group (dark gray head) breeds on Chatham Island. Remaining nine North American records are divided between those of *cauta/steadi* and those believed to be of *salvini* (including the Aleutians record).

YELLOW-NOSED ALBATROSS

Thalassarche chlororhynchos | L 32" (81 cm) WS 80" (203 cm)

Confused with Black-browed, but Yellow-nosed is smaller and slimmer, with longer neck. **Adult**'s bill appears black; at close range, yellow ridge on top and reddish tip are visible. Note light grayish wash on head and blackish triangular patch in front of eye of nominate race. Underwing extensively white with a narrow dark border; some **juveniles** show more dark on leading edge, which can cause confusion with adult Black-browed. Otherwise juvenile resembles adult, except for all-dark bill and reduced eye patch.

RANGE: Casual off Atlantic and Gulf coasts; a few inland sightings in East. Most of nominate subspecies breed on Tristan de Cunha and Gough Islands in south Atlantic, the probable source of North American records. The other subspecies, *bassi,* breeding on south Indian Ocean islands, has purer white head and reduced dark smudge in front of eye.

BLACK-BROWED ALBATROSS

Thalassarche melanophris | L 35" (89 cm) WS 88" (224 cm)

From similar Yellow-nosed, note larger size with thicker neck and chunkier body. **Adult** has broad dark leading edge to underwing and heavier orange bill with redder tip; black eyebrow. **Juveniles** have darker bills and gray shading about head and neck forming collar; underwings show limited white, can appear all-dark at a distance. **Subadults** have more yellowish bill with dark tip and some white in underwings.

RANGE: A circumpolar Southern Hemisphere species. Casual in North Atlantic, most recorded in northeast portion. One certain record of immature photographed off Virginia Beach on 6 Feb. 1999; about 20 other reports (some possibly correct).

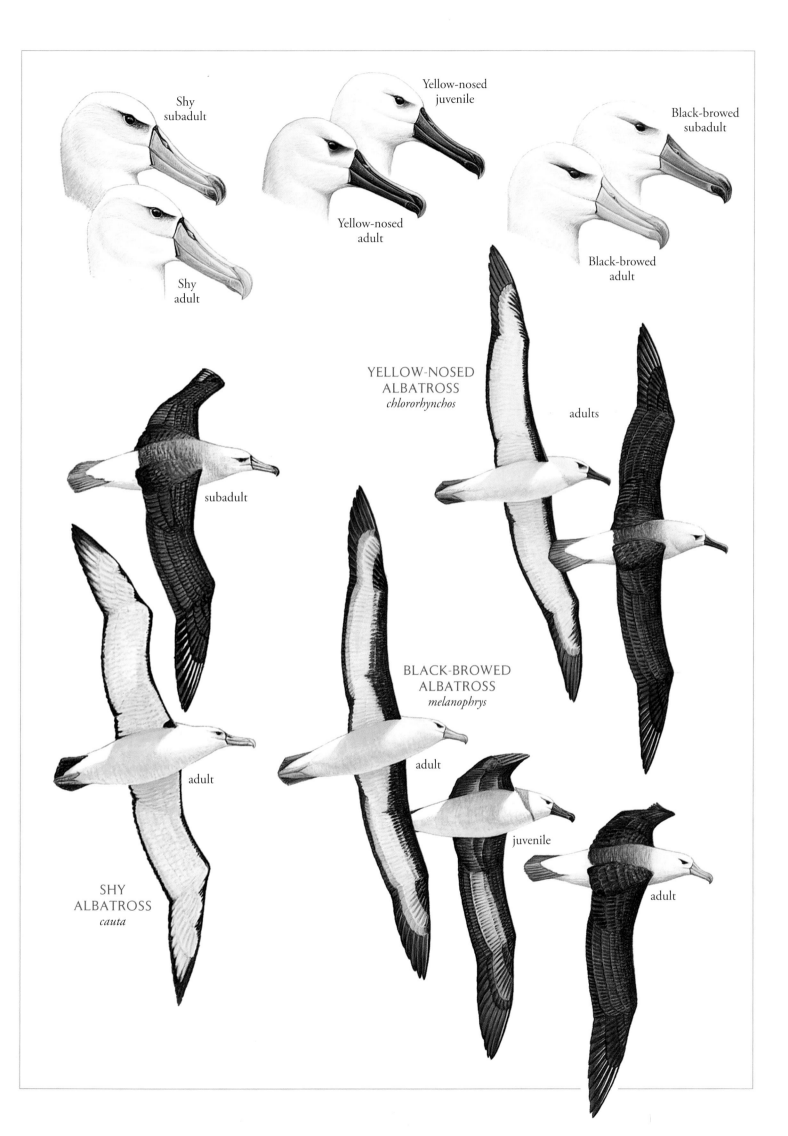

Shy
subadult

Shy
adult

Yellow-nosed
juvenile

Yellow-nosed
adult

Black-browed
subadult

Black-browed
adult

YELLOW-NOSED
ALBATROSS
chlororhynchos

adults

subadult

BLACK-BROWED
ALBATROSS
melanophrys

adult

adult

juvenile

adult

SHY
ALBATROSS
cauta

SHEARWATERS AND PETRELS (FAMILY PROCELLARIIDAE)

Pelagic seabirds rarely seen from shore; bills have nostril tubes. Fly with rapid wingbeats, stiff-winged glides.
Species: *75 World, 28 N.A.*

NORTHERN FULMAR
Fulmarus glacialis | L 19" (48 cm) WS 42" (107 cm)

Light morphs predominate over much of North Atlantic. In Pacific light morphs predominate in Bering Sea; **dark morphs** farther south. Pacific subspecies (*rodgersii*) has a more slender bill; shows greater variation in color morphs (darker dark morphs and paler lights), and darker tail contrasts with rump. **Intermediates** of all shades are frequent. Distinguished from gulls by short, thick bill with nostril tubes, and shearwater-like flight; from shearwaters by thick, yellow bill, stockier shape, and rounder wings.

RANGE: Northern Fulmars are common and increasing. Within winter range, numbers fluctuate annually; some summer.

PARKINSON'S PETREL
Procellaria parkinsoni | L 18" (46 cm) WS 45" (115 cm)

Almost wholly blackish petrel; rather thick yellowish bill, with black lines and tip; feet and legs dark. Compare to similarly colored Flesh-footed Shearwater (page 90), note bill shape and color. Westland Petrel (*P. westlandica*) from South Pacific is similar but is even larger and thicker billed; unrecorded north of Equator.

RANGE: Breeds on islands off New Zealand; ranges north in austral winter to east-central Pacific, north to southern Mexico. One certain record (1 Oct. 2005, about 18 miles off Point Reyes, California).

GADFLY PETRELS

Fast-flying petrels with arcing, acrobatic flight in high winds. Unlike shearwaters, they typically hold their wings slightly forward from the shoulder and bent sharply back at the "wrist."

GREAT-WINGED PETREL
Pterodroma macroptera | L 16" (41 cm) WS 38" (97 cm)

Similar to Murphy's, but browner (less gray); more uniform underwing; white more evenly distributed around bill; and dark legs and feet.

RANGE: Southern oceans species; two records (late summer, Oct.) off central California of presumed *gouldi* subspecies with whiter face.

MURPHY'S PETREL
Pterodroma ultima | L 16" (41 cm) WS 38" (97 cm)

Dark brownish gray color with faint dark M pattern on back, wedge-shaped tail, and white underwing flash; white most conspicuous below bill; legs and feet pink. The similar but larger Solander's Petrel (*P. solandri*), reported off California and Washington (no accepted records), has heavier bill with equal or more white above bill than below; prominent dark primary covert tips. Compare to dark morph Northern Fulmar.

RANGE: Breeds on remote central South Pacific islands. Fairly common spring visitor far off California coast; recorded at least north to Washington.

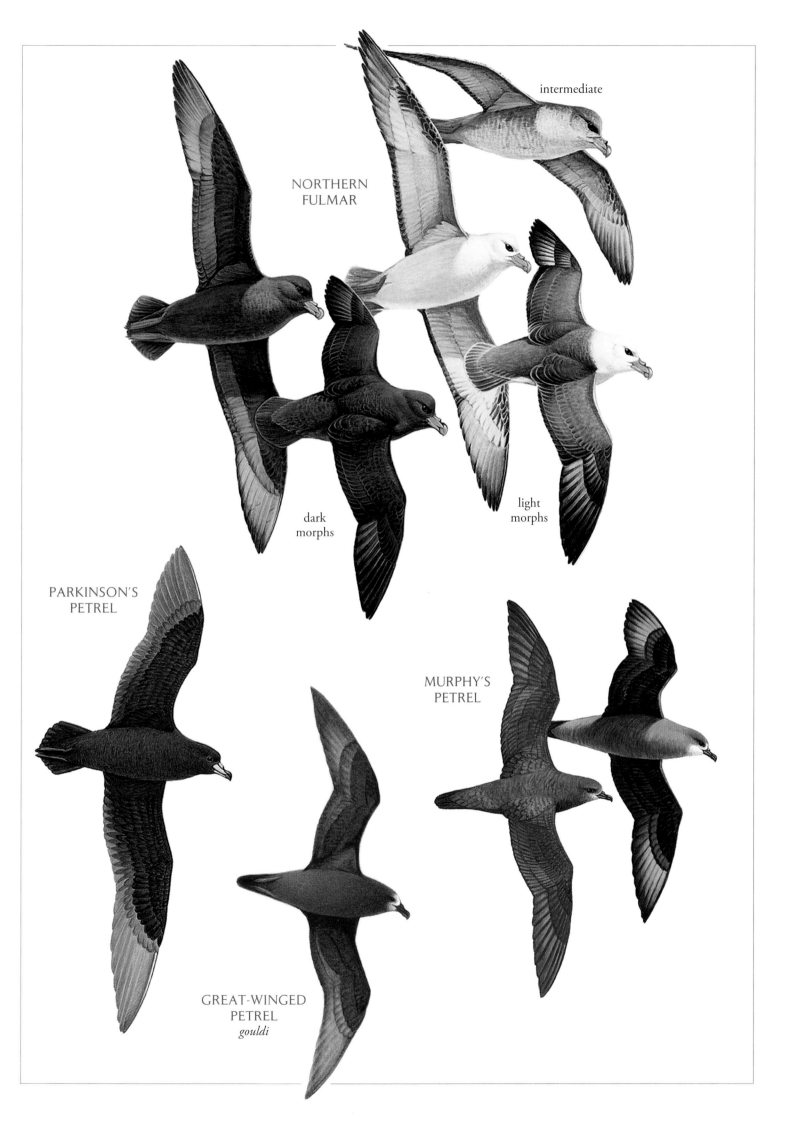

NORTHERN
FULMAR

intermediate

dark
morphs

light
morphs

PARKINSON'S
PETREL

MURPHY'S
PETREL

GREAT-WINGED
PETREL
gouldi

HAWAIIAN PETREL

Pterodroma sandwichensis | **E** | L 17" (43 cm) WS 39" (98 cm)

Recently split from Galapagos Petrel (*P. phaeopygia*), which together were formerly known as Dark-rumped Petrel. The two are not known to be separable with certainty in the field. Hawaiian is heavier than Galapagos Petrel, with shorter wings and shorter, deeper bill. Galapagos often has a freckled rather than pure white forehead. Both species have mostly black crown that extends down sides of neck forming a partial collar. Both uniformly dark above, but may show a slight M pattern and white on the upper tail coverts. Note underwing pattern. The endangered Hawaiian Petrel nests only on Hawaiian Islands.
RANGE: Casual, recorded off California and Oregon from May to Oct., over 20 records, but no specimens.

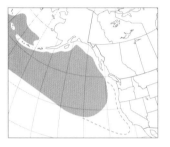

MOTTLED PETREL

Pterodroma inexpectata | L 14" (36 cm) WS 32" (81 cm)

White throat, breast, and vent contrast with rest of mostly gray underparts. Shows prominent black bar on otherwise white underwings; dark M across upperwings.
RANGE: Breeds on islands off New Zealand. Probably regular well off southern Alaska and Aleutians in summer; rare and irregular well off the West Coast, chiefly in late fall. Accidental in western New York (Livingston County, early Apr. 1880).

COOK'S PETREL

Pterodroma cookii | L 10¾" (27 cm) WS 26" (66 cm)

This and Stejneger's considered part of the *Cookilaria* petrel subgenus. They are smaller sized and more acrobatic than the larger *Pterodroma* petrels. All show a dark M across the dorsal surface. Cook's is small, with long wings and rather short tail. Crown and back uniformly gray; blackish eye patch. Note mainly white underwings and distinct dark M across upperwings. White on outer tail feathers and dark tip to central tail feathers can be hard to see. Two other *Cookilaria* petrels are very similar: the thicker-billed Masatierra or De Filippi's Petrel (*P. defilippiana*) from off west coast of South America and smaller Pycroft's Petrel (*P. pycrofti*) from New Zealand have not been recorded in North Pacific.
RANGE: Breeds on islands off New Zealand. Found well off California coast from spring through late fall, where it is the most numerous *Pterodroma* petrel. Casual off Aleutians and on the Salton Sea in summer.

STEJNEGER'S PETREL

Pterodroma longirostris | L 11" (28 cm) WS 23" (58 cm)

Resembles Cook's Petrel but distinct, dark half hood contrasts with grayish back and extensive white forehead; tail is longer and more uniformly colored, with less white in outer tail feathers; wings are shorter.
RANGE: Breeds on Juan Fernandez Islands off Chile. Casual well off the California coast, chiefly in fall. Accidental coastal Texas (Port Aransas, 15 Sept. 1995, tideline corpse).

HAWAIIAN
PETREL

MOTTLED
PETREL

COOK'S
PETREL

STEJNEGER'S
PETREL

BLACK-CAPPED PETREL

Pterodroma hasitata | L 16" (41 cm) WS 37" (94 cm)

Distinct dark cap; white collar; broad white band on uppertail coverts and base of tail. White wing lining, with variable dark diagonal bar on leading edge. Wing and bill shape, white forehead, broader band on tail, and languid, arching flight distinguish this species from Greater Shearwater (page 88). Some birds have less white at base of tail and duskier collar.

RANGE: Breeds on Hispaniola and Cuba. Common in Gulf Stream off North Carolina from late May to mid-Oct.; uncommon in winter; casual north to Nova Scotia and in Gulf of Mexico. Recorded inland in the East after hurricanes.

FEA'S PETREL

Pterodroma feae | L 14" (36 cm) WS 37" (94 cm)

Sightings (supported by excellent photos but no specimens) here believed to be Fea's, not Zino's Petrel (*P. madeira,* a highly endangered species breeding only on Madeira), based on larger bill size and likelihood. Fea's is brownish gray above with dark M pattern, pale uppertail coverts and tail. White below; partial breast band; mostly dark underwings.

RANGE: Breeds in the Madeira and Cape Verde Islands off West Africa. Rare visitor off North Carolina in late May and early June, casual into the fall; accidental to Nova Scotia.

BERMUDA PETREL

Pterodroma cahow | **E** | L 15" (38 cm) WS 35" (89 cm)

Endangered species. Believed extinct, but rediscovered in 1951; population increasing, now estimated at about 200. Note that larger Black-capped Petrel has heavier bill and disproportionately shorter wings. Bermuda Petrel's whitish rump, sometimes lacking, is restricted to base of uppertail coverts. Without Black-capped's white collar, dark on head is more like cowl than cap; Bermuda more buoyant in flight, with darker underwings than Black-capped.

RANGE: Nests only on islets off Bermuda; since mid-1990s, about 20 well-documented records, in late spring (mostly) and summer off Outer Banks, North Carolina.

HERALD PETREL

Pterodroma arminjoniana | L 15½" (39 cm) WS 37½" (95 cm)

Long-winged, slender-bodied species with languid wingbeats. Occurs in three morphs; **dark morph**, predominant in U.S., has pale-based flight feathers and greater primary coverts on underwing. Compare to chunkier Sooty Shearwater with its shorter tail, thinner bill, paler underwings, and faster wingbeats. **Light morph** has brownish gray head and chest, variably whitish throat, white belly, pale underwings. **Intermediate morphs** are variably mottled below.

RANGE: Tropical Southern Hemisphere species. South Atlantic birds of the nominate race nest on islands off Brazil; mid-Atlantic region sightings likely from these populations. Rare visitor late May to late Sept. in Gulf Stream off North Carolina. Accidental inland after hurricanes.

typical

darker
variant

BLACK-CAPPED
PETREL

Black-capped
Petrel

Greater
Shearwater

FEA'S
PETREL

BERMUDA
PETREL

darker
rump

intermediate
morph

HERALD PETREL

light
morph

dark
morph

Herald
Petrel

Sooty
Shearwater

dark
morph

CORY'S SHEARWATER

Calonectris diomedea | L 18" (46 cm) WS 46" (117 cm)

Large shearwater; grayish brown upperparts merge into white underparts without sharp contrast; bill is yellowish. Flight is more languid than most other shearwaters. Underside of primaries darker in *borealis*.

RANGE: Nominate *diomedea* breeds in the Mediterranean; slightly larger *borealis* breeds on the Azores, Canary, and Salvage Islands. Both occur (*borealis* much more common) off East Coast, mainly late spring through fall. Uncommon in Gulf of Mexico.

CAPE VERDE SHEARWATER

Calonectris edwardsii | L 15½" (39 cm) WS 39½" (101 cm)

Much smaller, slightly longer tailed than Cory's with darker gray upperparts and more contrasting face; much slimmer bill is olive-gray, not yellowish.

RANGE: Breeds Cape Verde archipelago, ranges to West African coast. Accidental off Hatteras Inlet, North Carolina (15 Aug. 2004).

GREATER SHEARWATER

Puffinus gravis | L 18" (46 cm) WS 44" (112 cm)

Dark brown cap contrasts with grayish brown upperparts and white cheeks. Rump usually shows a narrow, white, U-shaped band. Bill is dark; underparts white with indistinct dusky patch on belly. Many Greater Shearwaters have a white nape. Similar to Black-capped Petrel (page 86), but lacks white forehead and wide black bar on underwing; also has shorter, less wedge-shaped tail with less extensive white at base.

RANGE: Breeds in South Atlantic. Fairly common off East Coast during migration, chiefly in spring. Summers in large numbers from Gulf of Maine north. Casual off West Coast.

MANX SHEARWATER

Puffinus puffinus | L 13½" (34 cm) WS 33" (84 cm)

Blackish above, white below with white wing linings. Pure white undertail coverts extend to end of short tail. White wraps around dark ear coverts.

RANGE: Most breed on islands around United Kingdom; one small colony in Newfoundland. Winters off eastern South America. Fairly common off the northern Atlantic coast in summer. Rare in winter from Maryland south. Rare in summer and fall off West Coast to Alaska.

AUDUBON'S SHEARWATER

Puffinus lherminieri | L 12" (31 cm) WS 27" (69 cm)

Dark brown above, white below, with long tail, dark undertail coverts (a few with pale ones).

RANGE: Breeds on Caribbean islands. Common off the southern Atlantic coast, chiefly from May through Oct. Rare in winter.

LITTLE SHEARWATER

Puffinus assimilis | L 11" (28 cm) WS 25" (64 cm)

Similar to larger Audubon's Shearwater, but has shorter tail, whiter underwings, and grayer two-toned upperwing; white undertail coverts; face whiter. Flies with rapid, stiff, shallow wingbeats.

RANGE: Casual, two fall specimens from Nova Scotia and South Carolina (*baroli,* breeds Azores); sight records from off North Carolina and Nova Scotia; photo record from Monterey County, California (29 Oct. 2003) possibly Southern Hemisphere subspecies.

CORY'S
SHEARWATER
borealis

diomedea

GREATER
SHEARWATER

Cory's

CAPE VERDE
SHEARWATER

Greater

Black-capped
Petrel

MANX
SHEARWATER

LITTLE
SHEARWATER
baroli

AUDUBON'S
SHEARWATER
lherminieri

WEDGE-TAILED SHEARWATER
Puffinus pacificus | L 18" (46 cm) WS 40" (101 cm)

Polymorphic species. Long tail held in a point; wedge shape visible only when fanned. Slender, grayish bill has darker tip. Head and upperparts of **light morph** grayish brown; mostly white with mottled brown below. Most sightings are of wholly brown **dark morph**, which has paler base to flight feathers. Languid flight with prolonged soaring on bowed wings angled forward to "wrist," then swept back.
RANGE: Species from warm waters of Pacific and Indian Oceans. Casual in summer and fall to waters off central California; accidental on Salton Sea.

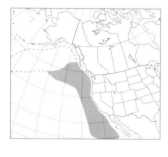

FLESH-FOOTED SHEARWATER
Puffinus carneipes | L 17" (43 cm) WS 41" (104 cm)

Dark above and below except for pale flight feathers, distinctive pale pink base of bill. Compare especially with Sooty Shearwater, which has whitish wing linings, all-dark bill, less languid flight.
RANGE: Breeds on islands off Australia and New Zealand. Winters (our summer) in North Pacific; rare off West Coast.

BULWER'S PETREL
Bulweria bulwerii | L 10" (26 cm) WS 26" (66 cm)

Sooty brown overall with pale diagonal bar across secondary coverts. Long tail usually held in a point; wedge shape visible only when fanned. Flight is buoyant and erratic, long wings slightly bowed and held forward. Flies within a few feet of the water; flight, when it is windy, is more like gadfly petrels.
RANGE: Bird of tropical and subtropical oceans; accidental summer visitor off Monterey, California, and Outer Banks, North Carolina.

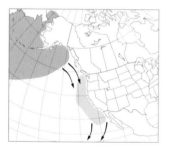

SHORT-TAILED SHEARWATER
Puffinus tenuirostris | L 17" (43 cm) WS 39" (99 cm)

Plumage variable. Usually dark overall, but often with pale wing linings like Sooty Shearwater; white is more evenly distributed, when present, forming a panel. Has shorter bill than Sooty, slightly steeper forehead, more rounded crown. Some birds, unlike Sooty, have pale throat and dark-capped appearance.
RANGE: Breeds off Australia. Winters (our summer) in North Pacific to Alaska. Seen along West Coast from British Columbia to California during southward migration in fall and winter.

SOOTY SHEARWATER
Puffinus griseus | L 18" (46 cm) WS 40" (101 cm)

Whitish underwing coverts contrast with overall dark plumage. Flies with fast wingbeats and, except when it is windy, short glides. Almost identical to Short-tailed Shearwater. White on underwings usually most prominent on primary coverts.
RANGE: Breeds in Southern Hemisphere. Fairly common off East Coast in spring, and in summer off New England and Canada. Abundant off West Coast, often seen from shore. Very rare in Gulf of Mexico.

dark
morph

dark
morph

WEDGE-TAILED
SHEARWATER

light
morph

FLESH-FOOTED
SHEARWATER

BULWER'S
PETREL

SHORT-TAILED
SHEARWATER

SOOTY
SHEARWATER

Short-tailed

Sooty

STORM-PETRELS (Family Hydrobatidae)

These small seabirds hover close to the water, pattering or hopping across the waves to pluck up small fish and plankton. Some species follow ships. Identification is often difficult. Flight behavior helps to distinguish the various species, but can vary deceptively depending on weather. **Species: 21 World, 12 N.A.**

EUROPEAN STORM-PETREL
Hydrobates pelagicus | L 5½–6½" (14–17 cm) WS 15" (38 cm)

Small, dark storm-petrel with white underwing bar; feet do not project beyond tail.
Range: Nests on islands in northeastern Atlantic and Mediterranean. Specimen record from Sable Island, Nova Scotia (Aug. 1970), and multiple recent records in late spring off Outer Banks, North Carolina.

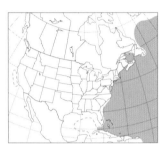

WILSON'S STORM-PETREL
Oceanites oceanicus | L 7¼" (18 cm) WS 16" (41 cm)

Flies with shallow, fluttery wingbeats. Wings short and rounded; long legs; in flight, feet trail behind tip of squarish or rounded tail. Often hovers to feed, pattering its yellow-webbed feet on the water. Bold white, U-shaped rump band extends onto undertail coverts; visible even on sitting bird. Smaller than Leach's and Band-rumped Storm-Petrels.
Range: Common off Atlantic coast; rare off eastern Gulf Coast and California (fall).

BAND-RUMPED STORM-PETREL
Oceanodroma castro | L 9" (23 cm) WS 17" (43 cm)

Rather shallow wingstrokes are followed by stiff-winged glides, like the flight of a shearwater but unlike the erratic flight of Leach's Storm-Petrel or the fluttery flight of Wilson's. White rump patch narrower and less extensive on undertail coverts than on Wilson's. Tail squarish or very slightly notched; no foot projection in flight. Larger, longer winged than Wilson's; darker than Leach's, with fainter carpal bar.
Range: Breeds on tropical islands. Fairly common from late May to late Aug. in Gulf Stream, especially off North Carolina; rare or casual farther north to waters off Massachusetts; uncommon well off Gulf Coast.

LEACH'S STORM-PETREL
Oceanodroma leucorhoa | L 8" (20 cm) WS 18" (46 cm)

Distinctive erratic flight, with deep strokes of long, pointed wings. In close view, note dusky line dividing white rump band on most birds. No white is visible on flanks of sitting bird. Brown overall, with pale wing stripes and forked tail. Amount of white on rump varies; a few birds seen off southern California have dark rumps.
Range: Fairly common well off Pacific coast; uncommon south of breeding range along Atlantic coast. Very rare in Gulf of Mexico; casual on Salton Sea.

WHITE-FACED STORM-PETREL
Pelagodroma marina | L 7½" (19 cm) WS 17" (43 cm)

Flies with stiff, shallow wingbeats and short glides. Habitually angles to water at 45 degrees, then bounces off the surface with long legs in a pogo-stick fashion. Distinctive white underparts, wing linings, and face. Dark eye stripe, crown, and upperparts; paler rump.
Range: Very rare off the Atlantic coast from North Carolina to Massachusetts in late summer.

EUROPEAN
STORM-PETREL

WILSON'S
STORM-PETREL

BAND-RUMPED
STORM-PETREL

LEACH'S
STORM-PETREL
leucorhoa

WHITE-FACED
STORM-PETREL

Leach's
West Coast

Wilson's

Band-rumped

northern

intermediate

southern

Wedge-rumped
kelsalli

BLACK STORM-PETREL
Oceanodroma melania | L 9" (23 cm) WS 19" (48 cm)

Deep, languid wingstrokes, graceful flight. Largest of the all-dark storm-petrels. Blackish brown overall with pale bar on upper surface of wing. Tail forked and fairly long. Slow, deep wingbeats and larger size distinguish Black Storm-Petrel from dark-rumped individuals of Leach's Storm-Petrel (preceding page).

RANGE: Breeds from May to Dec. off Baja California and in Gulf of California; small colony in vicinity of Santa Barbara Island. Fairly common off southern California coast from late spring; by late summer north to Monterey Bay. Casual in interior of southern California and in southwest Arizona after tropical storms.

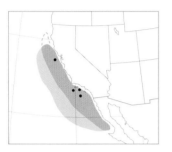

ASHY STORM-PETREL
Oceanodroma homochroa | L 8" (20 cm) WS 17" (43 cm)

Fluttery wingbeats, but flight fairly direct; not as swallowlike as Wilson's Storm-Petrel (preceding page). Gray-brown overall; pale mottling on underwing coverts may be visible at close range. Viewed from the side, Ashy appears to have long tail, unless in molt. Distinguished from Black Storm-Petrel by rapid, shallow wingbeats and overall paler, grayer appearance.

RANGE: Fairly common most of the year; rare in winter. Breeds on islands off central and southern California.

LEAST STORM-PETREL
Oceanodroma microsoma | L 5¾" (15 cm) WS 15" (38 cm)

Our smallest storm-petrel. Swift, indirect flight, low over the water, with deep wingbeats like the much larger Black Storm-Petrel. Blackish brown overall. Short tailed; appears almost tailless in flight; often confused with molting Ashies.

RANGE: Irregular; rare to common off the coast of southern California in late summer and fall; in peak years, small numbers occur to central California; after tropical storms, may be seen in southeast California and southwest Arizona.

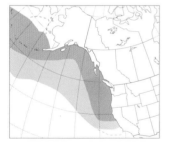

FORK-TAILED STORM-PETREL
Oceanodroma furcata | L 8½" (22 cm) WS 18" (46 cm)

Wingbeats shallow, rapid, often followed by glides. Looks fairly long-tailed in flight. Occasionally makes shallow dives for food. Distinctively bluish gray above, pearl gray below. Note also dark gray forehead and eye patch, dark wing linings. Nominate *furcata* from coastal northeast Asia and Aleutians is larger and paler than *plumbea* from off south-coastal to northern California.

RANGE: Found regularly south to central California; rare off southern California coast.

WEDGE-RUMPED STORM-PETREL
Oceanodroma tethys | L 6½" (17 cm) WS 13¼" (34 cm)

Distinctive bold white triangular patch of uppertail coverts gives the appearance of a white tail with dark corners. Compare with the rounded rump band and white flanks of Wilson's Storm-Petrel (preceding page). Wedge-rumped is almost as small as Least Storm-Petrel, with similar deep wingbeats.

RANGE: Breeds on the Galápagos Islands (nominate *tethys*) and on islets off Peru (much smaller *kelsalli*); disperses north to Panama. Casual off the California coast from Aug. to Jan.; single specimen (Jan.) is of *kelsalli*.

BLACK
STORM-PETREL

ASHY
STORM-PETREL

FORK-TAILED
STORM-PETREL

LEAST
STORM-PETREL

WEDGE-RUMPED
STORM-PETREL
kelsalli

FRIGATEBIRDS (Family Fregatidae)

These large, dark seabirds have the longest wingspan, in proportion to weight, of all birds.
Species: *5 World, 3 N.A.*

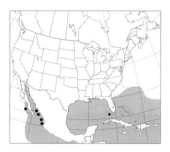

MAGNIFICENT FRIGATEBIRD
Fregata magnificens | L 40" (102 cm) WS 90" (229 cm)

Long, forked tail; long, narrow wings. **Male** is glossy black; orange throat pouch becomes bright red when inflated in courtship display. **Female** is blackish brown, with white at center of underparts. **Juveniles** show varying amount of white on head and underparts; require four to six years to reach adult plumage. Frigatebirds skim the sea, snatching up food from surface; also harass other birds in flight, forcing them to disgorge food.

RANGE: Generally seen along coast, but also casual inland, especially after storms. Breeds on Dry Tortugas off Florida. Very rare at Salton Sea and north along California coast, casual farther north; rare on East Coast north to North Carolina, casual farther north.

TROPICBIRDS (Family Phaethontidae)

Long central tail feathers identify adults. They are usually seen far out at sea.
Species: *3 World, 3 N.A.*

WHITE-TAILED TROPICBIRD
Phaethon lepturus | L 30" (76 cm) WS 37" (94 cm)

Smaller than Red-billed Tropicbird; distinctive black stripe on upperwing coverts; primaries show less black than in Red-billed. Bill orange in Atlantic subspecies (*catesbyi*) **adults**. **Juvenile** lacks tail streamers; upperparts are boldly barred, bill more yellowish.

RANGE: Tropical species, rare but regular in Gulf Stream off North Carolina in summer; casual on Dry Tortugas, Florida, and elsewhere off East Coast.

RED-BILLED TROPICBIRD
Phaethon aethereus | L 40" (102 cm) WS 44" (112 cm)

Flies with rapid, stiff, shallow wingbeats, unlike other tropicbirds, whose flight is more tern-like. **Adult** has red bill, black primaries, barring on back and wings, white tail streamers. **Juvenile** has black collar; lacks streamers; tail is tipped with black; barring on upperparts is finer than in other young tropicbirds. Also bill is yellowish, but soon becomes orange-red.

RANGE: Tropical species, rare well off southern California coast; very rare in Gulf of Mexico and off Atlantic coast to North Carolina; casual to Massachusetts.

RED-TAILED TROPICBIRD
Phaethon rubricauda | L 37" (94 cm) WS 44" (112 cm)

Broadest-winged tropicbird; flies with languid wingbeats. Flight feathers mostly white. **Adult** has red bill; red tail streamers, narrower than in other tropicbirds. **Juvenile**'s all-white tail lacks streamers; upperparts barred; bill black, gradually changing to yellow and then red. Note also lack of black collar on nape.

RANGE: Species from tropical and subtropical Pacific and Indian Oceans. Very rare, usually well off California coast.

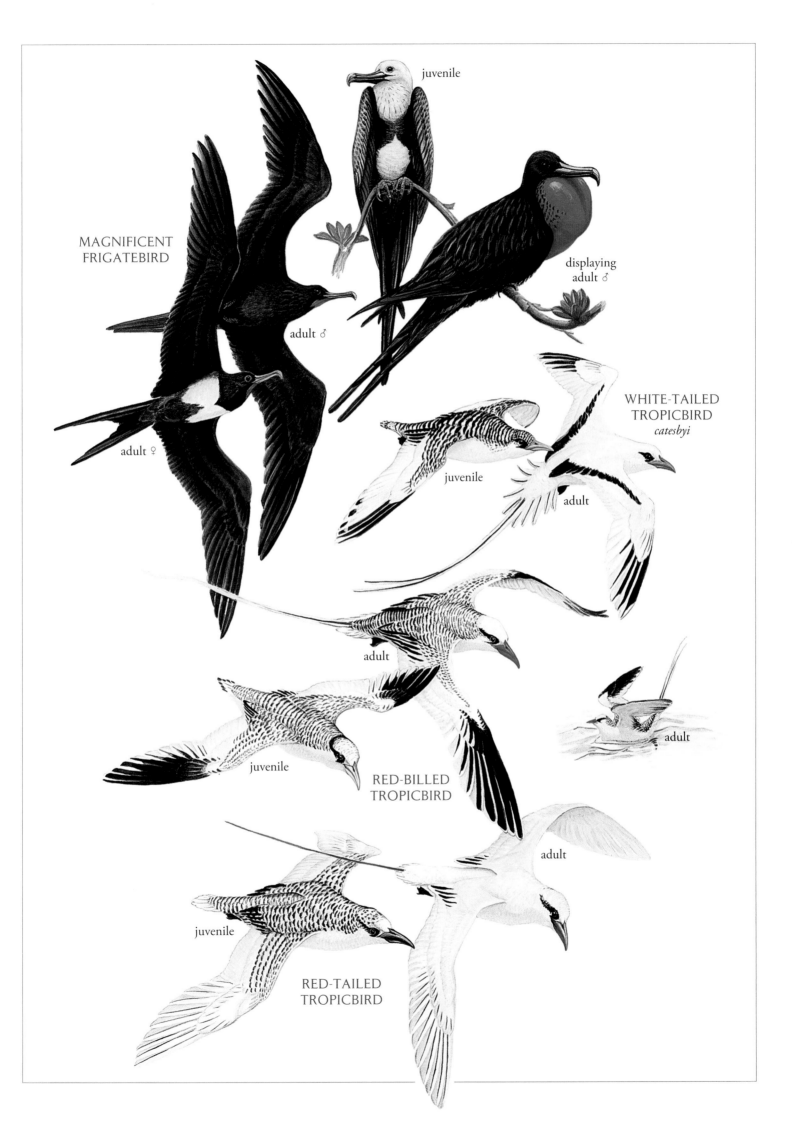

MAGNIFICENT
FRIGATEBIRD

juvenile

displaying
adult ♂

adult ♂

adult ♀

WHITE-TAILED
TROPICBIRD
catesbyi

juvenile

adult

adult

adult

juvenile

RED-BILLED
TROPICBIRD

adult

juvenile

RED-TAILED
TROPICBIRD

BOOBIES AND GANNETS (FAMILY SULIDAE)

High-diving seabirds that plunge into water. Gregarious, nesting in colonies on small islands. The rest of the year, gannets roost at sea, boobies primarily on land. **Species:** *10 World, 6 N.A.*

RED-FOOTED BOOBY
Sula sula | L 28" (71 cm) WS 60" (152 cm)

Smallest booby, with rounded head. All **adults** show bright coral red feet, and blue and pink at base of bill. Four principal morphs occur: **brown morph, white-tailed brown morph, white morph,** and **black-tailed white morph**; note that white morphs have black primaries, secondaries, and underwing median primary coverts. All **juveniles** and **subadults** are brownish overall with darker chest band, with mainly dark underwings and flesh pink legs and feet; dusky bill of juveniles becomes pinkish with black tip in subadult.
RANGE: Tropical species; casual on Florida's Dry Tortugas and California coast.

BROWN BOOBY
Sula leucogaster | L 30" (76 cm) WS 57" (145 cm)

Adults of nominate race and female *brewsteri* from western Mexico have dark brown heads and necks with sharply contrasting white bellies and underwing coverts; **adult male** *brewsteri* has white on head and neck. Adult female's bill, facial skin, legs, and feet are bright yellow; male's soft parts washed with grayish green, throat bluish. **Juveniles** are dark brown, with little or no contrast between breast and belly; underwing muted. **Subadults** show white on belly and sharp line of contrast with darker neck.
RANGE: A few found on Dry Tortugas and vicinity; rare north off both Florida coasts. Very rare through Gulf of Mexico; casual up Atlantic coast to Nova Scotia, and in California and the Southwest (*brewsteri*).

BLUE-FOOTED BOOBY
Sula nebouxii | L 32" (81 cm) WS 62" (158 cm)

Feet bright blue in adults, darker in young; long, attenuated bill is dark bluish gray. **Adults** have streaked heads, whitish patches on upper back and rump; central tail feathers mostly whitish. **Juvenile** has darker head and neck; compare with immature Masked Booby; note pale dorsal patches, all-dark underwing primary coverts.
RANGE: Breeds on islands in Gulf of California. Rare and irregular to inland California and southwest Arizona in late summer and fall (most to Salton Sea). Casual to California coast; accidental to Washington and Texas. Absent in U.S. most years.

MASKED BOOBY
Sula dactylatra | L 32" (81 cm) WS 62" (158 cm)

Proportionately, the shortest tailed booby. **Adult** distinguished from Northern Gannet (next page) by yellow bill and extensive black facial skin; black tail; and solid black trailing edge to wing. On **juvenile**, note more white on underwing, with contrasting dark median primary coverts and pale collar. **Subadult** has paler head and broader collar; note yellow on bill.
RANGE: Breeds on Dry Tortugas, Florida; uncommon in Gulf of Mexico in summer. Rare in Gulf Stream north to Outer Banks, North Carolina. Casual to coastal California, where it must be separated from Nazca Booby (page 467).

juvenile

white-morph
adults

RED-FOOTED
BOOBY

black-tailed
white-morph adult

white-tailed
brown-morph adult

brown-morph
subadult

brown-morph
adult

juvenile

adult ♂
brewsteri

BROWN
BOOBY
leucogaster

adult ♀

adult ♂

subadult ♀

juvenile

adult

BLUE-FOOTED
BOOBY
nebouxii

adults

juveniles

Northern Gannet
immature
(for comparison)

adult

juvenile

subadult

adult

MASKED
BOOBY
dactylatra

adult

subadult

juvenile

NORTHERN GANNET

Morus bassanus | L 37" (94 cm) WS 72" (183 cm)

Large, white seabird with long, black-tipped wings, pointed white tail. **Juvenile** is dark gray above, with pale speckling; grayish below. **First-year** birds are whiter below; distinguished from juvenile and immature Masked Booby (preceding page) by more uniformly dark underwings and, at close range, by different feathering pattern around the bill. Full **adult** plumage is acquired in three to four years.

RANGE: Common; breeds in large colonies on rocky cliffs; winters at sea. Often seen from shore during migration and winter. Casual in Great Lakes region in late fall.

PELICANS (FAMILY PELECANIDAE)

*These large, heavy waterbirds have massive bills and huge throat pouches used as dip nets to catch fish. In flight, pelicans hold their heads drawn back. **Species:** 7 World, 2 N.A.*

AMERICAN WHITE PELICAN

Pelecanus erythrorhynchos | L 62" (158 cm) WS 108" (274 cm)

White, with black primaries and outer secondaries. **Breeding adult** has pale yellow crest; bill is bright orange, usually with a fibrous plate on upper mandible. Plate is shed after eggs are laid; crown and nape become grayish. Juvenile is white with brownish wash on head, neck, and lesser coverts; soft parts more dully colored. White Pelicans do not dive for food but dip their bills into the water while swimming. Usually found in flocks.

RANGE: **Nonbreeding** birds are seen in summer throughout area enclosed by dashed line on map. In fall, vagrants may appear almost anywhere, increasingly in Northeast.

BROWN PELICAN

Pelecanus occidentalis | **E** | L 48" (122 cm) WS 84" (213 cm)

Nonbreeding adult has white head and neck, often washed with yellow; grayish brown body; blackish belly. In **breeding** bird, hindneck is dark chestnut; yellow patch appears at base of foreneck. On eastern race, *carolinensis* (shown here), gular pouch is grayish; breeding *californicus* from the West Coast has a bright red gular pouch. Molt during incubation and **chick feeding** produces speckled head and foreneck. Adult eye color is light except during chick feeding, when it darkens. Juvenile is grayish brown above, tipped with pale buff; underparts whitish. **Immatures** are browner; acquire adult plumage by third year. Dives from the air after prey, capturing fish in its pouch.

RANGE: Common on both coasts. Very rare, but widely recorded from interior U.S. and southern Ontario, inland except at Salton Sea, where may be common. Wanderers are seen mainly in spring and summer, to limit of dashed line on map, rarely to British Columbia and casually to Nova Scotia.

adult

NORTHERN
GANNET

juvenile

adult

2nd year

1st year

nonbreeding
adult

chick-feeding
adult

AMERICAN
WHITE PELICAN

immature

nonbreeding
adult

breeding
adult

chick-feeding
adult

BROWN
PELICAN
carolinensis

nonbreeding
adult

breeding
adult

subadult

immature

DARTERS (Family Anhingidae)

Long, slim neck helps to distinguish anhingas from cormorants. Anhingas often swim submerged to the neck. Sharply pointed bill is used to spear fish. **Species:** *2 World, 1 N.A.*

ANHINGA
Anhinga anhinga | L 35" (89 cm) WS 45" (114 cm)

Black above, with green gloss; silvery white spots and streaks on wings and upper back. During breeding season, **male** acquires pale, wispy plumes on upper neck; bill and bare facial skin become brightly colored. **Female** has buffy neck and breast. Immatures resemble adult female but are browner overall. Anhingas prefer freshwater habitats; often seen perched on branches or stumps with wings spread. In flight, profile looks headless. Flies with slow, regular wingbeats and circles like raptor on thermals.
Range: Casual wanderer north of breeding range.

CORMORANTS (Family Phaethontidae)

Dark birds with set-back legs; long, hooked bill; and colorful bare facial skin and throat pouch. Dive from the surface for fish. May briefly soar; may swim partially submerged. **Species:** *36 World, 6 N.A.*

NEOTROPIC CORMORANT
Phalacrocorax brasilianus | L 26" (66 cm) WS 40" (102 cm)

Small, long-tailed cormorant with white-bordered, yellow-brown or dull yellow throat pouch that tapers to a sharp point behind bill. In **breeding** plumage, **adult** acquires short white plumes on sides of neck. Distinguished from Double-crested Cormorant (next page) by smaller size, longer tail, and smaller, angled throat pouch that does not extend around eye. Neotropic **juveniles** are overall browner than adults, particularly on underparts.
Range: Fairly common; found at marshy ponds or shallow inlets near perching stumps and snags. Regular to Arizona; casual to southeast California, Colorado, and the western Midwest; accidental mid-Atlantic region. Formerly called Olivaceous Cormorant.

GREAT CORMORANT
Phalacrocorax carbo | L 36" (91 cm) WS 63" (160 cm)

Large, short-tailed cormorant with small, lemon yellow throat pouch broadly bordered with white feathering. **Breeding adult** shows white flank patches and wispy white plumes on head. Smaller Double-crested Cormorant (next page) has orange throat pouch; lacks flank patches; note also Great Cormorant's larger, blockier head and heavier bill. **First-year** birds are brown above; white belly contrasts with streaked brown neck, breast, and flanks. **Second-year** immatures resemble nonbreeding adults more closely but have a brown tinge above; compare with young Double-crested, which has a slimmer bill, deep orange facial skin, and, often, a darker belly.
Range: Winters in small numbers regularly south to South Carolina, very rarely as far south as Florida. Casual on Lake Ontario.

breeding
adult ♂

ANHINGA
leucogaster

♀

♀

Great adult

adult

Double-crested
Cormorants (for comparison)

eastern
breeding
adult

Neotropic
juvenile

NEOTROPIC
CORMORANT
mexicanus

nonbreeding
adult

juvenile

breeding
adult

2nd year

GREAT
CORMORANT
carbo

breeding
adult

1st year

PELAGIC CORMORANT
Phalacrocorax pelagicus | L 26" (66 cm) WS 39" (99 cm)

Adults are dark and glossy overall; bill dark. Smaller and slenderer than other western cormorants. **Breeding adult** has tufts on crown and nape; fine white plumes on sides of neck; white patches on flanks. Distinguished from Red-faced Cormorant by darker and less extensive red facial skin and lack of yellow in bill. **Juvenile** is uniformly dark brown; closely resembles young Red-faced, but note dark bill and smaller size. Pelagic Cormorant is distinguished in flight from Brandt's by smaller head, slimmer neck, smaller overall size, and disproportionately longer tail. Its bill is thinner and different in color than Red-faced Cormorant.
RANGE: Less gregarious than other species; breeds in smaller colonies. Despite name, rare out over open ocean.

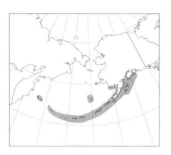

RED-FACED CORMORANT
Phalacrocorax urile | L 31" (79 cm) WS 46" (117 cm)

Heavier and paler bill distinguishes all ages from Pelagic Cormorant. In **adult**, dull brown wings contrast with glossy upperparts; more uniform in Pelagic. Extensive yellow on bill with bluish at base. Facial skin is red and becomes enlarged and brighter in the **breeding** season. **Juvenile** is uniformly dark brown, with pale yellowish gray bill.
RANGE: More gregarious than Pelagics, Red-faced Cormorants nest in colonies on the ledges of steep coastal cliffs and on rocky sea islands, alongside gulls, murres, and auklets.

BRANDT'S CORMORANT
Phalacrocorax penicillatus | L 35" (89 cm) WS 48" (122 cm)

A band of pale buffy feathers bordering the throat pouch identifies all ages. Throat pouch becomes bright blue in **breeding** plumage; head, neck, and scapulars acquire fine, white plumes. **Juvenile** is dark brown above, slightly paler below. In all ages, appears more uniformly dark above than Double-crested Cormorant; wings and tail are shorter. Head and bill are larger than in Pelagic Cormorant.
RANGE: Common and gregarious; often fishes in large flocks; flies in long lines between feeding and roosting grounds. Much more likely to be seen over open ocean than Pelagic Cormorant. A tiny breeding colony may still exist on Seal Rocks near Prince William Sound, Alaska; very rare in southeast Alaska.

DOUBLE-CRESTED CORMORANT
Phalacrocorax auritus | L 32" (81 cm) WS 52" (132 cm)

Large, rounded throat pouch is yellow-orange year-round. **Breeding adult** has a tuft curving back on either side of its head from behind eyes. Tufts are largely white in western birds, black and less conspicuous in eastern birds (see figure preceding page). **Juvenile** is brown above, variably pale below, but usually palest on upper breast and neck. Immatures sometimes have pouch edged with white, which can cause confusion with Neotropic Cormorant (preceding page). Among West Coast cormorants, Double-crested's kinked neck is distinctive in flight; its wings are also longer and more pointed than Brandt's and Pelagic.
RANGE: Common and widespread; found along coasts, inland lakes, and rivers. Breeding populations in the interior have greatly increased in the last three decades.

winter adult

juvenile

breeding adult

PELAGIC CORMORANT

RED-FACED CORMORANT

winter adult

juvenile

breeding adult

Red-faced

Pelagic

breeding adults

Brandt's

Double-crested

BRANDT'S CORMORANT

juvenile

winter adult

breeding adult

DOUBLE-CRESTED CORMORANT

juvenile

winter adult

western breeding adult

2nd year

1st year

1st year

HERONS, BITTERNS, AND ALLIES (Family Ardeidae)

Wading birds; most have long legs, neck, and bill for stalking food in shallow water.
Graceful crests and plumes adorn some species in breeding season. **Species:** *65 World, 18 N.A.*

AMERICAN BITTERN
Botaurus lentiginosus | L 28" (71 cm) WS 42" (107 cm)

Mottled brown upperparts and brownish neck streaks. Contrasting dark flight feathers are conspicuous in flight; note also that wings are longer, narrower, and more pointed, not rounded as in night-herons. **Juvenile** lacks neck patches. Distinctive spring and early summer **song**, *oonk-a-lunk*, is most often heard at dusk in dense marsh reeds. When alarmed, freezes with bill pointing up, or flushes with rapid wingbeats.
RANGE: Fairly common; casual breeder south of range.

LEAST BITTERN
Ixobrychus exilis | L 13" (33 cm) WS 17" (43 cm)

Buffy inner wing patches identify this small, rather secretive heron as it flushes briefly from dense marsh cover. When alarmed, it may freeze with bill pointing up. In **male** back and crown are black; in **female** they are browner. **Juvenile** resembles female but has more prominent streaking on back and breast. Rare dark morph, "**Cory's Least Bittern**" not documented since mid-20th century, is chestnut where typical plumage is pale. Least Bittern's **calls** include a series of harsh *kek* notes; **song**, a softer series of *ku* notes, is heard only on the breeding ground.
RANGE: Fairly common. May breed sporadically beyond mapped range in West.

YELLOW-CROWNED NIGHT-HERON
Nyctanassa violacea | L 24" (61 cm) WS 42" (107 cm)

Adult has buffy white crown, black face with white cheeks; acquires head plumes in breeding season. **Juvenile** told from young Black-crowned Night-Heron by grayer upperparts with less conspicuous white spotting; longer neck; stouter, mostly dark bill; and larger eyes. In flight, its legs extend well beyond its tail and it shows darker flight feathers and trailing edge on wings. Full adult plumage is acquired in third year. **Calls** include a short *woc*, higher and less harsh than in Black-crowned.
RANGE: Uncommon to fairly common; roosts in trees in wet woods and swamps. Casual in California and north to dashed line on map; casual also in Newfoundland.

BLACK-CROWNED NIGHT-HERON
Nycticorax nycticorax | L 25" (64 cm) WS 44" (112 cm)

Stocky heron with short neck and legs. **Adult** has black crown and back; white hindneck plumes are longest in breeding season. **Juvenile** distinguished from young Yellow-crowned Night-Heron by browner upperparts with bolder white spotting; thicker neck; paler, less contrasting face with smaller eyes; and longer, thinner bill with mostly pale lower mandible. In flight, legs barely extend beyond tail. Full adult plumage is not acquired until third year. **Calls** include a low, harsh *woc*, more guttural than in Yellow-crowned. Mainly nocturnal feeder. Typically roosts in trees.
RANGE: Fairly common to common. Occurs very rarely north of mapped range.

juvenile

adults

AMERICAN
BITTERN

high breeding
adult ♂

juvenile

"Cory's"
adult ♂

adult ♂

adult ♀

adult ♂

juvenile

LEAST
BITTERN
exilis

juvenile

1st spring

adult

adult

juvenile

high breeding
adult

YELLOW-CROWNED
NIGHT-HERON
violacea

adult

juvenile

1st spring

2nd
spring

juvenile

BLACK-CROWNED
NIGHT-HERON
hoactli

adult

high breeding
adult

GREEN HERON

Butorides virescens | L 18" (46 cm) WS 26" (66 cm)

Small, chunky heron with short legs. Back and sides of **adult**'s neck are deep chestnut; green on upperparts is mixed with blue-gray; center of throat and neck white. Greenish black crown feathers, sometimes raised to form shaggy crest. Legs are usually dull yellow but in male turn bright orange in high breeding plumage. **Juvenile** is browner above; white throat and underparts heavily streaked with brown. Common **call** is a loud, sharp *kyowk*. When alarmed, raises crest and flicks tail.

RANGE: Usually solitary; found in a variety of habitats, but prefers streams, ponds, and marshes with woodland cover; often perches in trees. Generally common; a few winter north of resident limit.

TRICOLORED HERON

Egretta tricolor | L 26" (66 cm) WS 36" (91 cm)

White belly and foreneck contrast with mainly dark blue upperparts; bill long and slender. **Juvenile** has chestnut hindneck and wing coverts.

RANGE: Common inhabitant of salt marshes and mangrove swamps of the East and Gulf Coasts. Rare inland, but has bred in North Dakota and Kansas. Rare but regular on southern California coast, chiefly in winter; casual in the Southwest.

LITTLE BLUE HERON

Egretta caerulea | L 24" (61 cm) WS 40" (102 cm)

Slate blue overall. During most of the year, plumage, head, and neck are dark purple; legs and feet dull green. In high **breeding** plumage, head and neck become reddish purple, legs and feet black. **Juvenile** is easily confused with immature Snowy Egret (next page); note Little Blue Heron's dull yellow legs and feet; two-toned bill with thicker, gray base and dark tip; mostly grayish lores; and, often, narrow, dusky primary tips. During first spring, juvenile's white plumage begins gradual **molt** to adult plumage.

RANGE: Slow, methodical feeders in freshwater ponds, lakes, and marshes and coastal saltwater wetlands. Common; disperses north in spring and during post-breeding dispersal. Casual north on the West Coast to southern British Columbia, and on the East Coast to Newfoundland.

REDDISH EGRET

Egretta rufescens | L 30" (76 cm) WS 46" (117 cm)

While feeding, this heron lurches, dashing about with wings spread in a canopy. **Dark-morph breeding adult** has shaggy plumes on rufous head, neck. Bill is pink with black tip; legs cobalt blue. Nonbreeding plumage varies, but in general duller, with shorter plumes, darker bill. **Dark-morph juvenile** is gray; some pale cinnamon on head, neck, inner wing; bill is dark. **White-morph adult** resembles immature Little Blue Heron or Snowy Egret (next page); note larger size, longer bill, dark legs and feet. A few dark-morph birds have much white on wings and resemble molting immature Little Blue Heron.

RANGE: Inhabits shallow, open salt pans. Wanders along Gulf Coast in post-breeding dispersal; casual inland to Midwest, Southwest, up the Atlantic coast to New England; rare to coastal southern California.

juvenile

adult

adult

GREEN
HERON
virescens

TRICOLORED
HERON
ruficollis

juvenile

breeding
adult

molting
immature

LITTLE BLUE
HERON

juvenile

breeding
adult

nonbreeding
adult

REDDISH
EGRET
rufescens

white-morph
breeding adult

canopy
feeding

dark-
morph
juvenile

dark-morph
breeding adult

CATTLE EGRET
Bubulcus ibis | L 20" (51 cm) WS 36" (91 cm)

Small, stocky white heron with large, rounded head; note throat feathering extends far out on bill. **Breeding adult** is adorned with orange-buff plumes on crown, back, and foreneck. At height of breeding season, bill is red-orange, lores purplish, legs dusky red. **Nonbreeding adult** has short yellow bill, yellowish legs. Juvenile's bill is black; begins to turn yellow in late summer; resembles nonbreeding adult. Often seen among livestock in fields, feeding on insects. In flight, resembles Snowy Egret but is smaller; bill and legs shorter; wingbeats faster.
RANGE: An Old World species, Cattle Egret came to South America from Africa, spread to Florida in the early 1950s, reached California by the mid-1960s. In spring, summer, and especially fall, wanders well north of breeding range.

LITTLE EGRET
Egretta garzetta | L 24" (60 cm) WS 36" (91 cm)

Closely resembles Snowy Egret, but often appears larger, with longer neck; longer, thicker bill and legs, the latter always entirely black; mostly grayish lores; and more extensive throat feathering out on lower mandible. Little Egret's crown is flatter; feet are yellow, like Snowy, but average slightly duller. In **breeding** plumage, lore color is variable, but can be yellow. Note the two or three long, tapering plumes on back, rather than Snowy's many curved plumes. Often feeds less frenetically than Snowy, with long neck bent over in a posture like Little Blue Heron.
RANGE: Old World species. Casual spring and summer visitor to East Coast, from Newfoundland to mid-Atlantic states.

SNOWY EGRET
Egretta thula | L 24" (61 cm) WS 41" (104 cm)

White heron with slender black bill, yellow eyes, black legs, and bright golden yellow feet. Graceful plumes on head, neck, and back (where they curve upward) are striking in **breeding adult**. In **high breeding** plumage, lores turn red, feet orange. Nonbreeding plumage is similar but plumes shorter; also note yellow on backs of legs. Compare to Little Egret. **Juvenile** resembles nonbreeding adult, but lacks plumes and shows some bluish gray at base of lower mandible. Can be confused with immature Little Blue Heron (preceding page); note young Snowy Egret's slimmer, mostly black bill; yellow lores; predominantly dark legs; and white wing tips. Snowy Egrets are active feeders.
RANGE: Common in various wetland habitats. Disperses north of mapped range in spring and after breeding season.

GREAT EGRET
Ardea alba | L 39" (99 cm) WS 51" (130 cm)

Large white heron with heavy yellow bill, blackish legs and feet. In **breeding** plumage, long plumes trail from back, extending beyond tail. In immature and **nonbreeding adult**, bill and leg colors are duller, plumes absent. Distinguished from most other white herons by large size; from white morph of the larger Great Blue Heron by black legs and feet.
RANGE: Common in wetlands. Partial to open habitats for feeding; stalks prey slowly, methodically. Occasionally breeds far north of usual range. Post-breeding wanderers reach far north of mapped breeding range.

CATTLE
EGRET
ibis

immature

high
breeding
adult

Snowy
Egret

Little Blue
Heron
juvenile

Cattle Egret
nonbreeding

LITTLE
EGRET
garzetta

nonbreeding
adult

breeding
adult

high
breeding
adult

juvenile

SNOWY
EGRET
thula

breeding
adult

high breeding
adult

nonbreeding

Great Blue Heron
white-morph adult,
"Great White Heron"
(for comparison)

GREAT EGRET
egretta

GREAT BLUE HERON *Ardea herodias* | L 46" (117 cm) WS 72" (183 cm)

Large, gray-blue heron; black stripe extends above eye; white foreneck is streaked with black. **Breeding adult** has yellowish bill and ornate plumes on head, neck, and back. Nonbreeding adult lacks plumes; bill is yellower. **Juvenile** has black crown, no plumes. All-white morph found in southern Florida formerly considered a separate species, "**Great White Heron**" (preceding page). "**Wurdemann's Heron**" morph, found chiefly in Florida Keys, has all-white head and pale neck with inconspicuous streaking underside.
RANGE: Common. A few winter far north into breeding range.

STORKS (FAMILY CICONIIDAE)

Large, long-legged birds that fly with slow beats of their long, broad wings, soaring and circling like hawks.
Species: 19 World, 2 N.A.

WOOD STORK *Mycteria americana* | **E** | L 40" (102 cm) WS 61" (155 cm)

Black flight feathers and tail contrast with white body. **Adult** has bald, blackish gray head; thick, dusky, downcurved bill. **Juvenile's** head is feathered largely with grayish brown; bill is yellow. Adult plumage attained in fourth year.
RANGE: Wood Storks inhabit wet meadows, swamps, ponds, and coastal shallows. A few wander north in late summer; accidental north to Maine and British Columbia.

JABIRU *Jabiru mycteria* | L 52" (132 cm) WS 90" (229 cm)

Distinguished from Wood Stork by larger size; large bill, slightly upturned; and all-white wings and tail. Red throat pouch brightens and inflates during breeding season. **Juvenile** is patchy brown-gray; head is blackish brown. Usually seen with flocks of Wood Storks.
RANGE: Huge stork of Central and South America, casual straggler in south Texas; recorded once in Oklahoma.

FLAMINGOS (FAMILY PHOENICOPTERIDAE)

Large waders with big, bent bills, used to strain food from the waters of shallow lakes and lagoons.
Species: 5 World, 1 N.A.

AMERICAN FLAMINGO *Phoenicopterus ruber* | L 46" (117 cm) WS 60" (152 cm)

Note pink legs, black flight feathers, tricolored bill. **Immature** is grayer, with pink wash below; paler bill. South Florida (rare) and south Texas (casual) sightings include wild birds; others, especially away from these regions, more likely escapes. In addition, escapes include a widespread Old World species, the Greater Flamingo (*P. roseus*), with pink-and-white plumage; the Chilean Flamingo (*P. chilensis*), with grayish legs with pink joints; and the Lesser Flamingo (*Phoeniconaias minor*), with dark red bill and blotchy red wing coverts and axillaries.
RANGE: Birds native to the New World, recently split from the Old World Greater Flamingo, breed as close to Florida as southern Bahamas, Cuba, and the Yucatan Peninsula.

breeding
adult

adult

juvenile

"Wurdemann's
Heron"
breeding adult

GREAT BLUE
HERON
herodias

adult

adult

WOOD
STORK

juvenile

juvenile

JABIRU

adult

AMERICAN
FLAMINGO

immature

breeding
adult

IBISES and SPOONBILLS (Family Threskiornithidae)

Gregarious, heronlike birds, these long-legged waders feed with long, specialized bills: slender and curved downward in ibises, wide and spatulate in spoonbills. **Species:** *32 World, 5 N.A.*

GLOSSY IBIS
Plegadis falcinellus | L 23" (58 cm) WS 36" (91 cm)

Breeding adult's chestnut plumage is glossed with green or purple; looks all-dark at a distance. Distinguished from White-faced Ibis by brown eye, gray-green legs with red joints, and lack of distinct white border to bare facial skin. Blue edge to gray facial skin does not extend behind eye or under chin. **Winter adult** closely resembles winter White-faced; look for gray facial skin partially bordered by blue line. **Juvenile** closely resembles juvenile White-faced Ibis, but note gray facial skin and at least trace of blue line on most birds. Adult breeding plumage is acquired in second spring.
Range: Glossy Ibises inhabit freshwater and saltwater marshes. Fairly common but local. Expanding north along the East Coast. Rare but annual in Texas. Rare inland wanderer, chiefly in spring; casual west to California.

WHITE-FACED IBIS
Plegadis chihi | L 23" (58 cm) WS 36" (91 cm)

Breeding adult distinguished from Glossy Ibis, with which it sometimes hybridizes, by bronzer tones in chestnut plumage, reddish bill, red eye, all-red legs, and white feathered border around red facial skin; border extends behind eye and under chin. **Winter adult** plumage is like Glossy, but lacks pale blue line from eye to bill; facial skin is pale pink. **Juvenile** closely resembles juvenile Glossy until winter, when facial skin turns pinkish; look for lack of blue line (or a hint of white border) and reddish tinge to eye.
Range: Breeds in freshwater marshes. Very rare in Midwest; casual on East Coast north to New England.

WHITE IBIS
Eudocimus albus | L 25" (64 cm) WS 38" (97 cm)

Adult's white plumage and pink facial skin are distinctive. In **breeding adult**, facial skin, bill, and legs turn scarlet. Dark tips of primaries most easily seen in flight. Immatures have white underparts and wing linings, pinkish bill; gradually molt into adult plumage by second fall.
Range: Locally common to abundant in coastal salt marshes, swamps, mangroves. Expanding north, now breeds to Virginia. Casual north to New Jersey, Midwest, Southwest. Closely related **Scarlet Ibis** (*E. ruber*), a South American species introduced or escaped in Florida, hybridizes with White Ibis; offspring are various shades of pink or scarlet.

ROSEATE SPOONBILL
Platalea ajaja | L 32" (81 cm) WS 50" (127 cm)

Adult has pink body with red highlights; long, spatulate bill; unfeathered greenish head. The head may become buffy during courtship. **Juvenile** has white feathering on head; body is mostly pale pink. Spoonbills feed in shallow waters, swinging their bills from side to side.
Range: Fairly common locally along the Gulf Coast; casual north to mid-Atlantic, Midwest, and Southwest regions.

breeding
adult

juvenile

juvenile

winter
adult

GLOSSY
IBIS

breeding
adult

winter
adult

juvenile

WHITE-FACED
IBIS

breeding
adult

juvenile

breeding
adult

breeding
adult

juvenile

WHITE IBIS

1st spring

Scarlet Ibis
adult

juvenile

ROSEATE
SPOONBILL

breeding
adult

breeding
adult

juvenile

NEW WORLD VULTURES (Family Cathartidae)

Small, unfeathered head and hooked bill aid in consuming carrion. **Species:** *7 World, 3 N.A.*

TURKEY VULTURE
Cathartes aura │ L 27" (69 cm) WS 69" (175 cm)

In flight, rocks side to side with little flapping and wings held upward in a shallow V; dark wing linings contrast with silvery flight feathers. Rather long tailed. **Adult** has red head, white bill, brown legs; **juvenile's** head and bill are dark, legs are paler. Feeds chiefly on carrion and refuse.

Range: Common in mapped range.

BLACK VULTURE
Coragyps atratus │ L 25" (64 cm) WS 57" (145 cm)

In flight, shows large white patches at base of primaries. Tail is shorter than Turkey Vulture; wings shorter and broader; legs white; feet usually extend to edge of tail or beyond. Flight includes rapid flapping and short glides, usually with wings flat. Gregarious and aggressive, but less efficient at spotting carrion than Turkey Vulture.

Range: Common in open country and near human settlements, often scavenge in garbage dumps. Range expanding in the Northeast; casual north to Ontario, Maritimes, and northern California; accidental to New Mexico.

CALIFORNIA CONDOR
Gymnogyps californianus │ **E** │ L 47" (119 cm) WS 108" (274 cm)

Huge size distinctive. **Adult** has white wing linings, orange head; **juvenile's** wing linings mottled, head dusky. Soars on flat wings without flapping, in search of carrion. Population in wild in 2006 about 100; 100 more in captivity.

Range: Last wild birds captured April 1987. Previously found in arid foothills, mountains of southern and central California. Decline to near extinction caused mostly by lead poisoning and illegal shooting. Recently reintroduced in California and introduced in northern Arizona and northern Baja California.

HAWKS, KITES, EAGLES, and ALLIES (Family Accipitridae)

Worldwide family of diurnal birds of prey, with hooked bills and strong talons. **Species:** *233 World, 28 N.A.*

OSPREY
Pandion haliaetus │ L 22–25" (56–64 cm) WS 58–72" (147–183 cm)

Dark brown above, white below, with white head, prominent dark eye stripe. Females average darker streaking on neck; juvenale plumage is fringed with pale buff above. In flight, long, narrow wings are bent back at "wrist," dark carpal patches conspicuous; wings slightly arched in soaring. Nests near fresh or salt water; eats mostly fish. Hovering over water, dives down, then plunges feetfirst to snatch prey. Bulky nests are built in trees, on sheds, poles, docks, and special platforms. **Call** is a series of loud, whistled *kyew* notes.

Range: Conservation programs successful and now fairly common.

adults

TURKEY
VULTURE

adult

juvenile

adult

BLACK
VULTURE

adult

juvenile

adult

CALIFORNIA
CONDOR

adult

adult

juvenile

OSPREY
carolinensis

adult

MISSISSIPPI KITE

Ictinia mississippiensis | L 14½" (37 cm) WS 35" (89 cm)

Long, pointed wings with first primary distinctly shorter; long, flared tail. Dark gray above, paler below, with pale gray head, averaging paler on **male**. **Female** with white shaft on outer tail feather and often whitish in vent region. White secondaries show in flight as white wing patch. Black tail readily distinguishes Mississippi from White-tailed Kite. Compare also with male Northern Harrier (next page); note Mississippi never hovers. **Juvenile** is heavily streaked and spotted, with pale bands on tail, but pattern and overall darkness highly variable on underparts, underwings, and tail. **First-summer** bird (page 142) more like adult but retains juvenal flight feathers. At all ages, may be confused with Peregrine Falcon (page 142); compare wing and tail shapes. Mississippi Kites capture and eat their prey, mainly insects, on the wing. Gregarious; often hunt in groups, nest in loose colonies.

RANGE: Found in woodlands, swamps, rangelands. Regular straggler (chiefly immatures in spring) to mid-Atlantic states. Casual north to Great Lakes region and California. Winters in South America.

WHITE-TAILED KITE

Elanus leucurus | L 16" (41 cm) WS 42" (107 cm)

Long, pointed wings; long tail. White underparts and mostly white tail distinguish **adults** from similar Mississippi Kite. Compare also with male Northern Harrier (next page). **Juvenile**'s underparts and head are lightly streaked with rufous, which rapidly fades. In all ages, black shoulders show in flight as black leading edge of inner wings from above, small black patches from below. Hovers while hunting, unlike any other North American kite. Eats mainly rodents, insects.

RANGE: Populations fluctuate. Fairly common in grasslands, farmlands, even highway median strips. Casual well north of mapped range to British Columbia, Wyoming, and New York. Often forms winter roosts of more than a hundred birds.

SWALLOW-TAILED KITE

Elanoides forficatus | L 23" (58 cm) WS 48" (122 cm)

Seen in flight, deeply forked tail and sharply defined pattern of black and white are like no other large bird except the young Magnificent Frigatebird (page 98). Perched, Swallow-tailed Kite's coloring more closely resembles White-tailed and Mississippi Kites; again, note long, forked tail. Juvenile is similar to **adult**, but tail is shorter, flight feathers and tail narrowly tipped with white. Agile and graceful, Swallow-tailed snatches flying insects; also drops down upon snakes, lizards, young birds; does not hover. Often eats prey in flight; also drinks in flight, skimming the water like a swallow. Found in open woods, bottomlands, and wetlands. Nests in the tops of tall trees. Somewhat social; several may hunt in the same territory.

RANGE: Casual in spring and summer as far north as Ontario and Nova Scotia; accidental as far west as Arizona. Most winter in South America.

MISSISSIPPI
KITE

adult ♂

adult ♀

juvenile

adult ♂

adults

WHITE-TAILED
KITE
majusculus

juvenile

adult

adults

SWALLOW-TAILED
KITE

SNAIL KITE
Rostrhamus sociabilis | **E** | L 17" (43 cm) WS 46" (117 cm)

This kite's wings are paddle shaped, bill thin and deeply hooked. **Male** is gray-black above and below, with white uppertail and undertail coverts; square, white tail with broad, dark band and paler terminal band; legs orange-red; eyes and facial skin reddish. **Female** is dark brown, with distinctive head pattern. **Juvenile** has dark brown eyes, duller facial skin and legs, streaked crown and underparts. Hunting flight is slow, with considerable flapping of wings, and head held down as the kite searches for apple snails, its chief and perhaps only food.

RANGE: A tropical species. Endangered; uncommon and local resident in southern Florida.

HOOK-BILLED KITE
Chondrohierax uncinatus | L 18" (46 cm) WS 36" (91 cm)

Plumage varies considerably, but look for large, heavy bill with long hook, white eyes, banded tail, and heavily barred underparts, including underwings. **Males** are generally gray overall. **Females** are brown, with a rufous collar and rufous, barred underparts. **Juveniles** have brown eyes, white collar, and whitish underparts with variable dark brown barring. In the **black morph**, rarely seen in the U.S., **adult** is all-black except for a single white or grayish tail band and whitish tail tip; **juveniles** are mostly brownish black, with two or more grayish tail bands. Hook-billed Kite flies with deep, languid wingbeats, its "wrists" slightly cocked upward and "hands" angled down. Wings are paddle shaped, slightly tapered in at the base. It eats insects and small amphibians, but prefers snails of various kinds. A pile of broken snail shells beneath a tree may indicate a favorite perch or a nest site above.

RANGE: Tropical species, uncommon over most of its range. Rare resident in lower Rio Grande Valley from Santa Ana to Falcon Dam. Found in dense woodlands, from which it thermals upward in late morning.

NORTHERN HARRIER
Circus cyaneus | L 16–20" (41–51 cm) WS 38–48" (97–122 cm)

White uppertail coverts and owl-like facial disk distinctive in all ages and both sexes. Body slim; wings long and narrow with somewhat rounded tips; tail long. **Adult male** is grayish above; mostly white below with variable chestnut spotting; has black wing tips and black tips to secondaries. **Female** is brown above, whitish below with heavy brown streaking on breast and flanks, lighter streaking and spotting on belly. **Juveniles** resemble adult female but are cinnamon below, fading to creamy buff by spring; streaked only on the breast; wing linings are cinnamon, distinctly darker on inner half. Harriers generally perch low and fly close to the ground, wings upraised, as they search for birds, mice, frogs, and other prey. Seldom soar high except during migration and in exuberant, acrobatic courtship display.

RANGE: Fairly common in wetlands and open fields. Adult males migrate later in fall and earlier in spring than females and immatures. In winter, harriers form communal ground roosts, sometimes with Short-eared Owls.

adult ♂

adult ♀

juvenile

adult ♂

SNAIL KITE
plumbeus

black-morph
adult

black-morph
juvenile

juvenile

adult ♂

**HOOK-BILLED
KITE**
uncinatus

adult ♀

juvenile

adult ♂

adult ♂

juvenile

**NORTHERN
HARRIER**
hudsonicus

adult ♂

juvenile

adult ♀

juvenile

adult ♂

GOLDEN EAGLE

Aquila chrysaetos | L 30–40" (76–102 cm) WS 80–88" (203–224 cm)

Brown, with variable yellow to tawny brown wash over back of head and neck; bill mostly horn colored; tail faintly banded. Tawny greater upperwing coverts form a bar. **Juveniles**, seen in flight from below, show well-defined white patches at base of primaries, white tail with distinct dark terminal band. Compare with juvenile Bald Eagle's larger head, shorter tail, blotchier tail and underwing pattern. **Adult** plumage is acquired in four years. Golden Eagle often soars with wings slightly uplifted.

RANGE: Inhabits mountainous or hilly terrain, hunting over open country for small mammals, snakes, birds, and carrion. Also found in valleys and western plains, especially in migration and winter. Nests on cliffs or in trees. Uncommon to rare in the East; uncommon to fairly common in the West.

WHITE-TAILED EAGLE

Haliaeetus albicilla | L 26–35" (66–89 cm) WS 72–94" (183–239 cm)

Note short, wedge-shaped white tail. Plumage mottled; head may be very pale and appear white at a distance; undertail coverts are dark, unlike subadult and adult Bald Eagles. **Juvenile**'s tail has variable dark mottling and tip and is less wedge-shaped, underwing darker, than Bald Eagle.

RANGE: Flies over northern Eurasia and Greenland in diminishing numbers. Very rare visitor to outer Aleutians, especially Attu Island, where it has nested.

STELLER'S SEA-EAGLE

Haliaeetus pelagicus | L 33–41" (84–104 cm) WS 87–96" (221–244 cm)

In flight, white shoulders show as white leading edge of wings; trailing edge of wing more curved than in White-tailed or Bald Eagles. Immense yellow-orange bill; long, white, wedge-shaped tail; white thighs. **Juvenile** lacks white shoulders; end of tail is dark.

RANGE: Nests in northeastern Asia; casual in Alaska; recorded on Aleutians, Pribilofs, Kodiak Island, near Juneau in southeast Alaska.

BALD EAGLE

Haliaeetus leucocephalus | L 31–37" (79–94 cm) WS 70–90" (178–229 cm)

Adults readily identified by white head and tail, large yellow bill. **Juveniles** are mostly dark, may be confused with juvenile Golden Eagle; compare blotchy white on underwing coverts, axillaries, and tail with Golden Eagle's more sharply defined pattern; note also Bald Eagle's disproportionately larger head, shorter tail. Neck is shorter and tail longer than White-tailed Eagle; Steller's Sea-Eagle has longer, wedge-shaped tail. Flat-winged soar distinguishes young Bald Eagle from Turkey Vulture (page 118). Bald Eagles require four or five years to reach full adult plumage.

RANGE: Seen most often on seacoasts or near rivers and lakes. Feed mainly on fish in breeding season, regularly on carrion, and on roadkill in winter, particularly in the Southwest. Nest in tall trees or on cliffs. Most abundant in Alaska; common in winter along Mississippi and Missouri Rivers, fairly common in the Northwest. Banning of pesticides and intense recovery programs have increased populations that had been seriously diminished in the East.

juvenile

adult

adult

GOLDEN
EAGLE
canadensis

WHITE-TAILED
EAGLE

juvenile

adult

adult

STELLER'S
SEA-EAGLE

juvenile

2nd year

juvenile

BALD EAGLE

adults

3rd
year

juvenile

ACCIPITERS

Comparatively long tails and short, rounded wings give these woodland hawks great agility. Flight is several quick wingbeats and a glide. The three species in North America are confusingly similar. Generally silent, except at nest site.

SHARP-SHINNED HAWK

Accipiter striatus | L 10–14" (25–36 cm) WS 20–28" (51–71 cm)

Distinguished from Cooper's Hawk by shorter, squared tail, often appearing notched when folded, thinner legs, and by smaller head and neck. **Adult** lacks Cooper's strong contrast between crown and back. **Juveniles** are whitish below, some streaked with brown (like Cooper's), others spotted with reddish brown. Note also the pale eyebrows, narrow white tip on tail, entirely white undertail coverts, less tawny head than other accipiters. In flight (see also page 143), again compare smaller head and proportionately shorter tail than Cooper's; wingbeats quick and choppy, slower on Cooper's.

Range: Sharp-shinned Hawk is fairly common over much of its range; found in mixed woodlands. Preys chiefly on small birds. Migrates singly or in loose groups.

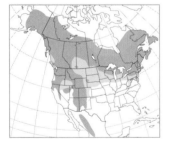

COOPER'S HAWK

Accipiter cooperii | L 14–20" (36–51 cm) WS 29–37" (74–94 cm)

Distinguished from Sharp-shinned Hawk by longer, rounded tail, larger head, and, in **adult**, stronger contrast between back and crown. **Juvenile** has whitish or buffy underparts with fine streaks on breast, streaking reduced or absent on belly; tawny rufous color on head is much richer, white tip on tail is broader, than in Sharp-shinned; undertail coverts entirely white. Note that some juveniles may have a pale eyebrow like Sharp-shinned. In flight (see also page 143), again compare larger head and longer tail.

Range: Inhabits broken woodlands or streamside groves, especially deciduous. Usually migrates singly. Rare, mainly in fall, in the Maritimes. Preys largely on songbirds, some small mammals. Often perches on telephone poles, unlike Sharp-shinned.

NORTHERN GOSHAWK

Accipiter gentilis | L 21–26" (53–66 cm) WS 40–46" (102–117 cm)

Conspicuous eyebrow, flaring behind eye, separates **adult**'s dark crown from blue-gray back. Underparts are white with dense gray barring; appear gray at a distance; has wedge-shaped tail with fluffy undertail coverts. Note disproportionately shorter tail, longer wings, than Cooper's Hawk. **Juvenile** is brown above, buffy below, with thick, blackish brown streaks, heaviest on flanks; tail has wavy dark bands bordered with white and a thin white tip; undertail coverts usually have dark streaks. In flight (see also page 143), note tawny bar on upperwing on greater secondary coverts. Juvenile also can be confused with Gyrfalcon (page 140) and Red-shouldered Hawk (page 130).

Range: Northern Goshawk inhabits deep, conifer-dominated, mixed woodlands; preys on birds and mammals as large as hares. Uncommon; winters irregularly south of mapped range in the East. Southward irruptions occur in some winters.

SHARP-SHINNED
HAWK
velox

adult ♂

juvenile ♀

juvenile

COOPER'S
HAWK

juvenile ♀

adult ♂

juvenile

NORTHERN
GOSHAWK
atricapillus

juvenile ♀

adult ♂

juvenile

BUTEOS

High-soaring hawks, among the easiest of birds of prey to spot.

COMMON BLACK-HAWK
Buteogallus anthracinus | L 21" (53 cm) WS 50" (127 cm)

Wings broad and rounded; tail short, broad. **Adult** blackish overall; tail has broad white band. Legs and cere orange-yellow. Distinguished from Zone-tailed Hawk by broader wings; broader, less banded tail; larger bill; more orange-yellow in lore region. **Juvenile** has strong face pattern; heavily streaked underparts; many-banded tail; buffy wing panel visible from above and below. **Call** is a series of loud whistles.
RANGE: Found along waterways. Rare, local, and declining; very rare in southwestern Utah and southern Texas; casual in California.

HARRIS'S HAWK
Parabuteo unicinctus | L 21" (53 cm) WS 46" (117 cm)

Chocolate brown overall, with chestnut shoulder patches, leggings, and wing linings; white at base and tip of long tail; rounded wingtips. **Juvenile** is heavily streaked below; chestnut shoulder patches are less distinct. Inhabits semiarid woodland, and brushland. Gregarious; sometimes hunts in small, cooperative groups.
RANGE: May straggle north and west of mapped range, but many may be escapes.

ZONE-TAILED HAWK
Buteo albonotatus | L 20" (51 cm) WS 51" (130 cm)

Grayish black overall, with barred flight feathers. Legs and cere yellow. Slimmer winged than Common Black-Hawk; longer tail, variably banded according to sex and age. Flies like Turkey Vulture; compare Zone-tailed's banded tail; smaller bill; yellow cere; larger, feathered head. **Juvenile** has grayish tail, some white flecking on breast. **Call**, a squealing whistle.
RANGE: Uncommon; found in mesa and mountain country, often near watercourses; drops from low glide on small birds, rodents, lizards, and fish. Rare in southern California and southern Texas.

SHORT-TAILED HAWK
Buteo brachyurus | L 15½" (39 cm) WS 35" (89 cm)

Small hawk with two color morphs. Secondaries seen from below are darker than primaries. **Light morph** has dark helmet and underwing resembling Swainson's Hawk (page 132), wings and tail are shorter, broader; lacks chest band. **Adults** have wide, dark subterminal tail band; of equal width on juveniles. Faint streaks on sides of light morph; spotted with white on wing linings and underparts on dark-morph juveniles. Most often seen in flight.
RANGE: Found in mixed woodland-grassland. Casual to south Texas and mountains of southeast Arizona (has bred) in spring and summer; accidental Michigan.

ROADSIDE HAWK
Buteo magnirostris | L 14" (36 cm) WS 30" (75 cm)

A small, slim, long-legged raptor with banded tail. **Adult** with brown bib, barred belly. **Juvenile** with some streaking on chest. Flies with stiff, rapid wingbeats; wingtips rounded with rufous patch on inner primaries.
RANGE: Tropical species, casual in winter in lower Rio Grande Valley, Texas.

COMMON
BLACK-HAWK
anthracinus

adult

juvenile

juvenile

adult

Turkey Vulture
(for comparison)

juvenile

adult ♂

HARRIS'S
HAWK
harrisi

adult

adult

juvenile

ZONE-TAILED
HAWK

adult

dark-
morph
adult

SHORT-TAILED
HAWK
fuliginosus

adult

light-
morph
adults

adult

ROADSIDE
HAWK

juvenile

BROAD-WINGED HAWK

Buteo platypterus | L 16" (41 cm) WS 34" (86 cm)

Pointed wing tips; white underwings have dark borders; tail has broad black and white bands, with last white band broader than the others. Wings broad but more pointed than in Red-shouldered Hawk; wing linings buffy or white; tail shorter, broader. Wingbeats are slower than in *elegans* race of Red-shouldered. **Juveniles** typically have black moustachial streak; dark-bordered underwings, indistinct bands on tail; very similar to juvenile eastern Red-shouldered but paler below; may have a pale area at base of primaries but lack the distinct pale crescent. Rare **dark morph** breeds in western and central Canada. Broad-winged is a woodland species. **Call**, heard on breeding and winter grounds, is a thin, shrill, slightly descending whistle: *pee-teee*.

RANGE: Often migrates in very large flocks. Rare migrant in the West. Most winter in South America; a few winter in southern Florida and very rarely in coastal California.

GRAY HAWK

Buteo nitidus | L 17" (43 cm) WS 35" (89 cm)

Gray upperparts, gray-barred underparts and wing linings, and rounded wing tips distinguish Gray from Broad-winged Hawk (see also page 143). Flight is accipiter-like: several rapid, shallow wingbeats and a glide. **Juvenile** resembles juvenile Broad-winged, but has much longer tail projection, stronger face pattern with outlined white cheek, and white, U-shaped rump band; dark trailing edge on wings is smaller or absent. **Calls** include a loud, descending whistle.

RANGE: Tropical species; local nester in southeastern Arizona. Rare in lower Rio Grande Valley year-round. Rare in summer upriver to Big Bend and in southwest New Mexico. Gray Hawk inhabits deciduous growth along streams with nearby open land.

RED-SHOULDERED HAWK

Buteo lineatus | L 15–19" (38–48 cm) WS 37–42" (94–107 cm)

Relatively long tailed and long legged. In flight, shows pale crescent at base of primaries. **Adult** has reddish shoulders and wing linings and extensive pale spotting above. Widespread eastern nominate race *lineatus* shows dark streaks on reddish chest. Southeastern *alleni* (not shown) is smaller, with grayish cast to head and back; usually lacks breast streaking. South Florida *extimus* is the smallest and palest race. California *elegans* and central Texas *texanus* (not shown) are decidedly more rufous below; *elegans* is often solidly rufous across the chest and has broader white tail bands. **Juveniles** show extensive variations; *lineatus* shows more finely streaked breast and more closely resembles juvenile Broad-winged Hawk; other eastern races show more coarsely marked underparts; *elegans* is quite dark and has more adultlike features, including some rufous on shoulders and wing linings. Flight of all ages of *elegans* is accipiter-like, with several quick wingbeats and a glide, while *lineatus* flies with slower wingbeats, more like Broad-winged. **Call** is an evenly spaced series of clear, high *kee-ah* or *kah* notes. Quite vocal, especially during Spring courtship.

RANGE: Found in moist, mixed woodlands; often seen near water.

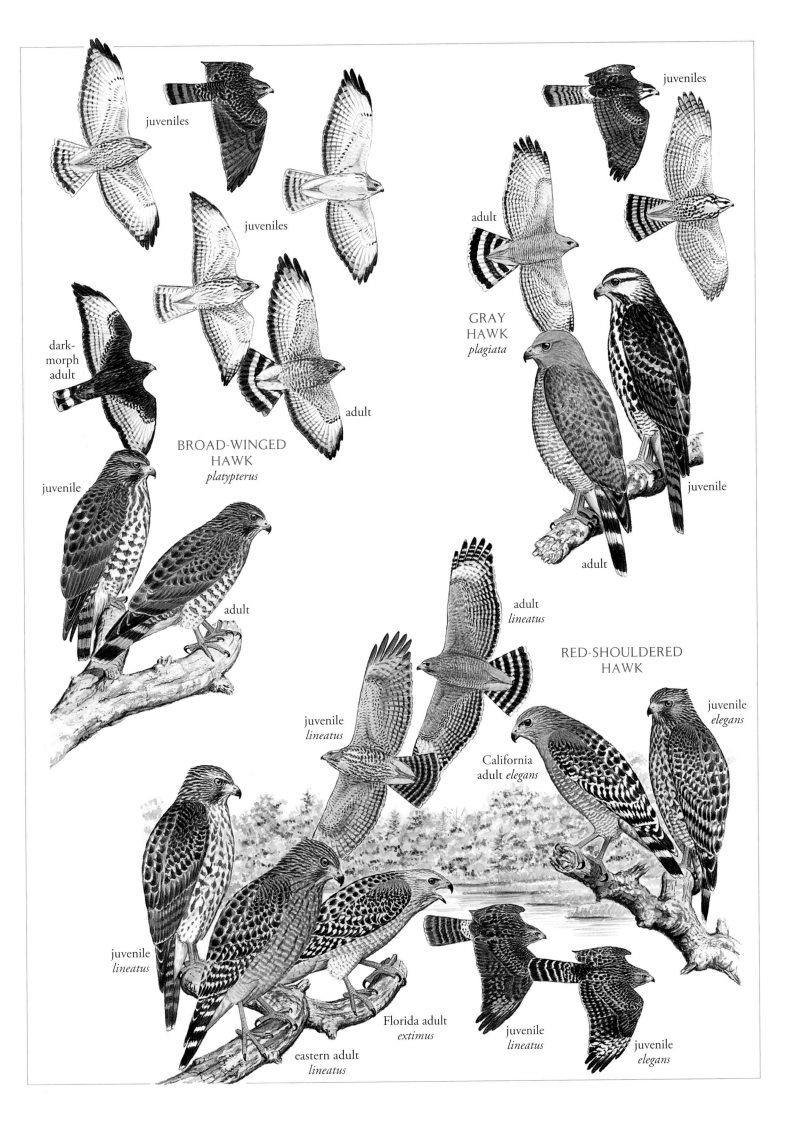

juveniles

juveniles

dark-
morph
adult

juvenile

BROAD-WINGED
HAWK
platypterus

adult

GRAY
HAWK
plagiata

adult

juvenile

adult

adult

adult
lineatus

juvenile
lineatus

RED-SHOULDERED
HAWK

California
adult *elegans*

juvenile
elegans

juvenile
lineatus

Florida adult
extimus

eastern adult
lineatus

juvenile
lineatus

juvenile
elegans

RED-TAILED HAWK

Buteo jamaicensis | L 22" (56 cm) WS 50" (127 cm)

Our most common buteo; wings broad and fairly rounded; plumage extremely variable. Looks heavy billed, unlike Rough-legged (next page) and Swainson's Hawks. Variable pale mottling on scapulars contrasts with dark mantle, often forming a broad-sided V on perched birds. Most **adults**, especially in the East, show a belly band of dark streaks on whitish underparts; dark bar on leading edge of underwing, contrasting with paler wing linings (see also page 144). Note reddish uppertail; paler red undertail. Great Plains race *krideri*, known as **"Krider's Red-tailed,"** has paler upperparts and whitish tail with pale reddish wash; in flight, shows pale rectangular patches at base of primaries on upperwing. Many southwestern birds of the *fuertesi* race (not shown) lack belly band and have entirely light underparts. Widespread dark and **rufous morphs** of western race, *calurus*, have dark wing linings and underparts, obscuring the bar on leading edge and belly band; tail is dark reddish above. In *harlani*, **"Harlan's Hawk,"** formerly considered a separate species, dark morph has dusky white tail, diffuse blackish terminal band; shows some white streaking on its dark breast; may lack scapular mottling; rare *harlani* light morph has typical tail pattern, but plumage resembles *krideri*. "Harlan's Hawk" breeds in Alaska and east to northwestern Canada; winters primarily in central U.S. **Juveniles** of all morphs except *harlani* have gray-brown tails with many blackish bands; otherwise heavily streaked and spotted with brown below. Distinctive **call**, a harsh, descending *keeeeer*.

RANGE: Habitat variable: woods with nearby open land; also plains, prairie groves, and desert. Preys on rodents.

SWAINSON'S HAWK

Buteo swainsoni | L 21" (53 cm) WS 52" (132 cm)

Distinguished from most other buteos by long, narrow, pointed wings; plumage is extremely variable. Lacks Red-tailed Hawk's pale mottling on scapulars; bill is smaller. All but darkest birds show contrast between paler wing linings and dark flight feathers; most show pale uppertail coverts. In **light morph**, whitish or buffy white wing linings contrast with darkly barred brown flight feathers (see also page 144); dark bib; underparts otherwise whitish to pale buff. **Dark-morph** bird is dark brown with white undertail coverts; shows less sharp contrast between wing linings and flight feathers; darkest birds show none. Compare with first-year White-tailed Hawk (next page). **Intermediate** colorations between light and dark morphs include a rufous morph. Intermediate and **light-morph juveniles** have dark moustachial stripe and conspicuous whitish eyebrows that meet on the forehead; variable streaking below, very heavy on dark morphs. Show less contrast between wing linings and flight feathers than adult birds (page 144). Swainson's soars over open plains and prairie with uptilted wings in teetering, vulturelike flight.

RANGE: Very rare spring and fall migrant in eastern North America. Gregarious; usually migrates in large flocks, often with Broad-winged Hawks (preceding page). Winters chiefly in South America; rarely in southern Florida and Central Valley of California.

RED-TAILED
HAWK

eastern adult
borealis

"Harlan's Hawk"
adult *harlani*

eastern juvenile
borealis

"Krider's Red-tailed"
adult *krideri*

eastern adult
borealis

rufous-morph
adult *calurus*

light-morph
juvenile

dark-morph
adult

SWAINSON'S
HAWK

light-morph
adult

light-morph
adult

intermediate-morph
adult

ROUGH-LEGGED HAWK
Buteo lagopus | L 21" (53 cm) WS 53" (135 cm)

Long, white tail with dark band or bands helps to identify this hawk in all plumages; bill small. Thin legs are feathered to the toes, the feathering barred in adults, unbarred in **juveniles**. **Adult male** has multibanded tail with a broad blackish subterminal band. **Adult female**'s tail is brown toward tip with a thin, black subterminal band. Juveniles have a single broad, brown tail band. Wings are long, fairly narrow. Seen in flight from above, white at base of tail is conspicuous; note also the small white patches at base of primaries on upperwings. In the common light morph, pale head contrasts with darker back and dark belly band, especially in females and immatures. Adult male has darker breast markings that may create a bib effect; belly is paler. Observe the square, black carpal patches at the "wrists" of the wings. **Dark morph** is less common. During breeding season gives a soft, plaintive courting whistle. Alarm **call** is a loud screech or squeal.
RANGE: A bird of the open country also seen in marshes in winter, Rough-legged often hovers while hunting.

FERRUGINOUS HAWK
Buteo regalis | L 23" (58 cm) WS 56" (142 cm)

This hawk has a pale head; extended "gape line" going back under eye; tail is a mixture of pale rust, white, and gray. Wings are long, broad, and pointed; note large, white, crescent-shaped patches on upperwing surface. Seen from below, flight feathers lack barring. **Adults** show rusty color on back and shoulders; rusty leggings form a conspicuous V against whitish underparts spotted with rufous. **Dark morph** is rare; varies from dark rufous to dark brown, with dark undertail coverts. Absence of dark tail bands separates it from dark-morph Rough-legged Hawk. **Juvenile** Ferruginous Hawk almost or entirely lacks rusty leggings and is less rufous above; resembles "Krider's" type of Red-tailed Hawk (preceding page), but wings longer and more pointed. Ferruginous gives harsh alarm **calls**, *kree-a* or *kaah*, chiefly in breeding season. Inhabits dry, open country. Often hovers when hunting or soars in a dihedral. Often sits on ground.
RANGE: Casual east to Wisconsin, Illinois, Arkansas, Louisiana, Florida in migration, winter. Very rare migrant to Minnesota; casual in summer.

WHITE-TAILED HAWK
Buteo albicaudatus | L 20" (51 cm) WS 51" (130 cm)

Legs longest of any North American buteo. Wings fairly long and pointed; at rest, **adult**'s wing tips project well beyond end of short tail; tail is white with single black band and other finer bands. Rusty shoulders are highly visible against dark gray upperparts. Underparts and wing linings vary from white on most to lightly barred. Females are darker above, more barred below. **Juveniles** brown above, variable below from mostly blackish to paler; most show a white patch on breast; tail is pale gray; undertail and uppertail coverts whitish, the latter forming a pale U at tail base. Compare with dark morphs of Swainson's (preceding page) and Ferruginous Hawks. Identifiable second-year plumage is intermediate.
RANGE: Rare to fairly common in open coastal grasslands and semiarid brush country. Casual to southwest Louisiana.

adult ♀

dark-morph
adult ♂

ROUGH-LEGGED
HAWK
sanctijohannis

juvenile

adult ♂

juvenile

dark-
morph
adult

FERRUGINOUS
HAWK

adults

adult

WHITE-TAILED
HAWK
hypospodius

juvenile

adult

adult

adult

juvenile

dark
juvenile

CARACARAS AND FALCONS (Family Falconidae)

These powerful hunters are distinguished from hawks by their long wings, which are bent back at the "wrist" and, except in the Crested Caracara, narrow and pointed. Females are larger than the males. Birds of the genus Falco *use their notched bills to kill prey by severing its spinal column at the neck.* **Species: 64 World, 11 N.A.**

EURASIAN HOBBY
Falco subbuteo | L 12¼" (31 cm) WS 30¼" (77 cm)

Small, short-tailed falcon with long, slender wings; in folded wing, wing tips extend well past tip of tail. Graceful and powerful flier. White cheeks; thin, pale eyebrow; thin, dark moustachial stripe; heavily streaked below. **Adult** has rufous-red undertail coverts and is dark gray above. **Juveniles** are blackish brown above with buffy feather fringes; lack rufous below. By following spring some look like adults, others intermediate in appearance. Compare all ages carefully to Merlin and Peregrine Falcon (next four pages).
RANGE: Old World species; long-distance migrant. Casual in late spring and summer in Bering Sea region and on western Aleutians. Record of a bird on a ship off Newfoundland and an Oct. record from Seattle.

APLOMADO FALCON
Falco femoralis | **E** | L 15–16½" (38–42 cm) WS 40–48" (102–122 cm)

In flight, often hovers; long, pointed wings and long, banded tail resemble young Mississippi Kite (page 120); underwings are dark, with pale trailing edge. Note slate gray crown, boldly marked head. Pale eyebrows join at back of head. Dark patches on sides sometimes extend across breast. **Juvenile** is cinnamon below with a streaked breast, and browner above.
RANGE: Once found in open grasslands and deserts from southern Texas to southeastern Arizona. Disappeared by the early 20th century; birds seen in New Mexico and west Texas from 1990s probably from a small extant population in northern Chihuahua, Mexico; recent nesting records southern New Mexico. A reintroduction project is now under way in south coastal Texas and is starting in New Mexico.

CRESTED CARACARA
Caracara cheriway | **T** | L 23" (58 cm) WS 50" (127 cm)

Large head, long neck, and long legs. Blackish brown overall, with white throat and neck and red-orange to yellow bare facial skin; underparts barred with black. **Juvenile** is browner; upperparts are edged and spotted with buff; underparts streaked with buff, unlike **adult** barring; second-year plumage closer to adult. In flight, shows whitish patches near ends of rounded wings. Flapping, ravenlike flight; soars with flat wings. **Calls** include a low rattle and a single *wuck* note. Inhabits open brushlands; often seen on the ground in company with vultures. Feeds chiefly on carrion; also hunts insects and small animals.
RANGE: Fairly common in Texas part of range. Rare in Louisiana and southern Arizona. Casual to southern New Mexico. Records from well outside known range are of debatable origin.

EURASIAN
HOBBY
subbuteo

adult

adult

juvenile

juvenile

adult

juvenile

adult ♀

adult ♂

APLOMADO
FALCON
septentrionalis

juvenile ♀

adult

CRESTED
CARACARA

juvenile

adults

AMERICAN KESTREL

Falco sparverius | L 10½" (27 cm) WS 23" (58 cm)

Smallest and most common of our falcons. Identified by russet back and tail, double black stripes on white face. Seen in flight from below, **adults** show pale underwings, and **males** a distinctive row of white, circular spots on trailing edge of wings. Male also has blue-gray wing coverts; compare with Merlin and much larger Peregrine Falcon (next page). **Juvenile male** is like adult male, but breast heavily streaked, back completely barred; by first fall looks more like adult, but some dark markings remain. **Call** is a shrill *killy killy killy*. Hovering over prey before plunging. Also eats small birds, chiefly in winter. Often perches on telephone wires; frequently bobs its tail.

RANGE: Found in open country and in cities, American Kestrel feeds on insects, reptiles, and small mammals.

EURASIAN KESTREL

Falco tinnunculus | L 13½" (34 cm) WS 29" (74 cm)

Resembles American Kestrel, but note larger size and single, not double, dark facial stripe. In flight, distinguished by wedge-shaped tail and two-toned upperwing, with back and inner wing paler. Hovers as it hunts. **Adult male** has russet wings, gray tail; **female** duller, often with gray rump. **Juvenile** similar to adult female, but dark barring heavier on upperparts and tail. Often hovers to spot prey.

RANGE: Old World species. Casual on western Aleutians and in Bering Sea region; accidental in fall, winter, and spring on the East Coast from New Brunswick and Nova Scotia to New Jersey and Florida; and on the West Coast to Washington.

MERLIN

Falco columbarius | L 12" (31 cm) WS 25" (64 cm)

Adult male is gray-blue above; **female** and juveniles usually dark brown. Merlins lack the strong facial markings and russet upperparts of kestrels, and have broader wings than American Kestrel. Plumage varies geographically from the very dark race, *suckleyi*, of the Pacific Northwest to the pale *richardsonii* that breeds on northern Great Plains from southern Canada to northern U.S. A few *suckleyi* winter to southern California; this race has dark cheeks and narrow, incomplete tail bands. All *richardsonii* have pale cheeks; male is paler blue-gray above; female and juvenile are pale brown, the latter with wide, pale tail bands. Winter to southern Great Plains, a few to the Great Basin and Pacific states. The widespread nominate race, *columbarius*, which breeds in the taiga region, is intermediate in plumage; western *columbarius* average slightly paler than eastern. In flight, strongly barred tail distinguishes Merlin from the much larger Peregrine and Prairie Falcons (next page). Underparts and underwings darker than in kestrels, particularly in *suckleyi* and *columbarius*, and head larger. Catches birds in flight by a sudden burst of speed rather than by diving. Also eats large insects and small rodents. Powerful flyer; does not hover.

RANGE: Fairly common to common on East Coast during fall migration. Many individuals in the Prairie Provinces do not migrate but winter in or near cities. Generally uncommon throughout U.S. in winter. Nests in open woods or wooded prairies; otherwise found in a variety of habitats.

adult ♀

adult ♂

adult ♂

EURASIAN
KESTREL

juvenile

adult ♀

adult ♂

juvenile ♂

AMERICAN
KESTREL

adult ♂

adult ♂

adult ♂
♀

MERLIN
columbarius

♂

♀ *suckleyi*

adult ♂
suckleyi

adult ♂

♀ *richardsonii*

adult ♂
richardsonii

PRAIRIE FALCON
Falco mexicanus | L 15½–19½" (39–50 cm) WS 35–43" (89–109 cm)

Pale brown above; creamy white and heavily spotted below. Brown crown, dark moustachial, and broad pale area below and behind eye; facial markings narrower and plumage paler overall than Peregrine Falcon. Compare also with female and juvenile male Merlin (preceding page), especially subspecies *richardsonii*. In flight (see also next page), all ages show distinctive dark axillaries and dark bar on wing lining, broader on **females**. Juvenile is streaked below (not spotted) and darker above; bluish cere. Preys chiefly on birds, especially flocking species in winter, and small mammals.

RANGE: Inhabits dry, open country, prairies. Uncommon to fairly common. Rare migrant and winter visitor in western Midwest. Casual elsewhere in Midwest and Southeast. Small numbers winter throughout breeding range.

PEREGRINE FALCON
Falco peregrinus | L 16–20" (41–51 cm) WS 36–44" (91–112 cm)

Crown and nape black; black wedge extends below eye, forming a distinctive helmet, absent in Prairie Falcon and smaller Merlin (preceding page). Tail is shorter than in Prairie; wing tips almost reach the end; also lacks dark bar and axillaries on underwings. Plumage varies from pale in subspecies *tundrius* of the North to very dark in *pealei*, found from Queen Charlotte Islands to the Aleutians. In *pealei*, the largest race, adult has heavy spotting on whitish breast, underparts very dark. Intermediate *anatum* race has thickest moustachial stripe; **adult** shows rufous wash below; **juvenile** is dark brownish above, and underparts are heavily streaked. Juvenile *tundrius* has a pale eyebrow and larger pale area on side of face; underparts more finely streaked.

RANGE: Peregrines inhabit open wetlands near cliffs; prey chiefly on birds. Now established also in cities; nest on bridges, tall buildings. Use of pesticides helped eliminate eastern *anatum* breeding populations; now reintroduced in parts of their former range, Peregrines are seen year-round. Most East Coast sightings in the fall are of *tundrius* birds. Uncommon to rare in winter in U.S.

GYRFALCON
Falco rusticolus | L 20–25" (51–64 cm) WS 50–64" (127–163 cm)

Heavily built; wings broader based than in other falcons. **Adult** has yellow-orange eye ring, cere, and legs (bluish gray in juveniles). Tail broad and tapered; may be barred or unbarred; in perched bird, tail extends far beyond wing tips, unlike other falcons. Compare also with Northern Goshawk (page 126). Plumages vary from **white morph** to **gray morph**, to very **dark morph**, with paler gray morphs intermediate between typical gray and white. Facial markings range from none on white morph to all-dark cheeks on dark morph. Juveniles of white and gray morphs are much browner above; **juveniles** of gray and dark morphs show darker wing linings and paler flight feathers. Flies with slow, powerful wingbeats. Preys chiefly on birds; capable of taking down flying geese..

RANGE: Gyrfalcon inhabits open tundra near rocky outcrops and cliffs. Uncommon; winters irregularly south to dashed line on map. Casual to central California, southern Great Plains, southern Great Lakes, and mid-Atlantic regions.

adult ♂

♀

PRAIRIE
FALCON

adult
pealei

juvenile
pealei

adult
pealei

PEREGRINE
FALCON

adult
anatum

adult
anatum

juvenile
anatum

adult
tundrius

juvenile
tundrius

gray-morph
adult

GYRFALCON

gray-morph
juvenile

dark-morph
juvenile

white-morph
adult

FEMALE HAWKS IN FLIGHT

adult

WHITE-TAILED KITE

1st summer

MISSISSIPPI KITE

adult

HOOK-BILLED KITE

adult

SNAIL KITE

MERLIN

columbarius

AMERICAN KESTREL

adult

PEREGRINE FALCON

gray-morph juvenile

GYRFALCON

adult

PRAIRIE FALCON

FEMALE HAWKS IN FLIGHT

adult — SHARP-SHINNED HAWK

adult — COOPER'S HAWK

adult — NORTHERN GOSHAWK

adult — NORTHERN HARRIER

adult — GRAY HAWK

adult — BROAD-WINGED HAWK

dark-morph adult — BROAD-WINGED HAWK

adult *lineatus* — RED-SHOULDERED HAWK

FEMALE HAWKS IN FLIGHT

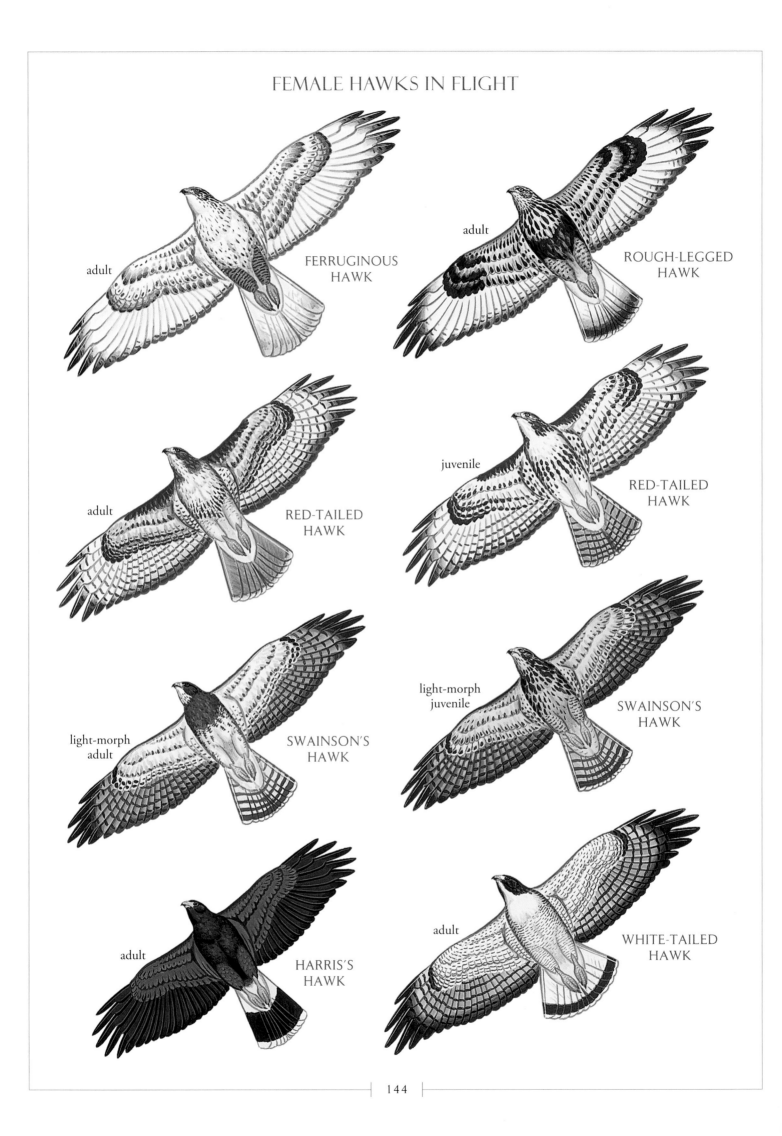

adult   FERRUGINOUS HAWK

adult   ROUGH-LEGGED HAWK

adult   RED-TAILED HAWK

juvenile   RED-TAILED HAWK

light-morph adult   SWAINSON'S HAWK

light-morph juvenile   SWAINSON'S HAWK

adult   HARRIS'S HAWK

adult   WHITE-TAILED HAWK

adult
CRESTED
CARACARA

adult
OSPREY

juvenile
ZONE-TAILED
HAWK

2nd year
BALD
EAGLE

adult
COMMON
BLACK-HAWK

adult
GOLDEN
EAGLE

adult
BLACK
VULTURE

adult
TURKEY
VULTURE

LIMPKINS (Family Aramidae)

Large, long-necked wading bird, named for its unusual limping gait.
Species: *1 World, 1 N.A.*

LIMPKIN

Aramus guarauna | L 26" (66 cm)

Chocolate brown overall, densely streaked with white above. Long bill, slightly downcurved. Long legs and large, webless feet are dull grayish green. Juvenile is paler than **adult. Call**, heard chiefly at night, is a wailing *krr-oww*.

RANGE: Generally uncommon in swamps and wetlands, where it wades or swims in search of snails, frogs, and insects. Rare to fairly common in Florida; casual in southern Georgia; accidental north to Maryland.

RAILS, GALLINULES, AND COOTS (Family Rallidae)

These marsh birds have short tails and short, rounded wings. Most species are local and secretive. Some, especially the rails, are identified chiefly by call and habitat. **Species:** *141 World, 14 N.A.*

KING RAIL

Rallus elegans | L 15" (38 cm)

Large freshwater rail with long, slightly downcurved bill. Much larger than similar Virginia Rail (next page). Adult distinguished from Clapper Rail by tawny edges on black-centered back feathers, tawny wing coverts. Head slate, with brown or grayish cheeks, buffy eyebrow; underparts cinnamon; flanks strongly barred black-and-white. **Juvenile** is darker above, paler below. King Rail favors freshwater and brackish swamps and marshes but hard to see. As with Clapper Rail, most often heard at dusk and dawn. Usually distinctive **call** is a series of fewer than ten *kek kek kek* notes, fairly evenly spaced.

RANGE: Fairly common to common in freshwater habitat near Gulf Coast; generally rather rare and local well inland in East. Rare in west Texas, where it may breed. Casual west to Colorado. Some birds winter in coastal marshes with Clapper Rails. Hybridizes with Clapper Rail in narrow zone of overlap; some calls of the two are identical.

CLAPPER RAIL

Rallus longirostris | L 14½" (37 cm)

Much larger than Virginia Rail (next page). Plumage variable but always has grayish edges on brown-centered back feathers, olive wing coverts. East Coast subspecies such as *crepitans* are much duller than King Rail: buffy below; cheeks gray; flanks less strongly barred than in King. Gulf Coast races such as *scottii* are brighter cinnamon below. West Coast races such as *levipes* (**E**) and inland *yumanensis* (**E**) are brighter below than East Coast birds; cheeks brownish gray. Clapper's **call** is a series of ten or more dry *kek kek kek* notes, like King but accelerating and then slowing.

RANGE: Common in coastal salt marshes except on West Coast, where populations have declined since introduction of the red fox; also found along lower Colorado River and at Salton Sea. In East, casual north to Maritimes.

LIMPKIN
pictus

adult

KING RAIL
elegans

adult

juvenile

crepitans

CLAPPER
RAIL

yumanensis

scottii

"Light-footed"
levipes

VIRGINIA RAIL

Rallus limicola | L 9½" (24 cm)

Similar to King Rail (preceding page) but smaller; cheeks grayer; wings richer chestnut; legs and bill often redder. **Juvenile** is blackish brown above, mottled black or gray below. **Song** a series of *kid kid kidick kidick* phrases, heard chiefly in breeding season; common **call**, heard year-round, is a descending series of *oink* notes.

RANGE: Common but a bit secretive; found in freshwater and brackish marshes and wetlands; also in coastal salt marshes.

SORA

Porzana carolina | L 8¾" (22 cm)

Short, thick bill, yellow or greenish yellow. **Breeding adult** is coarsely streaked above. Face and center of throat and breast are black. In **winter** plumage, black throat is somewhat obscured by gray edgings. **Juvenile** lacks black on face and throat; underparts are paler. Compare with Yellow Rail; juvenile Sora is not as black above; upperparts streaked, not barred, with white. **Calls** heard year-round are a descending whinny and a sharp, high-pitched *keek*; a whistled *ker-wheer* is heard on breeding grounds.

RANGE: Common in freshwater and brackish marshes, rice fields, grainfields; found in saltwater marshes during migration and winter.

YELLOW RAIL

Coturnicops noveboracensis | L 7¼" (18 cm)

A small, dark rail, deep tawny yellow above with wide dark stripes crossed by white bars. In flight, shows a large white patch on trailing edges of wings. Bill is short and thick; color varies from yellowish to greenish gray. **Juvenile** is darker than adult. Distinctive **call**, heard chiefly in breeding season, a four- or five-note *tick-tick, tick-tick-tick* in alternate twos or twos and threes, sounds like tapping two pebbles together.

RANGE: Uncommon and local; secretive. Breeds in grassy marshes, boggy swales, damp fields; not found in deepwater marshes or swamps. Rare in the West. Winters in fresh, brackish, or salt marshes, rice fields, dry fields.

BLACK RAIL

Laterallus jamaicensis | L 6" (15 cm)

Very small, extremely secretive. Blackish above, with white speckling; chestnut nape. Bill short and black. Underparts grayish black, with narrow white barring on flanks. Newly hatched juveniles of other rails resemble Black Rail. Unlike other rails, most vocal in the middle of the night. Distinctive **call**, heard chiefly in breeding season, is a repeated *kik-kee-do or kik-kee-derr*; sometimes four notes: *kik-kik-kee-do*.

RANGE: Uncommon and local; inhabits marshes, swamps, wet meadows. Very irregular inland; range speculative; declining in some coastal areas.

CORN CRAKE

Crex crex | L 10½" (27 cm)

Dull buffy yellow with short, thick, brownish bill; distinctive large chestnut wing patch.

RANGE: European species, formerly a very rare vagrant in fall along the East Coast; recently only three fall records: Saint-Pierre, Nova Scotia, and Avalon Peninsula, Newfoundland. Western European populations have seriously declined over last decades. Found in damp, grassy fields, croplands, not in marshes.

VIRGINIA RAIL
limicola

juvenile

SORA

winter
♀

juvenile

breeding
adult ♂

YELLOW RAIL
noveboracensis

juvenile

CORN
CRAKE

BLACK RAIL
jamaicensis

PURPLE GALLINULE
Porphyrio martinica | L 13" (33 cm)

Bright purplish blue head, neck, and underparts, with pale blue forehead shield, red-and-yellow bill. Legs and feet bright yellow. **Juvenile** is buffy brown overall, with brownish olive back, greenish wings; bill mostly dark olive, legs and feet dull olive. Molts into winter plumage after fall migration but may retain traces of juvenal plumage into first spring. In all ages, all-white undertail coverts are conspicuous.

RANGE: Fairly common in overgrown swamps, lagoons, and marshes. Highly migratory; winters from southern Florida to Argentina. Wanderers are seen in all seasons far north of mapped range; frequently breeds north of area shown.

COMMON MOORHEN
Gallinula chloropus | L 14" (36 cm)

Black head and neck, with red forehead shield, red bill with yellow tip. Back brownish olive; underparts slate; white along flanks is diagnostic. Outer undertail coverts white, inner ones black. Legs and feet yellow. **Juvenile** is paler, browner; throat whitish; bill and legs dusky. **Winter adult** has brownish facial shield and usually a brownish bill with dusky yellow tip.

RANGE: Common in freshwater marshes, ponds, and placid rivers; now uncommon to rare and declining from much of interior range.

PURPLE SWAMPHEN
Porphyrio porphyrio | L 18–20" (45-50 cm)

Resembles a huge Purple Gallinule with reddish bill and frontal shield, iris and legs. Florida birds appear to belong to the *poliocephalus* group subspecies (three included) from southern Asia with grayish blue neck, or possibly closely allied *viridis* (mainland Southeast Asia), but some bluer-headed swamphens likely represent other subspecies.

RANGE: Widespread species from southern Europe to island groups in tropical South Pacific. Various subspecies groups are recognized, some of which perhaps deserve recognition as separate species. Introduced into south Florida in 1996 and has slowly spread.

AMERICAN COOT
Fulica americana | L 15½" (39 cm)

Overall blackish, darker on head and neck; outer undertail coverts white. Whitish bill has dark subterminal band; note reddish brown forehead shield. Leg color ranges from greenish gray in **juvenile** birds to yellow or orangish in adults. Toes are lobed, unlike gallinules. Juvenile is quite pale; by first winter more like adult, but still paler, with whitish feather tips, especially below. In flight, white trailing edge on most of wing is distinctive. A few **variant** American Coots have extensively white facial shields like Caribbean Coot, *F. caribaea*, of the West Indies. Regarded by some as a subspecies of American, Caribbean has not yet been verified in Florida.

RANGE: Common to abundant. Nests in freshwater marshes, wetlands, or near lakes and ponds; winters in both fresh and salt water, usually in large flocks. Often dives to feed.

EURASIAN COOT
Fulica atra | L 15¾" (40 cm)

Slightly larger and darker than American Coot; undertail coverts all-black. Forehead shield and bill entirely white.

RANGE: Accidental straggler to Newfoundland, Labrador, Quebec, and the Pribilofs.

PURPLE
GALLINULE

juvenile

winter

COMMON
MOORHEN
cachinnans

breeding

juvenile

PURPLE
SWAMPHEN
poliocephalus

EURASIAN
COOT
atra

variant

AMERICAN
COOT
americana

juvenile

juvenile

CRANES (Family Gruidae)

Tall birds with long necks and legs. Tertials droop over the rump in a "bustle" that distinguishes cranes from herons. Cranes fly with their necks fully extended and circle in thermals like raptors. Courtship includes a frenzied, leaping dance. **Species:** *15 World, 3 N.A.*

SANDHILL CRANE
Grus canadensis | L 34–48" (86–122 cm) WS 73–90" (185–229 cm)

Races vary in size: northern nominate race smallest; more southerly *tabida* largest. Resident Florida race, *pratensis*, and endangered Gulf Coast race, *pulla* (**E**), are intermediate. **Adult** is gray, with dull red skin on crown and lores; whitish chin, cheek, and upper throat; and slaty primaries. **Juvenile** lacks red patch; head and neck vary from pale to tawny; gray body is irregularly mottled with brownish red; full adult plumage reached after two and a half years. Great Blue Heron (page 114), sometimes confused with Sandhill Crane, lacks bustle. Preening with muddy bills, cranes may stain feathers of upper back, lower neck, and breast with ferrous solution in mud. Common **call** is a trumpeting, rattling *gar-oo-oo*, audible for more than a mile.
Range: Locally common; breeds on tundra and in marshes and grasslands. In winter, regularly feeds in dry fields, returning to water at night. Resident near parts of the Gulf Coast, in Florida, and in Cuba; other North American subspecies migratory. Rare during fall and winter on East Coast from Massachusetts south. Migrating flocks fly at great heights, sometimes too high to be seen from the ground.

COMMON CRANE
Grus grus | L 44–51" (112–130 cm) WS 79–91" (202–231 cm)

Adult distinguished from Sandhill Crane by blackish head and neck marked by broad white stripe. **Juvenile** like juvenile Sandhill; may show trace of white head stripe by spring. In flight, in all ages, black primaries and secondaries show as a broad black trailing edge on gray wings. Gives a far-carrying trumpeting **call**.
Range: Eurasian species, casual vagrant on the Great Plains, accidental farther east; almost always with migrating flocks of Sandhill Cranes. Some records involve a mixed Common-Sandhill pair with accompanying hybrid young.

WHOOPING CRANE
Grus americana | **E** | L 52" (132 cm) WS 87" (221 cm)

Adult is white overall, with red facial skin; black primaries show in flight. **Juvenile** bird is whitish, with pale reddish brown head and neck and scattered reddish brown feathers over the rest of its body; begins to acquire adult plumage after first summer. A few abnormally colored Sandhill Cranes of *tabida* race ("Greater Sandhill Crane") have been taken for Whooping Cranes; check wingtip pattern. **Call** is a shrill, trumpeting *ker-loo ker-lee-loo*.
Range: Sparse wild population breeds in freshwater marshes of Wood Buffalo National Park, Canada, and winters in Aransas National Wildlife Refuge on Gulf Coast of Texas. A small population has been introduced in Florida, some migrate to Wisconsin. Wild population is now about 200, including introductions. Intensive management and protection seem to be slowly succeeding.

SANDHILL
CRANE
rowani

adult

juvenile

stained
adult

adult

COMMON
CRANE
lilfordi

juvenile

adult

adult

WHOOPING
CRANE

adult

adult

juvenile

LAPWINGS AND PLOVERS (Family Charadriidae)

These compact birds run and stop abruptly when foraging. Shape and behavior identify plovers in general.
Species: 66 World, 17 N.A.

BLACK-BELLIED PLOVER
Pluvialis squatarola | L 11½" (29 cm)

Black-and-white **breeding male** has frosty crown and nape, white belly region; **female** averages less black. **Winter** and **juvenile** birds distinguished from Pacific and American Golden-Plovers by larger size, larger bill, and grayer plumage (including crown); underparts streaked rather than softly barred, but note that juvenile can be speckled with gold above. In flight, shows black axillaries and white uppertail coverts, barred white tail, and bold white wing stripe. **Call** is a drawn-out, three-note whistle, the second note lower-pitched.
RANGE: Nests on Arctic tundra. Common migrant in Great Lakes region. Uncommon to rare elsewhere in interior.

AMERICAN GOLDEN-PLOVER
Pluvialis dominica | L 10¼" (26 cm)

Smaller, with a smaller bill than Black-bellied Plover; wing stripe is indistinct, and underwings are smoky gray with no black in axillaries; no contrasting white rump. Note the four evenly spaced primary tips. **Breeding male** shows broad white patches on sides of neck; underparts otherwise black. **Female** has less black. Flight **call** is a shrill *ku-wheep*.
RANGE: Mar. arrivals are in winter plumage; breeding plumage slowly acquired on migration north. Rare fall migrant in West (apparently all **juveniles**). Winters in South America.

PACIFIC GOLDEN-PLOVER
Pluvialis fulva | L 9¾" (25 cm)

Similar to American, but shorter primary tip projection with three, not four, staggered primary tips, the outer two close together; bill appears thicker, legs longer. **Breeding male** has less extensive white on sides of neck than American; white continues down sides and flanks; undertail coverts whiter; slightly larger gold markings above. Female has less black below. **Juveniles** and **winter** birds typically appear brighter than American. **Call**, a loud, rich *chu-wheet*.
RANGE: Breeds from northern Russia to western Alaska. Winters from southern Asia to Pacific islands; a few on West Coast and in central California. Some adults migrate earlier in fall than other golden-plovers. In Alaska, Americans favor less vegetated slopes; Pacifics, the coast and river valleys.

EUROPEAN GOLDEN-PLOVER
Pluvialis apricaria | L 11" (28 cm)

Similar to Pacific Golden-Plover; note larger size, plumper body shape, white underwings; also small bill and bolder wing bar. On **breeding males**, white nearly meets on front of breast; sides, flanks, and undertail coverts are more purely white than on Pacific; note dense pattern of smaller gold spots on upperparts, unlike coarser pattern of larger spots on other golden-plovers. **Call** is a mournful, drawn-out whistle.
RANGE: Breeds from Greenland and Iceland to northwestern Russia. Winters from Europe to North Africa. Irregular spring migrant to Newfoundland and Labrador; casually to eastern Quebec and Nova Scotia.

juveniles in flight

European   Pacific   American   Black-bellied

bright
juvenile

breeding ♀

juvenile

winter

BLACK-BELLIED
PLOVER

breeding ♂

bright
juvenile

breeding ♀

juvenile

April ♂

AMERICAN
GOLDEN-PLOVER

breeding ♂

juveniles

breeding ♂

winter

breeding
♀

PACIFIC
GOLDEN-PLOVER

breeding ♂

juveniles

breeding ♂

breeding ♀

EUROPEAN
GOLDEN-PLOVER
altifrons

SNOWY PLOVER
Charadrius alexandrinus | L 6¼" (16 cm)

Pale above, very pale in Gulf Coast birds; thin dark bill; dark or grayish legs; partial breast band; dark ear patch. **Females** and **juveniles** resemble Piping Plover; note Snowy Plover's thinner bill, darker legs. **Calls** include a low *krut* and a soft, whistled *ku-wheet*.
RANGE: Inhabits barren sandy beaches and flats. Uncommon and declining on Gulf Coast. Western *nivosus* is threatened (**T**).

PIPING PLOVER
Charadrius melodus | **E** | L 7¼" (18 cm)

Very pale above; orange legs; white rump conspicuous in flight (page 196). In **breeding** plumage, shows dark narrow breast band usually complete in *circumcinctus,* sometimes incomplete, especially in **females** and paler-faced East Coast birds, *melodus.* In **winter**, bill is all-dark. Distinguished from Snowy Plover by thicker bill, paler back; legs are brighter than in Semipalmated Plover. Distinctive **call**, a clear *peep-lo.*
RANGE: Found on sandy beaches, lakeshores, and dunes. Endangered: generally uncommon; rare and declining breeder in the Midwest; rare migrant in interior. In winter, *melodus* winters mainly on south Atlantic coast and northern Bahamas; *circumcinctus* on Gulf Coast. Casual in winter to coastal southern California.

WILSON'S PLOVER
Charadrius wilsonia | L 7¾" (20 cm)

Long, very heavy, black bill; broad neck band is black in **breeding male**, brown in **female** and winter male; legs grayish pink. **Juvenile** resembles adult female but note scaly-looking upperparts. Breeding male may have cinnamon-buff ear patch. **Call** is a sharp, whistled *whit.*
RANGE: Fairly common but declining on barrier islands, sandy beaches, mud flats. Recorded casually to California and the Maritimes and inland to Great Lakes region.

SEMIPALMATED PLOVER
Charadrius semipalmatus | L 7¼" (18 cm)

Dark back distinguishes this species from Piping and Snowy Plovers; bill much smaller than in Wilson's Plover (flight figures page 196). At very close range, Semipalmated shows partial webbing between toes. **Breeding male** often lacks white above eye; shows orangish eye ring. **Juvenile** has darker legs than adults'. Distinctive **call** is a whistled, upslurred *chu-weet;* **repeated in a series during breeding**.
RANGE: Common on beaches, lakeshores, and tidal flats; seen throughout the continent in migration, sometimes in large flocks.

COMMON RINGED PLOVER
Charadrius hiaticula | L 7½" (19 cm)

Almost identical to Semipalmated Plover; best distinguished by **call**, a soft, fluted *pooee;* **song**, delivered in display flight, is a series of these notes. Breast band averages slightly broader in center than in Semipalmated. White eyebrow is more distinct; eye ring is partial or lacking altogether; webbing between toes less extensive. Bill is slightly longer, of more even thickness, and shows more orange at base; black on face meets bill where mandibles join.
RANGE: Regular but rare spring migrant on western Alaska islands; occasionally breeds on St. Lawrence Island. Casual on East Coast south to Massachusetts.

SNOWY
PLOVER

Gulf Coast ♂
tenuirostris

western
nivosus

♀

♂

juvenile

PIPING
PLOVER
melodus

winter

breeding ♀

breeding ♂

interior
breeding ♂
circumcinctus

WILSON'S
PLOVER
wilsonia

juvenile

breeding ♂

♀

winter

breeding ♂

SEMIPALMATED
PLOVER

juvenile

breeding ♀

breeding ♂

breeding ♀

COMMON
RINGED PLOVER

LESSER SAND-PLOVER *Charadrius mongolus* | L 7½" (19 cm)

Bright rusty red breast; black-and-white facial pattern. **Females** are duller. **Juvenile** has broad buffy wash across breast; edged with buff above. In **winter** birds, underparts are white except for broad grayish patches on sides of breast.

RANGE: Asian species, rare migrant on Aleutians and on islands off western Alaska; casual along West Coast. Casual in summer in western and northwestern Alaska, where it has bred. Accidental in eastern North America.

LITTLE RINGED PLOVER *Charadrius dubius* | L 6" (15 cm)

A small, slim plover with conspicuous yellow eye ring; legs rather dull color. In flight (page 196), note lack of wing bar. In **breeding** adult, white line separates brown forecrown from rear of head. On winter birds and **juveniles**, brown replaces black on head and breast, and eye ring is slightly duller; juvenile often shows yellow-buff tint to pale areas on head and throat. Rather solitary. **Call** is a descending *pee-oo* that carries a long way.

RANGE: Old World species. Casual spring vagrant to western Aleutians.

KILLDEER *Charadrius vociferus* | L 10½" (27 cm)

Double breast bands distinctive, as is Killdeer's loud, piercing **call**, *kill-dee* or *dee-dee-dee*. Bright reddish orange rump is visible in flight (page 196). Downy young have only one breast band, but quickly grow out of this plumage.

RANGE: Common in grassy fields and on shores, but rarely along the ocean shore. Nests on open ground, usually on gravel. May form loose flocks and linger into early winter in summer range. Vagrant north of breeding range.

MOUNTAIN PLOVER *Charadrius montanus* | L 9" (23 cm)

In **breeding** plumage, unbanded white underparts separate this plover from all other brown-backed plovers. Buffy tinge on breast is more extensive in **winter** plumage; compare with winter American Golden-Plover (page 154). In flight, Mountain Plover shows white underwings; American Golden-Plover's are grayish. **Calls** heard on breeding grounds include low, drawn-out whistles and harsh notes. In migration and winter, gives a harsh *krrr* note.

RANGE: Inhabits plains; local and declining in many areas. Gregarious in winter; usually found on grassy or bare dirt fields.

NORTHERN LAPWING *Vanellus vanellus* | L 12½" (32 cm)

Most sightings of birds in **winter** plumage: dark, iridescent above, white below, with black breast; wispy but prominent crest. Wings broad and rounded, with white tips, white wing linings. Flight **call** is a whistled *pee-wit*.

RANGE: Eurasian species, casual primarily in late fall in northeast states and provinces; accidental elsewhere in East; recorded south to Florida.

EURASIAN DOTTEREL *Charadrius morinellus* | L 8¼" (21 cm)

Whitish band on lower breast is somewhat obscured in young and **winter** birds. Bold white eyebrow extends around entire head. Unlike other plovers, **females** are brighter than males. **Juvenile** is darker above, buffy below.

RANGE: Eurasian species, very approachable; uncommon, sporadic breeder in northwestern Alaska; casual along West Coast in fall.

breeding ♀

LESSER
SAND-PLOVER
stegmanni

breeding ♂

winter

juvenile

LITTLE RINGED
PLOVER
curonicus

juvenile

breeding ♂

KILLDEER

winter

MOUNTAIN
PLOVER

breeding

winter

juvenile

winter

NORTHERN
LAPWING

winter

winter

breeding ♂

breeding ♀

winter

EURASIAN
DOTTEREL

juvenile

JACANAS (Family Jacanidae)

Extremely long toes and claws allow these tropical birds to walk on lily pads and other floating plants.
***Species:** 8 World, 1 N.A.*

NORTHERN JACANA *Jacana spinosa* | L 9½" (24 cm)

Often raises its wings, revealing yellow flight feathers. **Immature** is white below.
RANGE: Mexican and Central American species, casual visitor to ponds and marshes in southern Texas, where it has probably bred. Accidental to west Texas and southern Arizona.

OYSTERCATCHERS (Family Haematopodidae)

These chunky shorebirds have laterally flattened, heavy bills that can reach into mollusks and pry the shells open; they also probe sand for worms and crabs. ***Species:** 11 World, 3 N.A.*

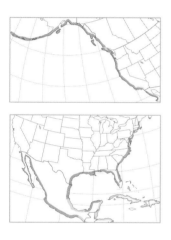

BLACK OYSTERCATCHER *Haematopus bachmani* | L 17½" (45 cm)

Large red-orange bill, all-dark body, pinkish legs. On immatures, outer half of bill is dusky during first year. **Call** is a loud, piercing *wheep* or *whee-ah*.
RANGE: Resident on rocky shores and islands along the Pacific coast.

AMERICAN OYSTERCATCHER *Haematopus palliatus* | L 18½" (47 cm)

Large red-orange bill. Black head and dark brown back; white wing and tail patches, white underparts. **Juvenile** appears scaly above; dark tip on bill is kept through first year.
RANGE: Expanding northward in the East; recently established as a breeder on Cape Sable Island, Nova Scotia. Casual in southern California, where has hybridized with Black Oystercatcher. Birds feed in small, noisy flocks on coastal beaches and mudflats.

STILTS AND AVOCETS (Family Recurvirostridae)

Sleek and graceful waders with long, slender bills and spindly legs. Two species inhabit North America.
***Species:** 7 World, 3 N.A.*

AMERICAN AVOCET *Recurvirostra americana* | L 18" (46 cm)

Black-and-white above, white below; head and neck rusty in breeding plumage, gray in winter. **Juveniles** have cinnamon wash on head and neck. Avocets feed by sweeping their bills through the water. **Male's** bill is longer, straighter than **female's**. **Call** is a loud *wheet*.
RANGE: Fairly common on shallow ponds, marshes, and lakeshores.

BLACK-NECKED STILT *Himantopus mexicanus* | L 14" (36 cm)

Male's glossy black back and bill contrast sharply with white underparts, long red or pink legs. Female is browner on back. **Juvenile** is brown above, with buffy edgings. Common **call** is a loud *kek kek kek*.
RANGE: Breeds and winters in a wide variety of wet habitats; breeding range is spreading north. Casual north to Great Lakes.

NORTHERN
JACANA
gymnostoma

adult

immature

BLACK
OYSTERCATCHER

adult

adults

AMERICAN
OYSTERCATCHER
palliatus

juvenile

AMERICAN
AVOCET

breeding ♀

juvenile ♂

breeding ♀

winter ♂

BLACK-NECKED
STILT

juvenile

♂

SANDPIPERS, PHALAROPES, and ALLIES (Family Scolopacidae)

The majority of these shorebirds have three distinct plumages. Most begin molting to winter plumage as they near or reach their winter grounds. **Species: *92 World, 65 N.A.***

WILLET

Tringa semipalmata | L 15" (38 cm)

Large, plump, and grayish overall with grayish legs. In flight, note black-and-white wing pattern. Two subspecies: eastern *semipalmata* is smaller, darker, browner, and thicker billed than western *inornata* in all plumages. **Breeding** *semipalmata* is more heavily barred below with more pinkish-based bill than *inornata*. **Juvenile** *semipalmata* has more contrasting scapulars than *inornata*. Separating winter birds to subspecies best done by structural features and range. Territorial **call** is *pill-will-willet*, distinctly faster and higher-pitched in nominate race. Other raucous calls are given year-round, all averaging higher-pitched and faster in the nominate eastern race.

RANGE: Nominate *semipalmata* nests in coastal Atlantic and Gulf salt marshes; *inornata* in interior marshes. In fall *semipalmata* is an early migrant, most departing by early Aug., and winters entirely outside North America (mainly South America). Winter-plumaged *semipalmata* typically not seen in North America, except for early Mar. arrivals on Gulf Coast.

LESSER YELLOWLEGS

Tringa flavipes | L 10½" (27 cm)

Legs yellow to rarely orange. Smaller than Greater Yellowlegs; all-dark bill is shorter, thinner, and straighter. In **breeding** plumage, not nearly as heavily as marked as Greater Yellowlegs, especially on sides and flanks. **Juvenile** and **winter** birds are overall darker than Greater; juvenile Lessers are washed with grayish brown across the neck and lack the streaks that are present in Greater. A more sedate feeder than Greater, leisurely picks at prey from a more vertical position. **Call** is one to three *tew* notes, a little higher than Greater Yellowlegs, the individual notes being more clipped and all on one pitch.

RANGE: Common in the East (often abundant on Great Plains and Midwest); uncommon in far West, especially in spring (more numerous in fall). Nests on tundra or in woodland. On nesting grounds, like Greater, often perches in trees and yelps at intruders. Most winter in South America, a few in U.S.

GREATER YELLOWLEGS

Tringa melanoleuca | L 14" (36 cm)

Legs yellow to orange. Larger than Lesser Yellowlegs; bill longer, stouter, often slightly upturned, and, in all plumages except breeding, two-toned. In **breeding** plumage, throat and breast are heavily streaked; sides and belly are spotted and barred with black; bill is all black. In **juvenile** birds the neck is distinctly streaked. Behavior is more active than Lesser Yellowlegs, often racing about with extended neck while pursuing prey (often very small fish). **Call** is a loud, slightly descending series of three or more *tew* notes.

RANGE: Fairly common; nests on muskeg, winters in wetland habitats. Compared to Lesser Yellowlegs, Greater is overall a later fall migrant and is much more capable of wintering at interior locations.

early March
in molt

western

eastern

breeding

eastern
semipalmata

juvenile

winter

WILLET

worn
breeding

western
inornata

breeding

winter

juvenile

breeding

LESSER
YELLOWLEGS

winter

winter

winter

juvenile

GREATER
YELLOWLEGS

breeding

winter

juvenile

COMMON GREENSHANK
Tringa nebularia | L 13½" (34 cm)

In plumage and structure resembles Greater Yellowlegs, but less heavily streaked; legs are greenish. In flight, white wedge extends up the middle of back. Typical flight **call** is a loud *tew-tew-tew*, very much like Greater Yellowlegs, but the notes are all on one pitch.

RANGE: Common Eurasian species that annually visits western Alaska in spring: rare to Aleutian and Pribilof Islands; casual to St. Lawrence Island; casual in western Alaska in fall. Accidental to northeast Canada and coastal northwest California.

MARSH SANDPIPER
Tringa stagnatilis | L 8½" (21 cm)

A small and slender *Tringa* with a long needle-like bill and disproportionately long, greenish legs; distinct supercilium in all plumages. Plumages overall suggestive of Common Greenshank. In **breeding** neck and sides are streaked and spotted with brown and mottled with black above. **Juvenile** is faintly streaked on sides of breast, brownish above with pale buff edges. **Winter** adult pale gray and uniform above and with long needle bill suggests a winter-plumaged Wilson's Phalarope; Marsh has much longer legs. In flight, white wedge extends up back; note long leg projection past tail. **Call** is a *tew* note, like call of Lesser Yellowlegs; often delivered in a series.

RANGE: Old World species. Common in Asia. Casual in central and western Aleutians (three fall records). One wintered recently on Oahu.

COMMON REDSHANK
Tringa totanus | L 11" (28 cm)

Bright orange legs, stout bill with reddish orange base, overall brownish plumage with distinct eye ring. **Juvenile** and **breeding** birds are extensively streaked below; winter birds (not illustrated) are diffusely mottled below. In flight, shows white dorsal wedge up back and distinctive broad white trailing edge to secondaries and inner primaries. **Calls** include a musical *tew* or more mournful *tew-hieu*. Alarm note is a series of *twek* notes.

RANGE: Eurasian species; breeds as close to North America as Iceland; numerous records for Greenland. Casual to Newfoundland in spring, once in winter.

SPOTTED REDSHANK
Tringa erythropus | L 12½" (32 cm)

Long bill with red-based lower mandible, droops at tip. **Breeding adult** is black overall with white spots above; legs dark red to blackish. Many underparts variably marked with white. **Juvenile** is brownish gray and is heavily barred and spotted below. **Winter** birds are pale gray above, spotted with white on coverts and tertials. Some fall North American sightings involve young birds in transitional plumage. Both plumages show brighter orange legs. In flight, shows white wedge on back and white wing linings. Distinctive **call**, a loud rising *chu-weet* closely resembles the call of Semipalmated Plover.

RANGE: Eurasian species, rare spring and fall visitor to Aleutian and Pribilof Islands; casual on both coasts during migration and winter; accidental elsewhere.

juvenile

breeding

winter

COMMON
GRENSHANK

juvenile

winter

MARSH
SANDPIPER

breeding

COMMON
REDSHANK

winter

juvenile

breeding

juvenile

breeding

SPOTTED
REDSHANK

winter

winter

WOOD SANDPIPER
Tringa glareola | L 8" (20 cm)

Dark upperparts are heavily spotted with buff; prominent whitish eyebrow. In flight (page 195), distinguished from Green Sandpiper by paler wing linings, smaller white rump patch, and more densely barred tail. Bill is straight like Green, but shorter. Common **call** is a loud, sharp whistling of three or more notes, similar to the call of Long-billed Dowitcher.
RANGE: Eurasian species, fairly common spring migrant and occasional breeder on the outer Aleutians; uncommon on the Pribilofs; rare on St. Lawrence Island. Casual to British Columbia and northeast North America.

GREEN SANDPIPER
Tringa ochropus | L 8¾" (22 cm)

Resembles Solitary Sandpiper in plumage, behavior, and calls. Structure also similar, but a little plumper, straighter billed. Note white rump and uppertail coverts, with less extensively barred tail; lacks solidly dark central tail feathers of Solitary; upperparts and wing linings are darker. Similar Wood Sandpiper has more spotting above, more barring on tail, and paler wing linings. **Call** is like Solitary, but louder.
RANGE: Eurasian species, casual in spring in Alaska on outer Aleutians, Pribilofs, and St. Lawrence Island.

SOLITARY SANDPIPER
Tringa solitaria | L 8½" (22 cm)

Dark brown above, heavily spotted with buffy white. White below; lower throat, breast, and sides streaked with blackish brown. Bolder white eye ring and shorter, olive legs distinguish Solitary Sandpiper from Lesser Yellowlegs (page 162). In flight (page 195), shows dark central tail feathers, white outer feathers barred with black. Underwing is dark. Two subspecies: nominate *solitaria* (illustrated) from East is spotted with white in breeding plumage; spots more cinnamon-buff in western *cinnamomea*. **Calls** include a shrill *peet-weet*, higher-pitched than calls of Spotted Sandpiper. Often keeps wings raised briefly after alighting; on the ground, often bobs its tail.
RANGE: Fairly common at shallow backwaters, pools, small estuaries, even rain puddles. Generally seen singly or in small flocks.

TEREK SANDPIPER
Xenus cinereus | L 9" (23 cm)

Note long, upturned bill and short orange-yellow legs. In **breeding adult**, dark-centered scapulars form two dark lines on back. In flight, shows distinctive wing pattern: dark leading edge, grayer median coverts, dark greater coverts, and white-tipped secondaries. Flight **call** is a series of shrill whistled notes on one pitch, usually in threes.
RANGE: Eurasian species, rare migrant on outer Aleutians; casual on Pribilofs, St. Lawrence Island, and in Anchorage area. Accidental in fall to coastal British Columbia, California, and Massachusetts.

WOOD
SANDPIPER

breeding

juvenile

breeding

breeding

juvenile

Solitary
Sandpiper
breeding

GREEN
SANDPIPER

Green
Sandpiper
breeding

breeding

juvenile

SOLITARY
SANDPIPER
solitaria

breeding

TEREK
SANDPIPER

breeding

juvenile

WANDERING TATTLER
Tringa incana | L 11" (28 cm)

Uniformly dark gray above; white eyebrow flecked with gray; bill dark, legs dull yellow. In **breeding** plumage, underparts are heavily barred. **Juvenile** and **winter** birds have only a dark gray wash over breast, sides, and flanks; juvenile has pale spots above. Closely resembles Gray-tailed Tattler; best distinguished by voice. Wandering Tattler's **call** is a rapid series of clear, hollow whistles, all on one pitch. Often teeters and bobs as it feeds.
RANGE: Breeds chiefly on gravelly stream banks. Winters on rocky coasts. Generally seen singly or in small groups. Casual inland during migration. Accidental to eastern North America.

GRAY-TAILED TATTLER
Tringa brevipes | L 10" (25 cm)

Closely resembles Wandering Tattler; upperparts are slightly paler; barring on underparts finer and less extensive; whitish eyebrows are more distinct and meet on forehead. Diagnostic shorter nasal groove hard to see in field. **Juvenile** has less extensive gray on underparts; white flanks and belly; more pale spotting above. Best distinction is voice. Gray-tailed Tattler's common **call** is a loud, ascending *too-weet*, similar to call of Common Ringed Plover.
RANGE: Asian breeding species, regular spring and fall migrant on outer Aleutians, Pribilofs, and St. Lawrence Island; casual visitor to northern Alaska. Accidental in fall to Washington and California.

COMMON SANDPIPER
Actitis hypoleucos | L 8" (20 cm)

Breeding adult is brown above with dark barring and streaking; white below; upper breast finely streaked. **Juvenile** and winter birds resemble Spotted Sandpiper. Note Common Sandpiper's longer tail; in juvenile, barring on edge of tertials extends along the entire feather. In flight, shows longer white wing stripe and longer white trailing edge; wingbeats not as shallow and rapid. **Call** in flight is a shrill, piping *twee-wee-wee*.
RANGE: Eurasian species, rare but regular migrant, usually in spring, on the outer Aleutians, Pribilofs, and St. Lawrence Island; casual to Seward Peninsula.

SPOTTED SANDPIPER
Actitis macularius | L 7½" (19 cm)

Striking in **breeding** plumage, with barred upperparts, spotted underparts. **Juvenile** and **winter** birds lack spotting below, resemble Common Sandpiper. Note Spotted Sandpiper's shorter tail; in flight, shows shorter white wing stripe, shorter white trailing edge. In juvenile and first-winter birds, barred wing coverts contrast with back; tertials have a black bar on tip; and barring, if any, on edge of tertials extends no farther than halfway along each feather. **Calls** include a shrill *peet-weet* and, in flight, a series of *weet* notes, lower-pitched than the calls of Solitary Sandpiper. Both Spotted and Common Sandpipers fly with stiff, rapid, fluttering wingbeats. On the ground, both nod and teeter constantly.
RANGE: Spotted is common and widespread, found at sheltered streams, ponds, lakes, or marshes. Generally seen singly; may form small flocks in migration. Most winter in Central and South America. Rare in winter to southern edge of breeding range.

WANDERING
TATTLER

breeding

breeding

winter

juvenile

GRAY-TAILED
TATTLER

breeding

juvenile

juvenile

COMMON
SANDPIPER

breeding

juvenile

1st winter

juvenile

SPOTTED
SANDPIPER

juvenile

breeding

LITTLE CURLEW
Numenius minutus | L 12" (30 cm)

Like a diminutive Whimbrel, with shorter and only slight curved bill; note mostly pale lores. **Calls** include a musical *quee-dlee* and a loud *tchew-tchew-tchew*.

RANGE: Breeds in Russian Far East; winters mainly northern Australia. Casual fall vagrant to coastal central California (four fall records involving both adult and juvenile birds) and one well-documented record in late spring for Gambell, St. Lawrence Island, Alaska. Also, an accepted spring record of a flyover in western Washington.

ESKIMO CURLEW
Numenius borealis | E | L 14" (36 cm)

Resembles a small Whimbrel, but upperparts darker, bill less curved; wing linings pale cinnamon. Closer in size to Little Curlew but slightly larger (though shorter legged), longer winged, with wingtips extending beyond tail tip at rest, overall more cinnamon coloration (pale buffy in Little Curlew), and more heavily barred breast and flanks with bold Y-shaped marks on the flanks (slight brown barring on Little). The lores are dark, like Whimbrel, not like Little Curlew. **Calls** are poorly known; one call reportedly a rippling *tr-tr-tr* and a soft whistle.

RANGE: Formerly bred on Arctic tundra. Only known nesting area was Anderson River region, Northwest Territories (nests found 1862-1866), but breeding range may have extended west to Alaska and perhaps Russian Far East. Wintered mainly on the Pampas of Argentina. Migrated up through Great Plains in spring; to northeast Arctic Canada and then over the Atlantic to South America in fall. Formerly common and found in large flocks, now probably extinct, the last certain record was of an adult female shot by a "sportsman" on Barbados on 4 Sept. 1963 (specimen eventually secured). Other more recent sightings not adequately documented. Main cause of likely extinction was unregulated market hunting, especially prevalent on central Great Plains in the two decades following the Civil War. Rare by 1900, thought possibly extinct by 1940; but a few persisted, as up to two were well photographed in Mar. and Apr. from 1959 to 1962 at Galveston Island, Texas, the last verified North American records.

WHIMBREL
Numenius phaeopus | L 17½" (45 cm)

Bold, dark-striped crown; dark eye line extending through lores; long downcurved bill. **Call**, a series of hollow whistles on one pitch. In flight, North American *hudsonicus* shows dark rump and underwings; European *phaeopus*, casual vagrant to east (mainly Atlantic coast), white rump and underwings; Asian *variegatus*, rare migrant on islands in western Alaska, casual elsewhere in Pacific region, has whitish, variably streaked rump and underwings. Some argue that based on distinct plumage and genetic differences, the Palearctic subspecies should be split from North American Whimbrels; however, all subspecies give similar vocalizations. In North America, the western breeding population (*rufiventris*; if recognized; migrating through Pacific states) is a little darker brown and larger than eastern birds.

RANGE: Fairly common; nests on open tundra; winters on beaches and coastal wetlands, including rocky shorelines of California; most winter south of U.S. Generally rare in interior, except Great Lakes and California.

adult

adult

LITTLE
CURLEW

ESKIMO
CURLEW

adult

adult

WHIMBREL
hudsonicus

phaeopus

variegatus

juvenile

adult

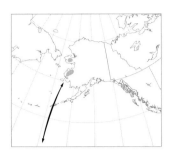

BRISTLE-THIGHED CURLEW
Numenius tahitiensis | L 18" (46 cm)

Bright buff rump and tail (paler when worn) and extensive pattern of large buff spots on upperparts distinguish this species from slightly smaller Whimbrel. The bill is also slightly more strongly decurved near tip. The stiff feathers on the sides and flanks are very hard to see in the field. Main **call**, a loud whistled *chu-a-whit*, and all other vocalizations are completely different from Whimbrel.

RANGE: Winters South Pacific islands. Migration chiefly over water to western Alaskan breeding grounds. Rare spring migrant on Middleton Island and on Pribilof Islands. Casual on St. Lawrence Island and on the Aleutians. Casual to West Coast in spring, after storms, as in May 1998 when over a dozen were found from coastal Washington to northern California.

LONG-BILLED CURLEW
Numenius americanus | L 23" (58 cm)

Cinnamon-brown above, buff below, with very long, strongly downcurved bill. Lacks dark head stripes of Whimbrel. Males and juveniles have shorter bills. Cinnamon-buff wing linings and flight feathers, visible in flight, distinctive in all plumages. At rest closely resembles the smaller Marbled Godwit, if bill is hidden; note paler legs. Two weakly differentiated subspecies; more southerly breeding *americanus* is larger and longer billed than more northerly *parvus*, but differences are clinal. Still, bill length extremes are significant and shorter-billed juveniles often look much more Whimbrel-like in structure. **Call**, a loud musical, ascending *cur-lee*.

RANGE: Fairly common; nests in wet and dry uplands; in migration and winter found on wetlands and agricultural fields. Rare on Gulf Coast east of Texas and on Atlantic coast from Virginia south. Casual farther north on Atlantic coast, and in Midwest.

EURASIAN CURLEW
Numenius arquata | L 22" (56 cm)

A large curlew that is heavily streaked below. The long bill is long and strongly decurved. Distinguished from Long-billed Curlew by paler overall coloration and by white rump and wing linings, readily visible in flight; from Eurasian races of Whimbrel by larger size, longer bill, and lack of dark stripes on head. Overall very similar to Far Eastern Curlew at rest, but paler, especially from lower belly to undertail; easily told in flight by white rump and wing linings.

RANGE: A widespread Eurasian species. Casual on East Coast in fall and winter. The seven records are from Newfoundland to Long Island, New York, except for one at Middle Cheyne Lake, Nunavut.

FAR EASTERN CURLEW
Numenius madagascariensis | L 25" (64 cm)

A large curlew with a very long, decurved bill. Closely resembles Eurasian Curlew, but overall browner, especially from lower belly to undertail coverts. In flight, underwings are heavily barred with dark; rump is the same color as remainder of upperparts.

RANGE: Despite scientific name, found nowhere near Madagascar! Breeds in Russian Far East, winters in the Sunda Isles, New Guinea, Australia, New Zealand; a few on mainland Southeast Asia. Casual in spring and early summer on Aleutians and Pribilofs; accidental (Sept. 1984) from coastal British Columbia.

BRISTLE-THIGHED
CURLEW

adult

adult

juvenile ♂

adult

LONG-BILLED
CURLEW

adult ♀

adult

EURASIAN
CURLEW
arquata

juvenile

FAR EASTERN
CURLEW

adult

adult

BLACK-TAILED GODWIT
Limosa limosa | L 16½" (42 cm)

Long, bicolored bill is straight or only slightly upcurved. Tail is mostly black, uppertail coverts white. In **breeding** plumage, shows chestnut head and neck and heavily barred sides and flanks. East Coast records are of *islandica*, which in the breeding male is of a deeper and more extensive reddish color below. **Winter** birds are gray above, whitish below. In all plumages, white wing linings and broad wing stripe are conspicuous in flight (page 194).
RANGE: Eurasian species, rare but regular spring migrant on western Aleutians; casual to Pribilofs; casual along Atlantic coast; accidental to eastern Ontario and Louisiana.

HUDSONIAN GODWIT
Limosa haemastica | L 15½" (39 cm)

Long, bicolored bill, slightly upcurved. Tail is black, uppertail coverts white. **Breeding male** is dark chestnut below, finely barred. **Female** is larger and much duller. **Juvenile**'s buff feather edges give upperparts a scaly look. Winter adult resembles Black-tailed Godwit; dark wing linings and narrower white wing stripe are distinctive in flight (page 194).
RANGE: Breeding range not fully known. Migrates through central and eastern Great Plains in spring, much farther east in fall. Casual to Pacific states.

BAR-TAILED GODWIT
Limosa lapponica | L 16" (41 cm)

Long, bicolored bill, slightly upcurved. **Breeding male** is reddish brown below; lacks heavy barring of Black-tailed Godwit. **Female** is larger and much paler than male. In **winter** plumage, Bar-tailed resembles Marbled Godwit but lacks cinnamon tones. Note also shorter bill and shorter legs. Black-and-white barred tail distinctive but hard to see at rest. **Juvenile** resembles winter adult but is buffier overall. Alaska-breeding *baueri* (shown opposite) appears casually in migration along Pacific coast and also recorded from Massachusetts; rump is heavily mottled, wing linings brown with white barring. European *lapponica*, very rare migrant along Atlantic coast, has a whiter rump, white wing linings, and brown-barred axillaries. (Both races are shown in flight on page 194.)
RANGE: Two subspecies of this Eurasian godwit occur in North America.

MARBLED GODWIT
Limosa fedoa | L 18" (46 cm)

Long, bicolored bill, slightly upcurved. Tawny brown; mottled with black above, barred below. Barring is much less extensive on **winter** birds and **juveniles**. The wing coverts are also less patterned on juveniles. Legs are longer than in Bar-tailed Godwit. In flight (page 194), cinnamon wing linings and cinnamon on primaries and secondaries are distinctive.
RANGE: Common on West Coast in winter, fairly common on Texas Gulf Coast and in Florida; rare but regular in the East. Nests in grassy meadows, near lakes and ponds.

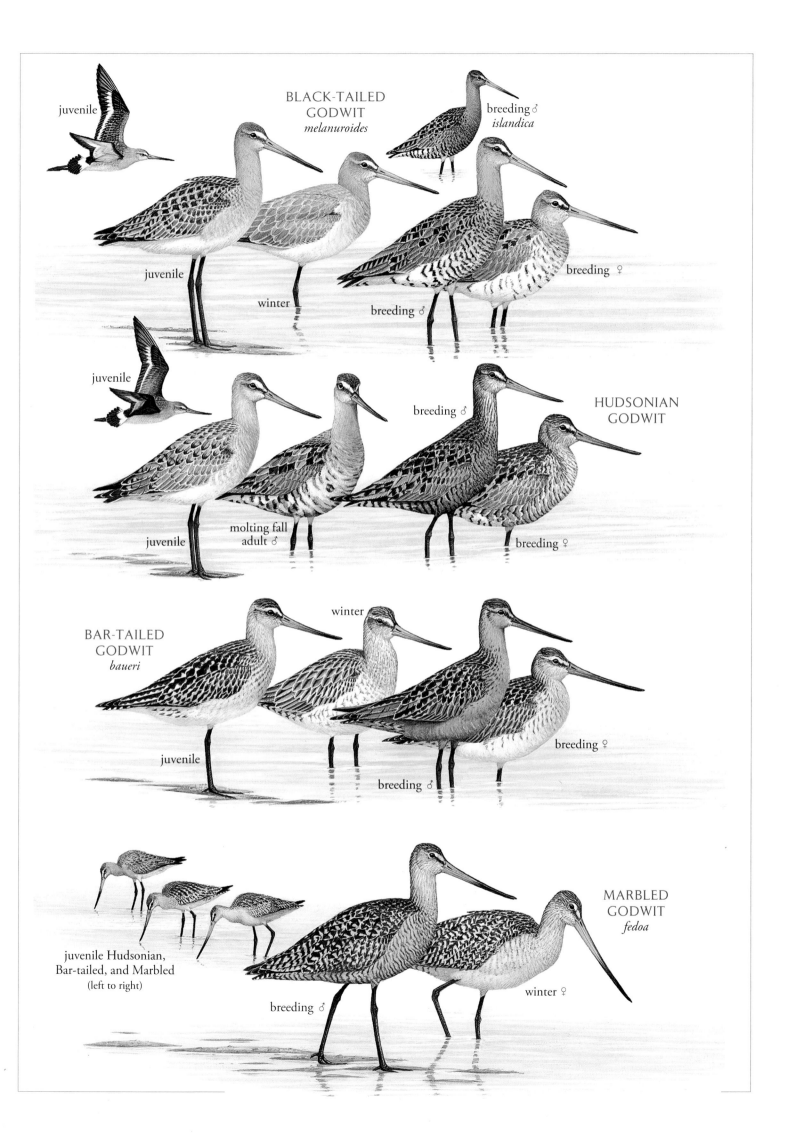

juvenile

BLACK-TAILED
GODWIT
melanuroides

breeding ♂
islandica

juvenile

winter

breeding ♀

breeding ♂

juvenile

HUDSONIAN
GODWIT

breeding ♂

juvenile

molting fall
adult ♂

breeding ♀

BAR-TAILED
GODWIT
baueri

winter

juvenile

breeding ♀

breeding ♂

juvenile Hudsonian,
Bar-tailed, and Marbled
(left to right)

MARBLED
GODWIT
fedoa

breeding ♂

winter ♀

RUDDY TURNSTONE *Arenaria interpres* | L 9½" (24 cm)

Striking black-and-white head and bib, black-and-chestnut back, and orange legs mark this stout bird in **breeding** plumage. Female is duller than **male**. Bib pattern and orange leg color are retained in **winter** plumage. **Juvenile** resembles winter adult but back has a scaly appearance. Distinctive **call** is a low-pitched, guttural rattle. Turnstones use their slender bills to flip aside shells and pebbles in search of food. In flight, complex pattern on back and wings identifies both Ruddy and Black Turnstones.

RANGE: Nests on coastal tundra. Rare inland migrant except in Great Lakes region, where much more numerous.

BLACK TURNSTONE *Arenaria melanocephala* | L 9¼" (24 cm)

Black upperparts. In **breeding** plumage head is marked by white eyebrow and lore spot; white spotting visible on sides of neck and breast. Legs dark reddish brown in all plumages. Juvenile and **winter adult** are slate gray, lack lore spot and mottling. **Calls** include a guttural rattle, higher than call of Ruddy Turnstone.

RANGE: Breeds in coastal Alaska. Winters on rocky coasts.

SURFBIRD *Aphriza virgata* | L 10" (25 cm)

Base of short, stout bill is yellow; legs yellowish green. **Breeding adult**'s head and underparts are heavily streaked and spotted with dusky black; upperparts edged with white and chestnut; scapulars mostly rufous. **Winter adult** has a solid dark gray head and breast. **Juvenile**'s head and breast are flecked with white; back appears scaly. In flight, all plumages show a conspicuous black band at end of white tail and rump.

RANGE: Nests on mountain tundra; winters along rocky beaches and reefs. Casual in spring on Texas coast, also at Salton Sea.

ROCK SANDPIPER *Calidris ptilocnemis* | L 9" (23 cm)

Black patch on lower breast in **breeding** plumage; compare with belly patch of Dunlin (next page). Crown and back are black, edged with chestnut. In flight (page 197), shows white wing stripe and all-dark tail. Nominate *ptilocnemis* breeding on Pribilofs is larger, has paler chestnut above, less black below, bolder white wing stripe. Long, slender bill is slightly downcurved, base greenish yellow. Legs greenish yellow. **Winter** bird separated from Purple Sandpiper by range, duller bill base and legs; from Surfbird by longer bill, smaller size, more patterned upperparts and breast.

RANGE: Nests on tundra; most winter on rocky shores, often with Black Turnstones and Surfbirds. Migrates late in fall.

PURPLE SANDPIPER *Calidris maritima* | L 9" (23 cm)

Breeding adult has tawny buff crown, streaked with black; back is edged with white and tawny buff; breast and flanks spotted with blackish brown. Long, slender bill is slightly downcurved, base orange-yellow. Legs orange-yellow. In flight (page 197) and in **winter**, adult resembles Rock Sandpiper.

RANGE: Migrates late in fall. Rare fall migrant on Great Lakes. Casual elsewhere inland in the East and in winter on the Gulf Coast. Winters on rocky shores, jetties, often with Ruddy Turnstones and Sanderlings.

breeding ♂

juvenile

winter

RUDDY
TURNSTONE

breeding ♂

BLACK
TURNSTONE

winter

winter

breeding

winter

juvenile

SURFBIRD

winter

breeding

winter

ROCK
SANDPIPER
tschuktschorum

juvenile

Pribilofs breeding
ptilocnemis

breeding

juvenile

breeding

PURPLE
SANDPIPER

winter

GREAT KNOT *Calidris tenuirostris* | L 11" (28 cm)

Larger than Red Knot, with longer bill. Compare to Surfbird and Rock Sandpiper (preceding page). In **breeding** plumage, shows black breast and black flank pattern. **Juvenile** has buffy wash and distinct spotting below; dark back feathers edged with rust. Resembles Red Knot in flight but primary coverts darker, wing bar fainter.
RANGE: Asian species, casual in spring to western Alaska.

RED KNOT *Calidris canutus* | L 10½" (27 cm)

Chunky and short legged. **Breeding adult** is dappled brown, black, and chestnut above, with buffy chestnut face and breast. In **winter**, back is pale gray; underparts white. Distinguished from dowitchers (page 188) by shorter bill, paler crown, and, in flight (page 195), by whitish rump finely barred with gray. **Juveniles** similar to winter adults but have distinct spotting below, scaly-looking upperparts.
RANGE: Feeds along sandy beaches and mudflats. Eastern populations declining; rare migrant in the interior.

SANDERLING *Calidris alba* | L 8" (20 cm)

Palest sandpiper of **winter**; pale gray above, white below. Bill, legs black. Bold white wing stripe shows in flight (page 197). In **breeding** plumage (acquired late Apr.), head, mantle, and breast are rusty. Feeds on sandy beaches, running to snatch mollusks and crustaceans exposed by retreating waves. **Juveniles** blackish above, with pale edges near tips of feathers. **Call**, a *kip*, often in a series.

DUNLIN *Calidris alpina* | L 8½" (22 cm)

Medium sized; long bill, curved at tip; in flight (page 197) shows dark center to rump. **Breeding** plumage with reddish upperparts, black belly. Subspecies differ in size, structure, and breeding plumage: more ventrally streaked in *hudsonia*, found in eastern North America; *pacifica* in western Alaska and Pacific region; similar *arcticola* (breeds northern Alaska) and *sakhalina* (migrant western Alaska) winter in Asia. Three races from Greenland and western Palearctic (*arctica*, *schinzii*, and *alpina*) are smaller, shorter billed, and darker above in breeding plumage; *arctica* and *alpina* accidental on East Coast. Compare to Rock Sandpiper (preceding page). In **winter** plumage, upperparts and chest brownish gray. **Juveniles** rusty above, spotted below. **Call**, a harsh, reedy *kree*.
RANGE: Rare on Great Plains. In North America, adults and juveniles stay north and molt into winter plumage, then migrate south.

CURLEW SANDPIPER *Calidris ferruginea* | L 8½" (22 cm)

Long, downcurved bill has whitish area at base. In **breeding** plumage, rich chestnut underparts, mottled chestnut back distinctive. Female slightly paler than male. Many sightings are of birds in patchy spring plumage or molting to winter plumage, showing grayer upperparts and partly white underparts. **Juvenile** appears scaly above; shows rich buff wash across breast; young birds seen in southern Canada and U.S. are in full juvenal plumage; compare with winter Dunlins. White rump and wing stripe are conspicuous in flight (page 195). **Call** is a soft, rippling *chirrup*.
RANGE: Eurasian species, rare on East Coast, casual elsewhere. Has bred in northern Alaska.

GREAT KNOT

breeding

juvenile

juvenile

RED KNOT

winter

breeding

SANDERLING

winter

breeding

juvenile

breeding
sakhalina

breeding
pacifica

breeding
schinzii

DUNLIN
hudsonia

juvenile

breeding

winter

CURLEW
SANDPIPER

juvenile

breeding ♂

winter

PEEPS

These are seven species of small *Calidris* sandpipers that are difficult to identify. Collectively known as stints by Old World English speakers, they can be roughly divided into four Old World and three New World species. Sometimes the larger Baird's and White-rumped Sandpipers are also included. Keys to identification include learning overall structure and feather topography, behavior, and the distribution patterns of each species. It is essential to thoroughly learn our three common species before attempting to identify one of the rare Eurasian ones.

SEMIPALMATED SANDPIPER
Calidris pusilla | L 6¼" (16 cm)

Black legs; tubular-looking, straight bill, of variable length. Easily confused with Western Sandpiper. In **breeding** birds, note that Semipalmated usually lacks spotting on flanks and shows only a tinge of rust on crown, ear patch, and scapulars. **Juveniles** are distinguished by stronger supercilium contrasting with darker crown and ear coverts and by more uniform upperparts. Some are brighter above than illustrated. **Winter** plumage (in North America seen most often on Gulf Coast in late Mar.) of these two species is very similar (see also page 197), but rounder-headed Semipalmated is plumper; note bill shape; face shows slightly more contrast; center of breast never shows the faint streaks visible on some winter Westerns. Semipalmateds tend to pick at the water surface, often just at the water's edge. **Call** is a short *churk*, very different from Western Sandpiper.
RANGE: Abundant. A common migrant in eastern half of continent; generally a rare migrant in the West, south of British Columbia; very rare in winter in south Florida; no valid winter records elsewhere.

WESTERN SANDPIPER
Calidris mauri | L 6½" (17 cm)

Black legs; tapered bill, of variable length (longer in females); distal portion usually slightly drooped. Easily confused with Semipalmated Sandpiper, but blockier-headed Western has more attenuated body. In **breeding** plumage, Western has arrow-shaped spots along sides, rufous at base of scapulars, and a bright rufous wash on crown and ear patch. **Juvenile** is distinguished from juvenile Semipalmated by less prominent supercilium, paler crown and face, and brighter rufous edges on back and inner scapulars. **Winter** plumage is very similar. Note structure (especially bill shape). **Call** is a raspy *jeet*.
RANGE: In North America, especially away from south Florida, any winter-plumaged individual in fall or winter is likely to be this species. Found in similar situations to Semipalmated Sandpipers, but often feeds a little farther out in the water away from the shore's edge.

LEAST SANDPIPER
Calidris minutilla | L 6" (15 cm)

Note small size and short, thin bill, slightly downcurved. Always darker above than Western and Semipalmated Sandpipers. Legs are yellowish, but can appear dark in poor light or when smeared with mud. In **winter** plumage, has prominent brown breast band. **Juvenile** has strong buffy wash across breast. **Call** is a high plaintive *kreee*.
RANGE: Feeds in a variety of wet habitats, but forages less in the water, often preferring to feed back from the shore's edge.

breeding

SEMIPALMATED
SANDPIPER

juvenile

winter

winter

juvenile

WESTERN
SANDPIPER

breeding

winter

LEAST
SANDPIPER

breeding

juvenile

RARE STINTS

These are the four Old World species of peeps that have been found in North America. Of these, Red-necked Stint is the most regular. Of the remaining three, both Temminck's and Long-toed Stints are almost unknown in North America away from Alaska. With any potential find of a rare stint it is essential to get documentation, preferably photographs. Even in migration, shorebirds hold feeding territories, and if flocks are disturbed, they soon return and sort themselves out. If the potential rarity cannot be refound, chances are it was a commoner species.

RED-NECKED STINT
Calidris ruficollis | L 6¼" (16 cm)

Rufous on throat and upper breast may be pale and indistinct; look for necklace of dark streaks on white lower breast. **Juvenile** distinguished from Little Stint by plainer wing coverts and tertials, more uniform crown pattern, and plainer breast sides.

RANGE: Asian species, regular migrant on western Alaska coast and islands. Breeding documented on Seward Peninsula coast; breeding range on map is conjectural. Casual migrant on both coasts; accidental in interior.

LITTLE STINT
Calidris minuta | L 6" (15 cm)

Breeding birds are brightly fringed with rufous above; throat and underparts white (more suffused with color in later summer adults), with bright buff wash and bold spotting on sides of breast. Redder above than Western and Semipalmated Sandpipers (preceding page); compare also with Red-necked Stint. **Juvenile** best distinguished from juvenile Red-necked by extensively black-centered wing coverts and tertials, usually edged with rufous; also note split supercilium and streaking on sides of chest.

RANGE: Eurasian species, very rare on western Alaska islands in spring and fall; casual on East and West Coasts, and accidental elsewhere.

LONG-TOED STINT
Calidris subminuta | L 6" (15 cm)

Distinguished in all plumages from Least Sandpiper by dark forehead, which pinches off prominent white supercilium before it meets the bill (dark forms J shape on its side); also shows a split supercilium. Median coverts are white edged; note also greenish base of lower mandible; yellowish legs. Gives a lower-pitched **call** than Least Sandpiper.

RANGE: Asian species, casual in spring on St. Lawrence Island and the Pribilofs; can be fairly common on the outer Aleutians. Accidental in fall from coastal Oregon and California.

TEMMINCK'S STINT
Calidris temminckii | L 6¼" (16 cm)

White outer tail feathers distinctive in all plumages. **Breeding adult** resembles the larger Baird's Sandpiper in plumage and shape (next page), but legs are yellow or greenish yellow; note Baird's dark legs and distinct primary tip projection past tertials. In Temminck's Stint **juvenile**, feathers of upperparts have a dark subterminal edge, buffy fringe. **Call** is a very distinctive, repeated, rapid dry rattle.

RANGE: Eurasian species; rare spring and fall migrant on Pribilofs, Aleutians, and St. Lawrence Island. Accidental northern Alaska and coastal Pacific Northwest.

juvenile

breeding

RED-NECKED
STINT

LITTLE
STINT

juvenile

juvenile

breeding

LONG-TOED
STINT

breeding

TEMMINCK'S
STINT

juvenile

breeding

WHITE-RUMPED SANDPIPER
Calidris fuscicollis | L 7½" (19 cm)

Long primary tip projection beyond tertials and tail on standing bird. Similar to Baird's Sandpiper structurally, but grayer overall and usually has an entirely white rump. In **breeding** plumage, streaking extends to flanks. **Juvenile** (see also page 197) shows rusty edges on crown and back. In winter, head and neck are dark gray, giving a hooded look. **Call** note, a very high-pitched insect-like *jeet*.
RANGE: Fairly common; common spring and rare fall migrant through Great Plains. Casual spring (mainly) and fall (all records are adults) in West. Spring migration late, extends to late June; juveniles don't migrate south until the end of Sept. Winters in southern South America; no valid midwinter records for North America. Feeds on mudflats.

BAIRD'S SANDPIPER
Calidris bairdii | L 7½" (19 cm)

Long primary tip projection beyond tertials and tail on standing bird gives the bird a horizontal profile. Buff-brown above and across breast. Pale fringing on **juvenile**'s back gives a scaly appearance (see also page 197). Distinguished from White-rumped Sandpiper by more buffy brown color and uniform plumage; in flight by dark rump. Distinguished from Least Sandpiper by much larger size, longer and straighter bill, and primary projection. **Call** is a low raspy *kreep*, similar to call of Pectoral Sandpiper, but less rich.
RANGE: Fairly common to common; found on upper beaches and inland on lakeshores, wet fields, even beaches. Migration is through Great Plains. Uncommon (usually juveniles) to both coasts in fall. Winters in South America; accidental in North America in midwinter.

SPOON-BILLED SANDPIPER
Eurynorhynchus pygmeus | L 6" (15 cm)

Spoon-shaped bill is diagnostic and gives bill a longer look, but spoon is sometimes hard to see with clarity at a distance; beware of other small *Calidris* with mud on bill tip. In **breeding** plumage is easily mistaken for Red-necked Stint, apart from bill. **Juvenile** has darker cheek and more contrasting supercilium than Red-necked Stint. On winter grounds in Southeast Asia, feeds farther out into the water than Red-necked Stints; probes like Western Sandpiper but often with a side-to-side motion.
RANGE: Globally threatened species. Recent estimated population on breeding grounds of between 600 and 1,000 individuals represents a greater than 50 percent reduction in the last 15 years. Breeds on coast of Russian Far East; winters coastal Southeast Asia; casual migrant (about six records) to western and northern Alaska. One record of a fall migrant, a breeding-plumaged adult, from Vancouver region, British Columbia (30 July to 3 Aug. 1978).

BROAD-BILLED SANDPIPER
Limicola falcinellus | L 7" (18 cm)

All sightings so far of juveniles. Plump body with short legs and long, broad-based bill with distinctive drooped tip give it a distinctive profile. Note also the distinctive split supercilium. **Call** is a dry and high-pitched buzzy trill; also shorter calls.
RANGE: Eurasian species, casual fall migrant on western and central Aleutians. Accidental in fall from coastal New York.

WHITE-RUMPED
SANDPIPER

juvenile

breeding

fall-molting
adult

breeding

juvenile

BAIRD'S
SANDPIPER

breeding

juvenile

SPOON-BILLED
SANDPIPER

BROAD-BILLED
SANDPIPER

breeding

juvenile

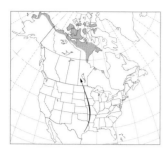

PECTORAL SANDPIPER
Calidris melanotos | L 8¾" (22 cm)

Prominent streaking on breast, darker in **male**, contrasts sharply with clear white belly. Male is larger than **female**. **Juvenile** has buffy wash on streaked breast. Compare especially with juvenile Sharp-tailed Sandpiper (see also page 197). **Call** is a rich, low *churk*.
RANGE: Often feeds in wet meadows, marshes, pond edges. Common in Midwest; fairly common on East Coast; scarcer from Great Plains to West Coast, where mainly juveniles are seen in fall.

SHARP-TAILED SANDPIPER
Calidris acuminata | L 8½" (22 cm)

Most sightings are **juveniles**, distinguished from juvenile Pectoral Sandpiper by white eyebrow that broadens behind the eye; bright buffy breast lightly streaked on upper breast and sides; streaked undertail coverts; and brighter rufous cap and edging on upperparts (see also page 197). **Adult** in **breeding** plumage is similar to juvenile, but more spotted below with dark chevrons on flanks; also distinct white eye ring. **Call** is a mellow, two-note whistle.
RANGE: Breeds in Russian Far East; casual spring and fairly common fall migrant in western Alaska; rare fall migrant along entire Pacific coast. Casual in fall across rest of continent.

UPLAND SANDPIPER
Bartramia longicauda | L 12" (31 cm)

Small head, with large, dark, prominent eyes; long, thin neck, long tail, long wings. Legs yellow. Prefers fields, where often only its head and neck are visible above the grass. Also perches on posts on breeding grounds. In flight (page 195), blackish primaries contrast with mottled brown upperparts. **Calls**, a rolling *pulip pulip*, and a call like a wolf whistle, heard in flight on breeding grounds.
RANGE: Fairly common except in eastern range, where declining. Casual on West Coast and in the Southwest in migration.

BUFF-BREASTED SANDPIPER
Tryngites subruficollis | L 8¼" (21 cm)

Dark eye stands out prominently on buffy face; underparts paler buff; legs orange-yellow. In flight, shows flashy white wing linings. **Juveniles** are paler below, with scaly white fringing to feathers above. (see also page 195)
RANGE: Prefers plowed fields, turf farms, and wet rice fields. Migrates through the interior of the continent. In fall, rare on the West Coast, uncommon in the East; most sightings west or east of eastern Great Plains are of juveniles.

RUFF
Philomachus pugnax | ♂ L 12" (31 cm) ♀ L 10" (25 cm)

Breeding males acquire dramatic ruffs in colors that range from black to rufous to white. **Female** lacks ruff, is smaller, and has a variable amount of black below. Both sexes have a plump body, small head, and white underwings. Leg color may be yellow, orange, or red. **Juvenile** is buffy below, has prominently fringed feathers above. In flight (page 197), the U-shaped white band on rump is distinctive in all plumages. Ruff and Buff-breasted Sandpiper are both silent away from breeding grounds.
RANGE: Old World species. Rare migrant in western Alaska and along West and East Coasts and in Great Lakes region. Casual elsewhere. Very rare in winter in California; has bred in Alaska.

breeding ♂

breeding ♀

PECTORAL
SANDPIPER

juvenile

juvenile Sharp-tailed (center)
with juvenile Pectorals

breeding
adult

SHARP-TAILED
SANDPIPER

juvenile

adult

UPLAND
SANDPIPER

juvenile

displaying
adult

juvenile

BUFF-BREASTED
SANDPIPER

breeding
adult

summer
molting ♂

breeding
males

♀

juvenile ♀

RUFF

summer
molting ♀

winter ♂

DOWITCHERS

Medium-size, chunky, dark shorebirds, dowitchers have long, straight bills and distinct pale eyebrows. Feeding in mud or shallow water, they probe with a rapid jabbing motion. Dowitchers in flight show a white wedge from barred tail to middle of back. Separating the two species is easiest with juveniles, more difficult with breeding adults, and very difficult in winter except when distinctive calls are heard. Both species give the same song, a rapid *di di da doo*, year-round.

SHORT-BILLED DOWITCHER
Limnodromus griseus | L 11" (28 cm)

In flight, tail usually looks paler than in Long-billed Dowitcher. **Breeding** plumage varies among the three subspecies: *griseus* (which breeds in northeast Canada), *hendersoni* (central and western Canada), and *caurinus* (Alaska). Unlike Long-billed, most Short-billed show some white on the belly, especially *griseus*, which also has a heavily spotted breast and may have densely barred flanks; *caurinus* is variable but similar to *griseus*. In *hendersoni*, which may be mostly reddish below, foreneck is much less heavily spotted than in Long-billed; sides have less or no barring; upperparts are brighter. In all subspecies, **juvenile** is brighter above, buffier and more spotted below than juvenile Long-billed; tertials and visible greater wing coverts have broad reddish-buff edges and internal bars, loops, or stripes. **Winter** birds are brownish gray above, white below, with gray breast; at close range note fine dark speckling on and below the breast on many birds. **Call** is a mellow *tu tu tu*, repeated in a rapid series as an alarm call.

RANGE: Common in migration along the Atlantic coast (*griseus*); from the eastern plains to Atlantic coast from New Jersey south (*hendersoni*); and along the Pacific coast (*caurinus*); a few *griseus* are seen on eastern Great Lakes in late spring. Fall migration begins earlier than Long-billed, usually in late June or early July for adults; early Aug. for juveniles. Migrant juveniles are seen through early Oct.

LONG-BILLED DOWITCHER
Limnodromus scolopaceus | L 11½" (29 cm)

Male's bill is no longer than on Short-billed Dowitcher, **female**'s is. In flight, tail usually looks darker than in Short-billed. **Breeding adult** is entirely reddish below; foreneck heavily spotted; sides usually barred. Bold white scapular tips in spring help separate this species from Short-billed. **Juvenile** is darker above, grayer below than Short-billed; tertials and greater wing coverts are plain, with thin gray edges and rufous tips; some birds show two pale spots near the tips. In **winter** birds, breast is unspotted and more extensively dark than on most Short-billed. **Call** is a sharp, high-pitched *keek*, given singly or in a rapid series. Adult Long-billeds go to favored locations in late summer to molt; Short-billeds molt when they reach winter grounds.

RANGE: Common in migration in western half of continent; less common in the East in fall, rare in spring. Fall migration begins later than Short-billed, in mid-July (West) or late July (East). Juveniles migrate later than adults; rare before Sept. Dowitchers seen inland after mid-Oct. are almost certainly Long-billed.

breeding
caurinus

worn breeding
griseus

breeding
griseus

breeding
hendersoni

molting
juvenile

juvenile tertials

juvenile

winter

SHORT-BILLED
DOWITCHER

griseus

winter
hendersoni

LONG-BILLED
DOWITCHER

molting
juvenile

winter

juvenile

juvenile tertials

winter

worn
breeding ♀

fresh breeding ♂

STILT SANDPIPER *Calidris himantopus* | L 8½" (22 cm)

Breeding adult has pale eyebrow, chestnut on head, slender, slightly downcurved bill, and heavily barred underparts. **Winter adult** is grayer above, whiter below; **juvenile** has more sharply patterned upperparts; the two resemble Curlew Sandpiper (page 178), but note straighter bill, yellow-green legs, and, in flight (page 195), lack of prominent wing stripe, paler tail; early juvenile has a buffy wash on breast. Feeds like dowitchers, with which it often associates, but note smaller size and disproportionately longer legs. **Call** is a low, hoarse *querp*, but often silent.

RANGE: Common to East in fall.

WILSON'S SNIPE *Gallinago delicata* | L 10¼" (26 cm)

Stocky, with very long bill; boldly striped head; barred flanks. Usually seen when flushed, as it gives a harsh two-syllable *ski-ape* **call** in rapid, twisting flight. Where breeding, males deliver loud *wheet* notes from perches. In swooping display flight, vibrating outer tail feathers make quavering hoots, like song of Boreal Owl.

RANGE: Widespread. Winter birds favor wet fields, pond edges, and muddy patches.

COMMON SNIPE *Gallinago gallinago* | L 10½" (27 cm)

Paler, buffier color overall than Wilson's Snipe; broader white trailing edge to secondaries, fainter flank markings; paler white-striped underwings. Flight-display notes of the male are lower-pitched; **calls** similar to Wilson's.

RANGE: Eurasian species, migrant to the western Aleutians where it has bred and been found casually in winter; rarer to central Aleutians, Pribilofs; casual to St. Lawrence Island.

PIN-TAILED SNIPE *Gallinago stenura* | L 10" (26 cm).

Chunkier, shorter billed, and shorter tailed than Common Snipe. On ground, note barred secondary coverts and even-width pale edges on inner and outer webs of scapulars for a scalloped look. In flight, note buffy secondary covert panel, uniformly dark underwings, no pale edge to secondaries, distinct foot projection past tail. Larger Swinhoe's Snipe (*G. megala*) from Asia (unrecorded in North America) perhaps not separable in field. In hand, razor-thin outer tail feathers are diagnostic. **Call** is high, ducklike *squak*.

RANGE: Breeds in Siberia and Russian Far East; winters chiefly in Southeast Asia. Two certain records from western Aleutians.

AMERICAN WOODCOCK *Scolopax minor* | L 11" (28 cm)

Chunky, with long bill, barred crown, large eyes set high in head. Rounded wings. Nocturnal, secretive. Flies up abruptly; wings make a twittering sound. Males give an elaborate flight display; visible at dusk and dawn. **Call**, heard mainly in spring, a nasal *peent*.

RANGE: Common; nests in moist woodlands. In mild winters, a few are found farther north. Casual to Colorado, New Mexico.

JACK SNIPE *Lymnocryptes minimus* | L 7" (18 cm)

Secretive, reluctant to flush. Flight is low, short, fluttery, on rounded wings. Bobs while feeding. Pale base to short bill, pale split eyebrow stripes with no median crown stripe, broad buffy back stripes, streaked flanks, pale underparts.

RANGE: Small, chunky Eurasian species. Three late fall records for California and Labrador; one, possibly two, spring records for Pribilofs.

STILT
SANDPIPER

breeding

molting
juvenile

juvenile

winter

displaying

underwing

WILSON'S
SNIPE

underwing

COMMON SNIPE
gallinago

AMERICAN
WOODCOCK

underwing

PIN-TAILED
SNIPE

tail

JACK SNIPE

PHALAROPES

These elegant shorebirds have partially lobed feet and dense, soft plumage. Feeding on the water, phalaropes often spin like tops, stirring up larvae, crustaceans, and insects. Females, larger and more brightly colored than the males, do the courting; males incubate the eggs and care for the chicks. In fall, adults and juveniles (particularly Wilson's and Red Phalaropes) rapidly molt to winter plumage; many are seen in transitional plumage farther south.

WILSON'S PHALAROPE
Phalaropus tricolor | L 9¼" (24 cm)

Long, thin bill; bold blackish stripe on face and neck. In **winter** plumage, upperparts are gray, underparts white; note also lack of distinct dark ear patch; legs yellowish. Briefly held **juvenal** plumage resembles winter adult but back is browner with buffy edge to feathers, breast buffy. In flight, white uppertail coverts, whitish tail, and absence of white wing stripe distinguish juvenile and winter birds from other phalaropes. **Calls** include a hoarse *wurk* and other low, croaking notes.
RANGE: Wilson's is chiefly an inland phalarope, nesting on grassy borders of shallow lakes, marshes, and reservoirs. Feeds as often on land as on water. Common to abundant in western North America; uncommon to rare in the East; rare in southern California in winter.

RED-NECKED PHALAROPE
Phalaropus lobatus | L 7¾" (20 cm)

Chestnut on front and sides of neck distinctive in **breeding female**, less prominent in **male**. Both have dark back with bright buff stripes along sides; bill shorter than in Wilson's Phalarope, thinner than in Red Phalarope. **Winter** birds are blue-gray above with whitish stripes; underparts and front of crown white; dark patch extends back from eye. In flight, show white wing stripe, whitish stripes on back, dark central tail coverts. Fresh **juvenile** resembles winter adult but is blacker above, with bright buff stripes. **Call**, a high, sharp *kit*, often given in a series.
RANGE: Breeds on Arctic and subarctic tundra; winters chiefly at sea in Southern Hemisphere. Common inland in West and off West Coast during migration; rare in Midwest and East; uncommon off East Coast; more numerous off Maine and Maritimes.

RED PHALAROPE
Phalaropus fulicarius | L 8½" (22 cm)

Bill shorter and much thicker than in other phalaropes; yellow with black tip in breeding adult, usually all-dark in juvenile and winter adult. **Female** in breeding plumage has black crown, white face, chestnut red underparts. **Male** is duller. **Juvenile** resembles male but is much paler below; juveniles seen in southern Canada and the U.S. are **molting** to winter plumage; more closely resemble Red-necked Phalaropes. **Winter** bird is pale gray above. In flight, shows a bolder white wing stripe than Red-necked and dark central tail coverts. **Call** is a sharp *keip*, is higher-pitched than Red-necked's.
RANGE: Breeds on Arctic shores; winters at sea. Irregularly common off West Coast during fall migration; rare to very rare inland, chiefly seen in fall. Generally uncommon off East Coast in spring and fall.

juvenile

winter

molting
juvenile

WILSON'S
PHALAROPE

breeding ♀

winter

breeding ♂

breeding adults with juvenile

juvenile

RED-NECKED
PHALAROPE

winter

molting
juvenile

breeding ♀

winter

breeding ♂

juvenile

molting fall
adults

winter

molting
juvenile

RED
PHALAROPE

breeding ♀

winter

breeding ♂

SHOREBIRDS IN FLIGHT

MARBLED
GODWIT

winter

BAR-TAILED
GODWIT

winter
baueri

HUDSONIAN
GODWIT

winter

BAR-TAILED
GODWIT

winter
lapponica

BLACK-TAILED
GODWIT

winter

FAR EASTERN
CURLEW

EURASIAN
CURLEW

LONG-BILLED
CURLEW

juvenile

phaeopus

variegatus

hudsonicus

WHIMBREL

BRISTLE-THIGHED
CURLEW

ESKIMO
CURLEW

LITTLE
CURLEW

SHOREBIRDS IN FLIGHT

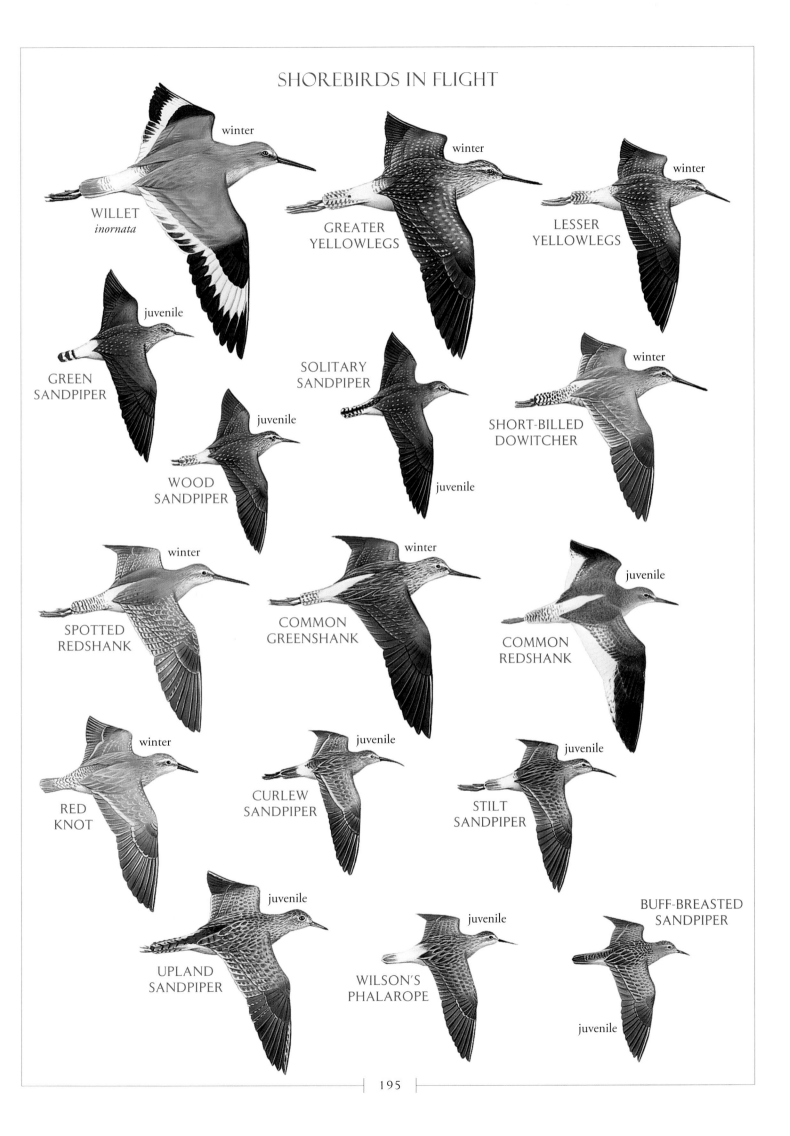

WILLET
inornata
winter

GREATER
YELLOWLEGS
winter

LESSER
YELLOWLEGS
winter

GREEN
SANDPIPER
juvenile

SOLITARY
SANDPIPER
juvenile

SHORT-BILLED
DOWITCHER
winter

WOOD
SANDPIPER
juvenile

SPOTTED
REDSHANK
winter

COMMON
GREENSHANK
winter

COMMON
REDSHANK
juvenile

RED
KNOT
winter

CURLEW
SANDPIPER
juvenile

STILT
SANDPIPER
juvenile

UPLAND
SANDPIPER
juvenile

WILSON'S
PHALAROPE
juvenile

BUFF-BREASTED
SANDPIPER
juvenile

SHOREBIRDS IN FLIGHT

winter

breeding

SNOWY
PLOVER

PIPING
PLOVER

LITTLE RINGED
PLOVER

juvenile

winter

juvenile

COMMON
RINGED
PLOVER

LESSER
SAND-PLOVER

SEMIPALMATED
PLOVER

winter

KILLDEER

MOUNTAIN
PLOVER

WILSON'S
PLOVER

juvenile

juvenile

juvenile

EURASIAN
DOTTEREL

PACIFIC
GOLDEN-PLOVER

AMERICAN
GOLDEN-PLOVER

juvenile

juvenile

EUROPEAN
GOLDEN-PLOVER

BLACK-BELLIED
PLOVER

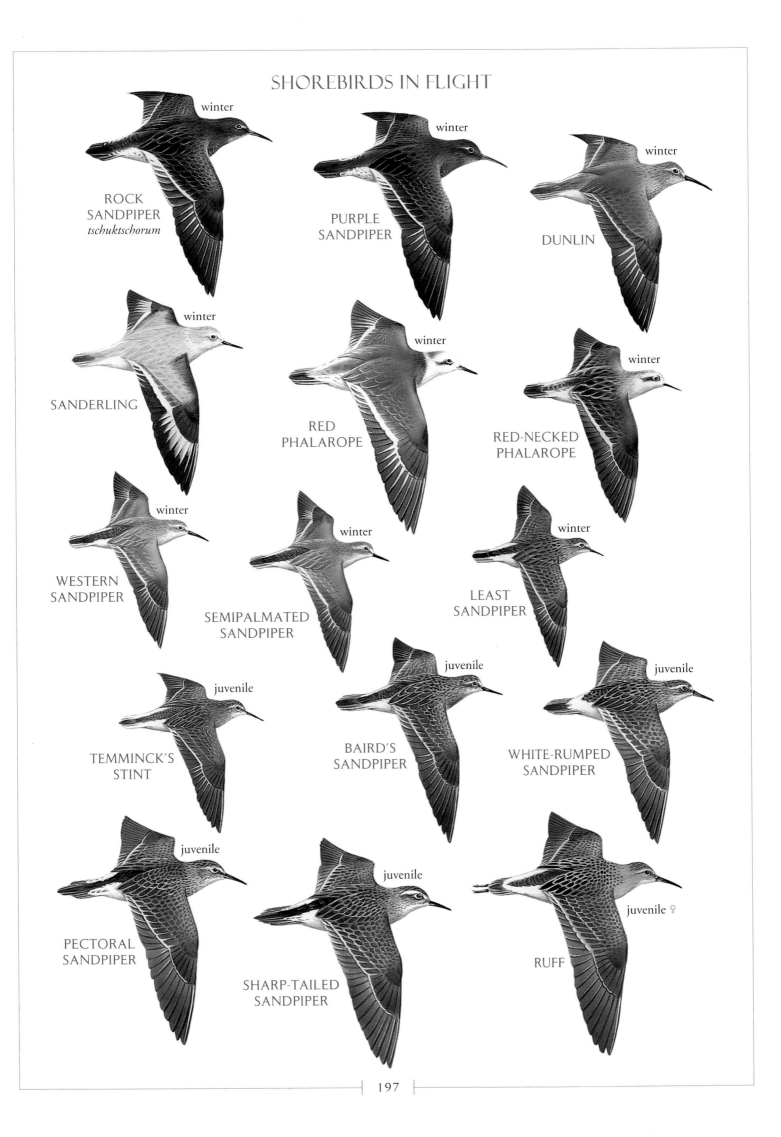

SHOREBIRDS IN FLIGHT

ROCK
SANDPIPER
tschuktschorum
winter

PURPLE
SANDPIPER
winter

DUNLIN
winter

SANDERLING
winter

RED
PHALAROPE
winter

RED-NECKED
PHALAROPE
winter

WESTERN
SANDPIPER
winter

SEMIPALMATED
SANDPIPER
winter

LEAST
SANDPIPER
winter

TEMMINCK'S
STINT
juvenile

BAIRD'S
SANDPIPER
juvenile

WHITE-RUMPED
SANDPIPER
juvenile

PECTORAL
SANDPIPER
juvenile

SHARP-TAILED
SANDPIPER
juvenile

RUFF
juvenile ♀

GULLS, TERNS, AND SKIMMERS (Family Laridae)

A large, diverse family with strong wings and powerful flight. Some species are largely pelagic; others frequent coastal waters or inland lakes and wetlands. Gulls take from about two to four years to reach adult plumage; immatures are often variable and hard to identify. In general, male gulls are larger than females.
Species: *97 World, 49 N.A.*

HEERMANN'S GULL
Larus heermanni | L 19" (48 cm) WS 51" (130 cm)

Four-year gull. **Adult** distinctive with white head, streaked gray-brown in winter; red bill; dark gray body; black tail with white terminal band; white trailing edge on wings. **Second-winter** bird is browner, with two-toned bill and buff tail band. **First-winter** bird has dark brown body, lacks contrasting tail tip and trailing edge on wing. Wings are fairly long, flight buoyant.
RANGE: Common post-breeding visitor along the West Coast; rare at Salton Sea. Casual elsewhere inland in California and the Southwest; accidental to west Texas, Great Lakes, southeast Alaska, Florida, and Virginia.

FRANKLIN'S GULL
Leucophaeus pipixcan | L 14½" (37 cm) WS 36" (91 cm)

Three-year gull. **Breeding adult** has black hood, white underparts variably tinged with pink, slate gray wings with white bar and black-and-white tips on primaries. Distinguished from Laughing Gull by white bar and large white tips on primaries, pale gray central tail feathers, and broader white eye crescents. All **winter** birds have a dark half-hood, more extensive than in any winter Laughing Gull. Second-summer Franklin's has partial or no bar on primaries. **First-summer** bird is like winter adult but lacks white primary bar; bill and legs black. **First-winter** bird resembles first-winter Laughing; note white outer tail feathers, half-hood, broader eye crescents, white underparts, and, in flight (page 216), pale inner primaries. Juvenile is like first-winter bird but back is brown. At all ages, distinguished from Laughing Gull by smaller size, smaller bill with less prominent hook, rounder forehead, less extensive dark on underside of primaries; shorter legs and wings give a stocky look when standing.
RANGE: Mainly a bird of the midcontinent. Rare migrant along both coasts; very rare in winter along Gulf Coast and in southern California.

LAUGHING GULL
Leucophaeus atricilla | L 16½" (42 cm) WS 40" (102 cm)

Three-year gull. **Breeding adult** has black hood, white underparts, slate gray wings with black outer primaries. In **winter**, shows gray wash on nape; compare with the half-hood of Franklin's Gull. Second-summer bird has partial hood, some spotting on tip of tail. **Second-winter** bird (see also page 216) is similar to second-summer but has gray wash on sides of breast, lacks hood. **First-winter** bird has extensively gray sides, complete tail band, gray wash on nape, slate gray back, dark brown wings; compare with first-winter Franklin's Gull. **Juvenile** is like first-winter bird but brown on head and body.
RANGE: Common along Gulf and Atlantic coasts; rare inland except at Salton Sea where it is fairly common, chiefly as a post-breeding visitor.

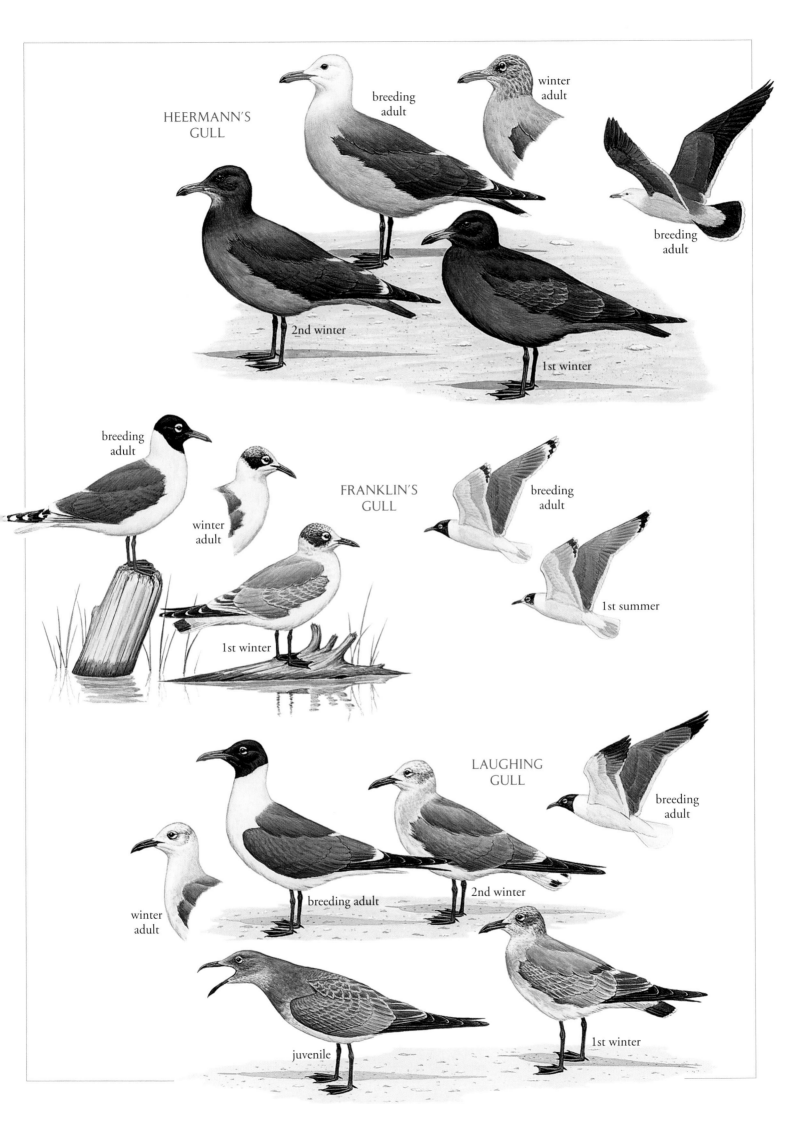

HEERMANN'S GULL

breeding adult

winter adult

breeding adult

2nd winter

1st winter

FRANKLIN'S GULL

breeding adult

winter adult

1st winter

breeding adult

1st summer

LAUGHING GULL

breeding adult

winter adult

breeding adult

2nd winter

breeding adult

juvenile

1st winter

BONAPARTE'S GULL

Chroicocephalus philadelphia | L 13½" (34 cm) WS 33" (84 cm)

Two-year gull. **Breeding adult** has slate black hood, black bill, gray mantle with black wing tips that are pale on underside; white underparts, orange-red legs. In flight, shows white wedge on wing. **Winter** bird lacks hood. **First-winter** bird has a dark brown carpal bar on leading edge of wing, dark band on secondaries, black tail band; compare with juvenile Black-legged Kittiwake (pages 214, 216). First-summer bird may show partial hood; wings and tail are like first-winter. Flight is buoyant, wingbeats rapid.

RANGE: Uncommon inland migrant in West; common to abundant on Great Lakes.

BLACK-HEADED GULL

Chroicocephalus ridibundus | L 16" (41 cm) WS 40" (102 cm)

Two-year gull. **Breeding adult** has dark brown hood; maroon-red bill and legs; mantle slightly paler gray than Bonaparte's Gull; black wing tips; white underparts. **Winter adult** lacks hood; bill brighter red. **First-winter** birds have orange-red bill, pale legs, dark tail band, dark brown carpal bar (page 216). Distinguished from Bonaparte's by larger size and bill color. **First-summer** bird's hood varies from minimal, like first-winter's, to nearly complete; wings and tail like first-winter. In all plumages, shows dark underside of primaries in flight, darker on adults; compare with Bonaparte's.

RANGE: Colonizer from Europe. Fairly common in winter in Newfoundland, where a few breed; rare off western Alaska, Maritimes, and coastal New England south to North Carolina; casual elsewhere in North America.

LITTLE GULL

Hydrocoloeus minutus | L 11" (28 cm) WS 24" (61 cm)

Two- to three-year gull. **Breeding adult** has black hood, black bill, pale gray mantle, white wing tips, white underparts, red legs. **Winter adult** has dusky cap, dark spot behind eye. Wings uniformly pale gray above, dark gray to black below, with white trailing edge. Some second-winter birds are like adult but underwing pattern is incomplete; show some dusky slate in primaries (page 216). **First-winter** is like Bonaparte's but primaries blackish above, lack white wedge; wings show strong blackish W; crown shows more black. In all plumages, note short, rounded wings, fluttery wingbeat.

RANGE: Western Palearctic species, has bred irregularly from Great Lakes to Hudson Bay. Generally rare on Great Lakes in migration and in winter on East Coast; very rare to casual elsewhere in North America.

ROSS'S GULL

Rhodostethia rosea | L 13½" (34 cm) WS 33" (84 cm)

Two-year gull. Variably pink below; upperwing pale gray; underwing pale to dark gray. Black collar in summer, partial or absent in winter. **First-winter** bird has black at tip of tail, dark spot behind eye; acquires black collar by first summer; in flight (page 216), shows W pattern like Little Gull. In all plumages, note long, pointed wings; long, wedge-shaped tail; and broad, white trailing edge to wings.

RANGE: Arctic species of the Russian Far East, has bred in northern Canada and Greenland in last two decades. Common fall migrant along northern coast of Alaska; presumably winters at sea. Casual south to northern U.S.

breeding
adult

winter
adult

1st winter

winter
adult

BONAPARTE'S
GULL

breeding
adult

1st
summer

winter
adult

1st
winter

winter
adult

BLACK-HEADED
GULL

breeding
adult

LITTLE
GULL

1st winter

breeding
adult

winter
adult

winter
adult

ROSS'S
GULL

1st winter

breeding
adult

RING-BILLED GULL

Larus delawarensis | L 17½" (45 cm) WS 48" (122 cm)

Three-year gull. **Adult** has pale gray mantle; yellow bill with black subterminal ring; pale eyes; yellowish legs; head streaked with brown in **winter. Second-winter** birds are like winter adult but bill has broader band, black on primaries is more extensive, tail usually has some blackish terminal spots. **First-winter** bird has gray back, brown wings with dark blackish-brown primaries, brown-streaked head and nape; underparts white with brown spots and scalloping on breast and throat; tail has medium-wide but variable brown band and extensive mottling above band; uppertail and undertail coverts are lightly barred; secondary coverts medium gray; wing linings mostly white, with some barring (page 216). Distinguished from first-winter Mew Gull (*brachyrhynchus*) by white underparts spotted on breast and throat, tail pattern, darker primaries, heavier bill, and paler back. **Juvenal** plumage may be largely kept into early winter; resembles first-winter but back is brown, spotting below more extensive, bill has more black.

RANGE: Abundant and widespread; winters uncommonly outside mapped range.

MEW GULL

Larus canus | L 16" (41 cm) WS 43" (109 cm)

Three-year gull. North American race *brachyrhynchus* is the smallest of three subspecies found here; has least black on wing tips; in flight, shows much more white on primaries than Ring-billed Gull. European nominate race, *canus* ("Common Gull"), and East Asian *kamtschatschensis* (the largest race) have more extensive black on wing tips. All **adults** have white head, washed with brown in **winter**; dark gray mantle; thin yellow bill; most have large, dark eyes. **Second-winter** bird is like adult but has two-toned bill; has less white on primaries; variably spotted tail band. **First-winter** *brachyrhynchus* is heavily washed with brown below, almost solid brown on belly; spotted with white on breast. The head and nape are washed with soft brown; mantle dark gray; primaries light brown with pale edges. Tail is almost entirely brown, with heavily barred tail coverts; wing linings evenly pale brown (page 216). **Juvenile** is like first-winter, but brown on the back and head, darker below. Second-winter and adult *kamtschatschensis* have pale eyes; first-winter birds are more like *brachyrhynchus*, but note dark tail band rather than all-dark tail, and paler tail coverts. In flight, first-winter's underwing coverts show intermediate coloration between *brachyrhynchus* (dark) and *canus* (pale). European *canus* resembles Ring-billed Gull in first winter; but note *canus*'s mostly white tail with dark subterminal band, unbarred white tail coverts, darker gray back, white wing linings mottled with brown. Mew Gulls average smaller than Ring-billed Gulls, especially *brachyrhynchus*, with rounder heads, larger eyes, thinner bills.

RANGE: American race, rare inland in winter in Pacific states, casual east to Great Lakes region. Asian race, rare on the Aleutians and islands in the Bering Sea. European race, casual on East Coast in winter, annual in Newfoundland.

winter adult

breeding adult

2nd winter

breeding adult

RING-BILLED GULL

juvenile

1st winter tail

1st winter

1st winter tail

winter adult

2nd winter

breeding adults

juvenile

1st winter

MEW GULL
brachyrhynchus

kamtschatschensis

winter adult

1st winter

canus

winter adult

1st winter

adult
kamtschatschensis

adult
canus

CALIFORNIA GULL
Larus californicus | L 21" (53 cm) WS 54" (137 cm)

Four-year gull. **Adult** has darker gray mantle than Herring Gull (next page), paler than Lesser Black-backed Gull (page 212); white head, heavily streaked with brown in winter; dark eyes; yellow bill with black and red spots; black spot often smaller in breeding season; gray-green or greenish-yellow legs. In flight, shows dusky trailing edge on underwing. Third-winter plumage is like adult but bill is more extensively smudged with black; wings show some brown; tail has some brown spotting. **Second-winter** has gray back, brown wings, grayish legs, two-toned bill; compare to first-winter Ring-billed Gull (preceding page). **First-winter** is brown overall with veiled gray on scapulars; usually palest on throat, breast, and upper belly; legs pinkish; bill two-toned, the colors sharply defined. In flight (page 217), first-winter birds show double dark bar on inner half of wing, caused by darker secondaries and greater secondary covert bases. Distinctly smaller than Western Gull (page 210), with thinner bill. Compare first-winter birds to first- and second-winter Herring and Lesser Black-backed. **Juveniles** are variably pale below, lack pale bill base.
RANGE: Casual on East Coast and Gulf Coast in winter.

BLACK-TAILED GULL
Larus crassirostris | L 18½" (47 cm) WS 47¼" (120 cm)

Three-year or four-year gull, about size of Ring-billed Gull (preceding page); bill and wings long; legs short. Distinctive white eye crescents except on **breeding adult** and third-winter bird. Adult has black ring near red tip of bill; yellow iris, red orbital ring. Mantle dark slate gray; tail has broad subterminal band. Head of **winter adult** heavily streaked. **First-winter** bird has white on face, otherwise heavily washed with brown.
RANGE: East Asian species, casual in coastal Alaska, Washington, and eastern North America; accidental in California.

BELCHER'S GULL
Larus belcheri | L 20" (51 cm) WS 49" (124 cm)

Medium-size, three-year gull. Plumages and bill color similar to Black-tailed Gull, but Belcher's has dark eyes, longer legs, and thicker bill. **Winter adults** and **second-winter** birds have dark hood, red only on tip of bill. Adult has yellow orbital ring. **First-winter**'s head and breast are smoky brown; belly white; mottled above.
RANGE: Resident on west coast of South America; accidental to California and Florida.

KELP GULL
Larus dominicanus | L 23" (58 cm) WS 53" (135 cm)

Three-year gull; size, structure, and bill shape suggest Western Gull (page 210). **Adult** Kelp has black back and dull greenish legs; head streaking in winter indistinct. Eye color variable. Note restricted white in outer primaries, unlike Great Black-backed (page 212). Change to adult plumage rapid; mantle blackish by **second summer**.
RANGE: Widespread Southern Hemisphere species; casual to Gulf Coast. A few Kelps nested on Chandeleur Islands off southeast Louisiana for about a decade starting in early 1990s. This resulted in pure Kelp pairings, and mixed pairings with Herring Gull that produced **hybrids.** Accidental elsewhere.

breeding
adult

winter
adult

2nd
winter

breeding
adult

CALIFORNIA
GULL

1st winter

pale juvenile

dark juvenile

breeding
adult

winter
adult

BLACK-TAILED
GULL

1st winter

breeding
adult

2nd winter

breeding
adult

BELCHER'S
GULL

1st winter

breeding
adult

2nd winter

winter adult

molting
juvenile

Kelp x Herring hybrid
winter adult

KELP GULL

2nd summer

winter
adult

breeding
adult

HERRING GULL

Larus argentatus | L 25" (64 cm) WS 58" (147 cm)

Highly variable four-year gull. **Adult** has pale gray mantle; white head streaked with brown in winter; legs and feet pink; bill yellow with red spot. **Third-winter** plumage is like winter adult but with black smudge on bill, some brown on body and wing coverts. **Second-winter** bird has pale gray back; brown wings; pale eyes; two-toned bill. **First-winter** birds are brown overall, with dark brownish black primaries and tail band, dark eyes, dark bill, with variable pink at base; some may have bill like first-winter California Gull (preceding page); but usually distinguished by darker bill, paler face and throat, and, in flight (page 217), by pale panel at base of primaries and single dark bar on secondaries. Distinguished from first-winter Western Gull (page 210) by smaller bill, paler and more mottled body plumage, and, in flight, by paler wings with pale panel, and lack of contrast between back and rump. Distinguished from first-winter Lesser Black-backed Gull (page 212) by browner, less-contrasting body plumage, usually darker belly, and, in flight, by pale primary and outer secondary coverts and less-contrasting rump pattern. Widespread North American race is *smithsonianus*; Bering Sea region *vegae* has darker mantle in adult plumage and similar wing-tip pattern to Slaty-backed Gull (page 212); usually has dark eyes; head heavily streaked in winter. First-year birds are paler and more checkered; tail base is white. West European race, *argenteus*, noted casually from Newfoundland, is similar in adult plumage to *smithsonianus*, but first-winter bird is paler.

RANGE: Herring Gull is generally local in the West; common in some regions, uncommon to rare in most. Common in eastern North America.

YELLOW-LEGGED GULL

Larus michahellis | L 24" (61 cm) WS 57" (144 cm)

Size similar to Herring Gull, but Yellow-legged has squarer head, peaked at rear of crown; bill stouter and shorter. Adult mantle darker gray than *smithonianus* Herring. From above, wing tip darker than Herring; from below more gray, less black, on outermost primaries. Red gonys spot often extends onto upper mandible; orbital ring is redder than on Herring. Yellow legs distinctive, but some *smithsonianus* Herrings may show some yellow during winter. In Yellow-legged, fainter winter head streaking is restricted to nape and crown, making white head stand out; by midwinter most **adults** are white headed. **First-winter** birds are much paler on head and underparts than first-winter Herring; blocky head and extensive white on uppertail coverts and base of tail suggests same-age Great Black-backed Gull (page 212). Compare also to first-winter Lesser Black-backed (page 212). Molt to first-winter occurs earlier than in Herring; by fall, young Yellow-legged often appear worn.

RANGE: Palearctic species. Casual winter visitor to northeastern coast from Newfoundland to mid-Atlantic, possibly south Texas.

winter
adult

juvenile

breeding
adult

3rd
winter

1st
winter

2nd
winter

1st
winter

breeding adult
vegae

1st winter
smithsonianus

1st winter
argenteus

HERRING
GULL
smithsonianus

1st winter
vegae

winter adult
vegae

1st winter

YELLOW-LEGGED
GULL

winter adult

breeding
adult

GLAUCOUS GULL
Larus hyperboreus | L 27" (69 cm) WS 60" (152 cm)

Heavy-bodied, four-year gull. All have translucent tips to white primaries. **Adult** has very pale gray mantle, yellow eye. Head is streaked with brown in winter. Late **second-winter** bird has pale gray back and pale eye. **First-winter** birds may be buffy or almost all-white; bill is bicolored. Distinguished from Iceland Gull by size; heavier, longer bill; flatter crown; slightly paler mantle of adults; disproportionately shorter wings, barely extending beyond tail. At all ages, distinguished from Glaucous-winged Gull (pages 210, 217) by more buffy white color, contrasting pale primaries; in first-winter birds by sharply two-toned bill.
RANGE: Rare in winter south to Gulf states and southern California. Northern Alaskan birds are slightly smaller, and adults have slightly darker mantles, than birds from eastern Canada and Bering Sea. Occasionally hybridizes with Herring Gull.

ICELAND GULL
Larus glaucoides | L 22" (56 cm) WS 54" (137 cm)

Highly variable four-year gull. **Adults** have white heads, suffused with brown in winter; most have yellow eyes, a few brown. Late **second-winter** birds have pale eyes, gray back, two-toned bill. **First-winter** birds are buffy to mostly white; chiefly dark bill is short; eyes dark; wing tips white or irregularly washed with brown. Canadian-breeding adult *kumlieni* have wing tips variably marked with gray; a few are pure white. Greenland breeding *glaucoides* is slightly smaller and paler overall in all plumages; adults are slightly paler mantled and have pure white wing tips; most migrate southeast to Iceland, a few to Europe; rare or casual to northeast North America. First-winter birds distinguished from Thayer's by paler primaries, checkered tertials, usually paler body plumage and on some by checkered tail (see also page 217); from Glaucous by usually darker bill and structural features (smaller size, rounder head, and longer wings that extend beyond tail at rest).
RANGE: Casual south to Gulf Coast states; uncommon to rare on Great Lakes; casual to Pacific region.

THAYER'S GULL
Larus thayeri | L 23" (58 cm) WS 55" (140 cm)

Variable four-year gull. In most **adults**, eye is dark brown, mantle slightly darker than Iceland Gull or Herring Gull (preceding page); bill yellow with dark red spot; legs darker pink than Herring. Primaries pale gray below, with thin, dark trailing edge; show some black or slaty gray from above. Many have paler eyes. **Second-winter** has gray mantle, contrasting gray-brown tail band, dark eye. **First-winter** variable but primaries always entirely pale below, darker than mantle above. Distinguished from Herring Gull by smaller size, paler checkered markings in plumage (see also page 217), and paler primaries with whitish edges; from Iceland Gull by generally darker plumage, primaries darker than mantle, and usually by unspeckled tail. Compare to larger-sized and bigger-billed Glaucous-winged Gull (next page), in which the immatures are less mottled and the wing tips are uniform with the mantle.
RANGE: Rare winter visitor to Great Lakes region and through the interior, but identification difficult. Casual on East Coast. Considered by some a race of Iceland Gull.

GLAUCOUS
GULL

2nd winter

winter adult

winter
adult

breeding
adult

1st winter

1st winter

breeding adult ♂
glaucoides

1st winter ♂

winter
adult ♀

ICELAND
GULL
kumlieni

winter
adult

1st winter ♀
glaucoides

2nd winter ♀

2nd
winter

winter
adult

breeding
adult

THAYER'S
GULL

winter
adult

1st winter

1st winter

YELLOW-FOOTED GULL
Larus livens | L 27" (69 cm) WS 60" (152 cm)

Three-year gull. **Adult** is like Western Gull but has yellow legs and feet; note also thicker yellow bill with red spot; dark slate gray wings; yellow eyes. **Second-winter** bird is like adult but tail looks entirely black, bill two-toned. In first-winter plumage (shown in flight on page 217), head and body are mostly white, back and wings brown, eyes dark, bill mostly dark, legs pinkish. **Juvenile** resembles first-winter Western but white belly contrasts sharply with streaked breast; upperparts are more boldly patterned; rump whiter. Yellow-footed Gull's **calls** are lower-pitched than Western.

RANGE: Breeds on small islands in the Gulf of California. Fairly common post-breeding visitor in summer to Salton Sea; a few usually linger into winter; rarest in spring. Casual to coastal southern California.

WESTERN GULL
Larus occidentalis | L 25" (64 cm) WS 58" (147 cm)

Four-year gull. **Adults** north of Monterey, California, have paler backs and darker eyes than southern birds. All adults have white head, dark gray back, pink legs, very large bill. In **winter**, head is moderately streaked with brown in northern birds, faintly streaked in southern. **Third-winter** plumage resembles second-winter Yellow-footed Gull but tail is mostly white. **Second-winter** bird has a dark gray back, yellow eyes, two-toned bill, dark brown wings. **First-winter** bird is one of the darkest young gulls; bill is black; in flight (page 217), distinguished from young Herring Gull by contrast of dark back with paler rump. Note also the often sootier underparts and head, heavier bill. **Juvenile** is like first-winter but darker. Western Gulls hybridize extensively with Glaucous-winged Gulls in the Northwest; **hybrids** are seen all along the West Coast in winter; two ages are shown here. These are easily confused with Thayer's Gull (page 208); note large bill, pattern of wing tips.

RANGE: Western Gulls are very rare to casual well inland; definite records as far east as Illinois and Texas. Casual to southeastern Alaska.

GLAUCOUS-WINGED GULL
Larus glaucescens | L 26" (66 cm) WS 58" (147 cm)

Four-year gull. **Adult** has white head, moderately streaked with brown in winter. Body is white, mantle pale gray; primaries are the same color as rest of wing above, paler below. Eyes dark; large bill yellow with red spot; legs pink. Third-winter bird is like adult but has some buff on body, bill is smudged black; some have a partial tail band. **Second-winter**'s back is gray, rest of body and wings are pale buff to white with little mottling; tail evenly gray; bill mostly dark. **First-winter** bird (shown in flight on page 217) is uniformly pale gray-brown to whitish with subtle mottling; primaries are the same color as the mantle; note young Glaucous Gull (pages 208, 217) has sharply two-toned bill, pale primaries; and young Thayer's Gull (previous page) is smaller, with smaller bill, more speckled body plumage, and darker primaries. Glaucous-winged Gull hybridizes with Western Gull and with Herring Gull (page 206) in south-central Alaska. Hybrids are extremely variable.

RANGE: Rare well inland in Pacific states; casual east to Great Lakes region.

YELLOW-FOOTED
GULL

2nd winter

adult

juvenile

adult

1st summer

southern 3rd
winter *wymani*

southern winter
adult *wymani*

WESTERN GULL

juvenile
occidentalis

southern breeding
adult *wymani*

southern 2nd
winter *wymani*

northern
breeding adult
occidentalis

1st winter
occidentalis

GLAUCOUS-WINGED x
WESTERN HYBRID

1st winter

breeding
adult

breeding
adult

2nd
winter

GLAUCOUS-
WINGED GULL

1st winter

breeding
adult

RING-BILLED GULL

1st winter

2nd winter

FRANKLIN'S GULL

1st winter

LAUGHING GULL

2nd winter

1st winter

MEW GULL

2nd winter

1st winter

BONAPARTE'S GULL

1st winter

BLACK-HEADED GULL

1st winter

BLACK-LEGGED KITTIWAKE

juvenile

ROSS'S GULL

1st winter

LITTLE GULL

2nd winter

RED-LEGGED KITTIWAKE

juvenile

SABINE'S GULL

juvenile

1st winter

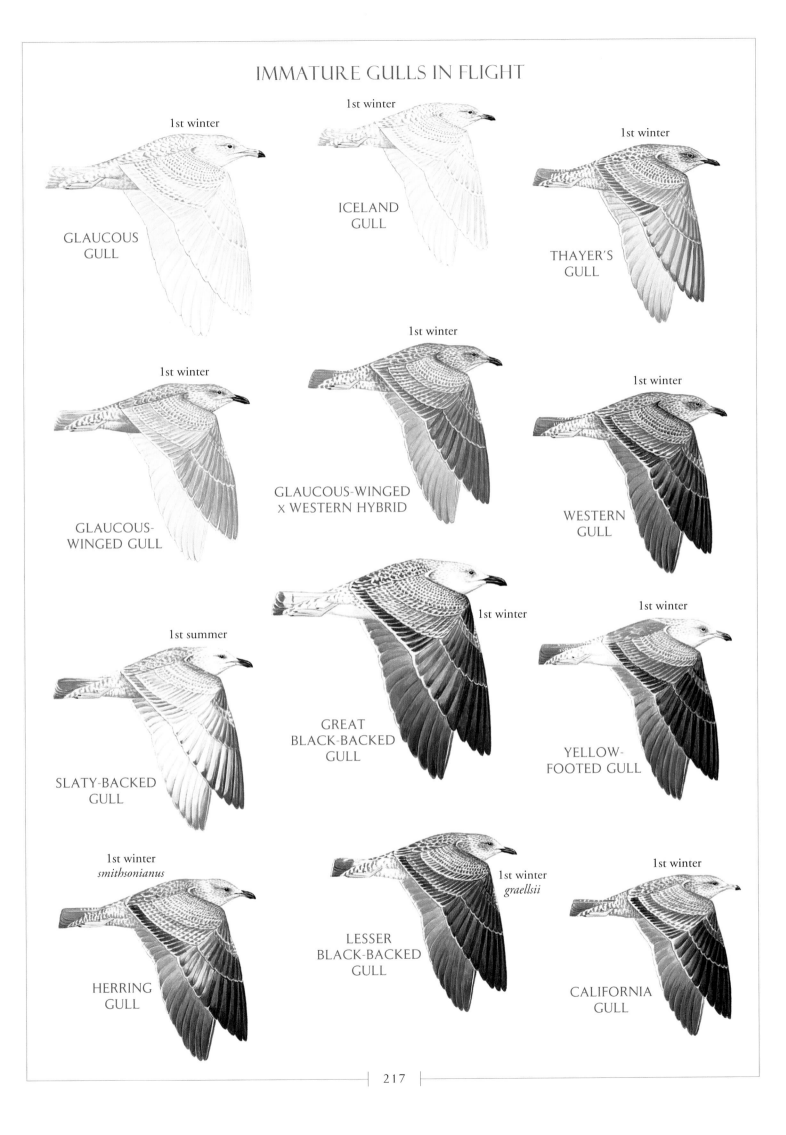

IMMATURE GULLS IN FLIGHT

1st winter

GLAUCOUS
GULL

1st winter

ICELAND
GULL

1st winter

THAYER'S
GULL

1st winter

GLAUCOUS-WINGED
x WESTERN HYBRID

1st winter

GLAUCOUS-
WINGED GULL

1st winter

WESTERN
GULL

1st summer

SLATY-BACKED
GULL

1st winter

GREAT
BLACK-BACKED
GULL

1st winter

YELLOW-
FOOTED GULL

1st winter
smithsonianus

HERRING
GULL

1st winter
graellsii

LESSER
BLACK-BACKED
GULL

1st winter

CALIFORNIA
GULL

TERNS

Distinguished from gulls by long, pointed wings and bill and by feeding technique. Most terns plunge-dive into the water after prey. Most species have a forked tail.

SANDWICH TERN
Thalasseus sandvicensis | L 15" (38 cm) WS 34" (86 cm)

Slender, black bill, tipped with yellow. **Breeding adult** is pale gray above with black crown, short black crest. In flight, shows some dark in outer primaries. White tail is deeply forked, comparatively short. Adult in **winter** plumage, seen as early as July, has a white forehead, streaked crown, grayer tail. **Juvenile**'s tail less deeply forked; bill often lacks yellow tip, in a few it is entirely yellow. By late summer, juvenile loses dark V-shaped markings and spots on back and scapulars. **Calls** include abrupt *gwit gwit* and *skee-rick* notes, like calls of Elegant Tern.

RANGE: Nests on coastal beaches and islands. Casual spring and summer visitor to coastal southern California.

ELEGANT TERN
Thalasseus elegans | L 17" (43 cm) WS 34" (86 cm)

Bill longer, thinner than in larger Royal Tern; reddish orange in adults, yellow in some juveniles. In flight, shows mostly pale underside of primaries; compare to Caspian Tern. **Breeding adult** is pale gray above with black crown and nape, black crest; white below, often with pinkish tinge. **Winter adult** and **juvenile** have white forehead; black over top of crown extends forward around eye; compare with Royal. Juveniles variably mottled above, may have orange legs; some have less black on crown, like juvenile Royal. Sharp *kee-rick* **call** very similar to Sandwich Tern.

RANGE: Post-breeding dispersal to northern Washington; casual to British Columbia; very rare spring and summer at Salton Sea; accidental in Southwest, Gulf and Atlantic coasts.

ROYAL TERN
Thalasseus maximus | L 20" (51 cm) WS 41" (104 cm)

Orange-red bill, thinner than Caspian. In flight, shows mostly pale underside of primaries; tail more deeply forked than Caspian. **Adult** shows white crown most of year; black cap acquired briefly early in **breeding** season. In **winter adult** and **juvenile**, black on nape does not usually extend to encompass eye; compare Elegant Tern. **Calls** include a bleating *kee-rer* and ploverlike whistled *tourreee*.

RANGE: Nests in dense colonies. Fairly common in winter on southern California coast, where a few breed; uncommon to rare north of breeding range along Atlantic coast in late summer; casual in North American interior.

CASPIAN TERN
Hydroprogne caspia | L 21" (53 cm) WS 50" (127 cm)

Large, stocky; bill orange to coral red, much thicker than in Royal Tern. In flight, shows dark underside of primaries; tail less deeply forked than Royal. **Adult** acquires black cap in **breeding** season; in **winter adult** and **juvenile**, crown streaked; never shows fully white forehead of Royal. Adult's **calls** include a harsh *kowk* and *ca-arr*; immature's a distinctive, whistled *whee-you*.

RANGE: Small colonies nest on coasts, in wetlands.

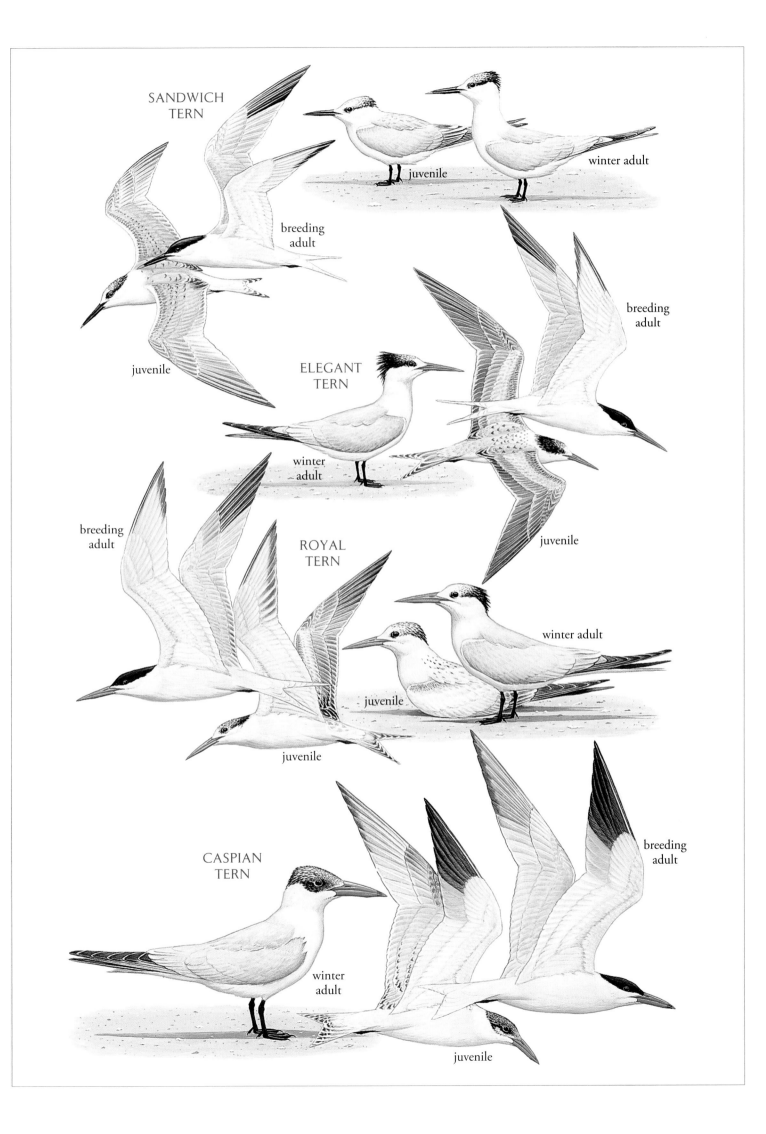

SANDWICH
TERN

juvenile

winter adult

breeding
adult

juvenile

ELEGANT
TERN

winter
adult

breeding
adult

juvenile

breeding
adult

ROYAL
TERN

winter adult

juvenile

juvenile

CASPIAN
TERN

winter
adult

breeding
adult

juvenile

ROSEATE TERN

Sterna dougallii | **E** | L 15½" (39 cm) WS 29" (74 cm)

Breeding adult is white below with slight, variable pinkish cast visible in good light; pale gray above with black cap and nape. Much paler overall than Common and Arctic Terns (next page). Lacks dark trailing edge on underside of outer wing. Bill mostly black; during summer more red appears at base. Wings shorter than in Common and Arctic; flies with rapid wingbeats suggestive of Least Tern (page 224). Deeply forked all-white tail extends well beyond wings in standing bird. Legs and feet bright red-orange. **Juvenile**'s brownish cap extends over forehead; mantle looks coarsely scaled, lower back barred with black; bill and legs black. **First-summer** bird has white forehead; lacks dark secondaries of immature Common. Full adult plumage is attained by second spring. **Call** is a soft *chi-weep*; alarm signal a drawn-out *zra-ap*, like ripping cloth.

RANGE: Uncommon and highly maritime, Roseate Terns usually come ashore only to nest. Rare on mid-Atlantic coast in late spring.

FORSTER'S TERN

Sterna forsteri | L 14½" (37 cm) WS 31" (79 cm)

Breeding adult is snow white below, pale gray above, with black cap and nape; mostly orange bill, orange legs and feet. Wingbeat much slower than in Roseate Tern. Legs and bill longer than in Common and Arctic Terns (next page). Long, deeply forked gray tail has white outer edges. In flight, shows pale upperwing area formed by silvery primaries; white rump contrasts with gray back, gray tail. **Winter** plumage resembles Common and Arctic but is acquired by mid- to late Aug., much earlier than those species, which molt chiefly after migration out of U.S. Note also lack of dark shoulder bars; most have dark eye patches not joined at nape as in Common, but many have dark streaks on nape. **Juvenile** and **first-winter** bird have shorter tails than adults and more dark color in wings. Juvenile has ginger brown cap and dark eye patch; shoulder bar is faint or absent. **Calls** include a hoarse *kyarr*, lower and shorter than in Common.

RANGE: Forster's Terns nest in widely scattered colonies in marshes.

GULL-BILLED TERN

Gelochelidon nilotica | L 14" (36 cm) WS 34" (86 cm)

Breeding adult is white below, pale gray above, with black crown and nape, stout black bill, black legs and feet. Stockier and paler than Common Tern (next page); wings broader; tail shorter and only moderately forked. **Juveniles** and **winter** birds appear largely white headed apart from some fine streaking. Juvenile has pale edgings on upperparts, bill is brownish. Adult **call** is a raspy, sharp *kay-wack*; call of juvenile is a faint, high-pitched *peep peep*.

RANGE: Fairly common but local; nests in salt marshes and on beaches; often seen hunting for insects over fields and marshes in direct powerful flight. Does not hover or dive in water. Casual in interior of North America, New England, and Atlantic Canada, except at Salton Sea, where it nests. Breeds also in California in coastal San Diego County.

ROSEATE TERN
dougallii

juvenile

breeding
adult

juvenile

1st summer

breeding
adult

FORSTER'S
TERN

breeding
adult

juvenile

1st winter

winter adult

GULL-BILLED
TERN

breeding
adult

juvenile

winter
adult

COMMON TERN

Sterna hirundo | L 14½" (37 cm) WS 30" (76 cm)

Medium gray above, with black cap and nape; paler below. Bill red, usually tipped with black. Slightly stockier than the Arctic Tern, with flatter crown, longer neck and bill. In flight, usually displays a dark wedge, variably shaped, near tip of upperwing; in late summer all outer primaries can appear dark. Note also that head projects farther than in Arctic. Common Tern's shorter tail gives it a chunkier look. Early **juvenile** shows some brown above, white below, with mostly dark bill. Juvenile's forehead is white, crown and nape blackish, secondaries dark gray; compare with juvenile Forster's Tern (preceding page). All immature and winter plumages have a dark shoulder bar. Full **adult breeding** plumage is acquired by third spring. Some **calls** are similar to Arctic Tern; distinctive in Common, a low, piercing, drawn-out *kee-ar-r-r-r*.

RANGE: Common Terns nest in large colonies. Common throughout breeding range. Uncommon and declining migrant on Pacific coast from Washington south; uncommon in fall in western interior. An East Asian subspecies, *longipennis*, seen regularly on the islands of western Alaska, is darker overall; bill and legs black.

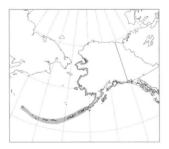

ARCTIC TERN

Sterna paradisaea | L 15½" (39 cm) WS 31" (79 cm)

Medium gray above, with black cap and nape; paler below. Bill deep red. Slightly slimmer than Common Tern, with rounder head, shorter neck and bill. In flight, upperwing appears uniformly gray, lacking dark wedge of Common; underwing shows very narrow black line on trailing edge of primaries; all flight feathers appear translucent. Note also that tail is longer and head does not project as far as in Common. **Juvenile** largely lacks brownish wash of early juvenile Common; shoulder bar less distinct; secondaries whitish and a portion of coverts whitish too, creating an effect like Sabine's Gull (page 214). Forehead is white, crown and nape blackish; compare juvenile Forster's Tern (preceding page). Full adult breeding plumage is acquired by third spring. **Calls** include a raspy *tr-tee-ar*, higher than Common Tern's call.

RANGE: Holarctic breeder. Arctic Terns migrate well offshore; casual inland during migration, especially in late spring.

ALEUTIAN TERN

Onychoprion aleuticus | L 13½" (34 cm) WS 29" (74 cm)

Dark gray above and below, with white forehead, black cap, black bill, black legs. In flight, distinguished from Common and Arctic Terns by shorter tail, white forehead, and dark, white-edged bar on secondaries, most visible from below. **Juvenile** is buff and brown above; legs and lower mandible reddish. **Call** is a squeaky *twee-ee-ee*, unlike any other tern.

RANGE: Aleutian Terns nest in loose colonies, sometimes with Arctic Terns. Casual in spring off British Columbia. Migration routes uncertain; has been seen in fall and winter off Hong Kong, the Philippines, and Sumatra. Main winter grounds still unknown, probably southwest Pacific and Indian Oceans.

breeding
adult

1st summer

2nd
summer

juvenile

breeding adult
longipennis

1st fall

COMMON
TERN
hirundo

breeding adult

breeding
adult

juvenile

ARCTIC
TERN

1st summer

breeding
adult

juvenile

ALEUTIAN
TERN

breeding
adult

juvenile

LEAST TERN
Sternula antillarum | **E** | L 9" (23 cm) WS 20" (51 cm)

Smallest North American tern. **Breeding adult** is gray above, with black cap and nape, white forehead, orange-yellow bill with dark tip; underparts are white; legs orange-yellow. By late summer, bill base is more greenish. In flight, black wedge on outer primaries is conspicuous; note also the short, deeply forked tail. **Juvenile** shows brownish, U-shaped markings; crown is dusky; wings show dark shoulder bar. By first fall, upperparts are gray, crown whiter, but dark shoulder bar is retained. **First-summer** birds are more like adults but have dark bill and legs, shoulder bar, black line through eye, dusky primaries. Flight is rapid and buoyant. **Calls** include high-pitched *kip* notes and a harsh *chir-ee-eep*.
RANGE: Nests in colonies on beaches and sandbars; also on rooftops. Fairly common but local on East and Gulf Coasts; declining inland and on the California coast. Winters from Central America south.

BLACK TERN
Chlidonias niger | L 9¾" (25 cm) WS 24" (61 cm)

Breeding adult is mostly black, with dark gray back, wings, and tail; white undertail coverts. In flight, shows uniformly pale gray underwing and fairly short tail, slightly forked. Bill is black in all plumages. **Juvenile** and winter birds are white below, with dark gray mantle and tail; dark ear patch extends from dark crown; flying birds show dark bar on side of breast. Some juveniles show a contrastingly paler rump. Shoulder bar on upper wing is much darker than in juvenile White-winged Tern. First-summer birds can be like winter adults or may have some dark feathers on head and underparts; second-summer birds are like breeding adults, but show some whitish on head; full breeding plumage is acquired by third spring. **Molting fall adults** appear patchy black-and-white as they acquire winter plumage in late summer; these birds are easily confused with the White-winged Tern. **Calls** include a metallic *kik* and a slurred *k-seek*.
RANGE: Black Terns nest on lakeshores and in marshes; declining over part of range inland and on East Coast; rare migrant on West Coast.

WHITE-WINGED TERN
Chlidonias leucopterus | L 9½" (24 cm) WS 23" (58 cm)

Bill and tail shorter than in Black Tern; tail less deeply notched. In **breeding** plumage, bill usually black, but sometimes red; white tail, whitish upperwing coverts, and black wing linings are distinctive; upperwing shows black outer primaries. **Molting** birds are patchy black-and-white, but whitish tail and rump are distinctive; black wing linings often last until late summer. **Winter adult** has white wing linings; lacks dark breast bar of Black Tern; crown, speckled rather than solid black, not usually connected to dark ear patch. First-summer bird resembles winter adult; second-summer usually like breeding adult; adult plumage reached by third spring. **Juvenile**'s head pattern resembles Black, but browner back shows greater contrast with grayish wing coverts and whitish rump than juvenile Black.
RANGE: Eurasian species, very rare vagrant to East Coast; casual to Great Lakes region; accidental in West, where recorded in western Aleutians, south coastal Alaska, coastal California.

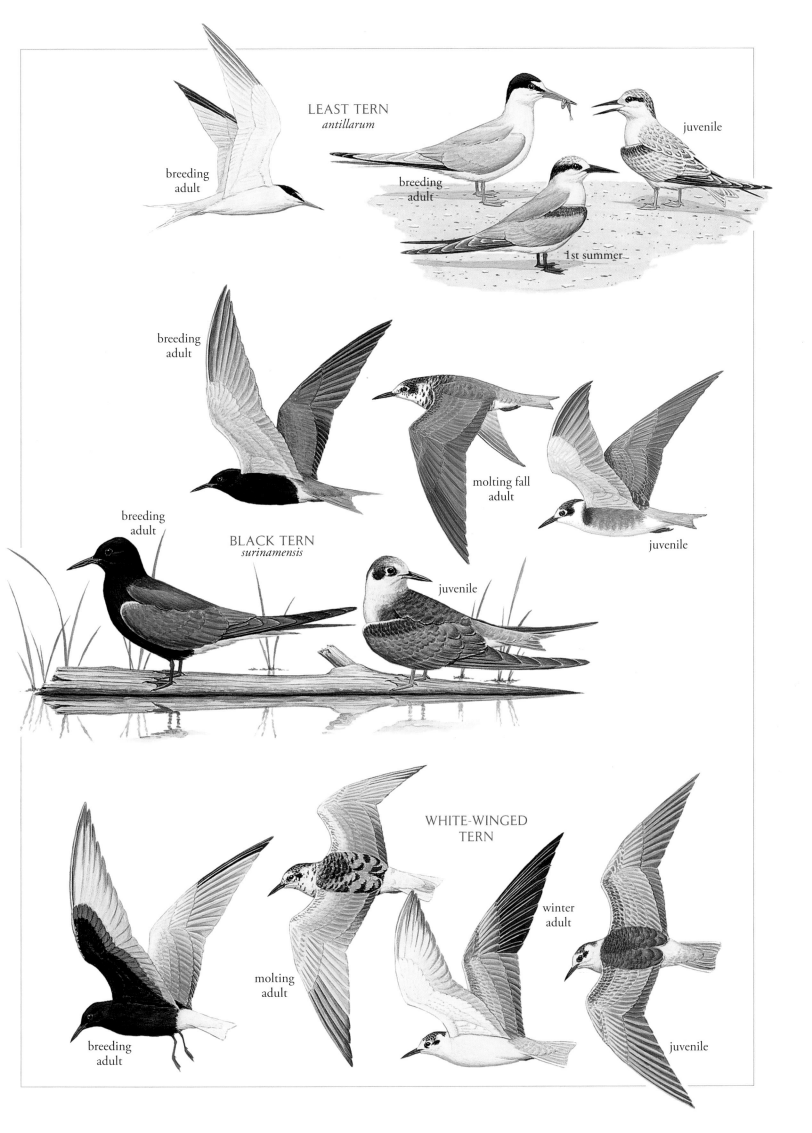

LEAST TERN
antillarum

breeding
adult

breeding
adult

juvenile

1st summer

breeding
adult

BLACK TERN
surinamensis

molting fall
adult

juvenile

breeding
adult

juvenile

WHITE-WINGED
TERN

winter
adult

molting
adult

breeding
adult

juvenile

BRIDLED TERN *Onychoprion anaethetus* | L 15" (38 cm) WS 30" (76 cm)

Note white collar between brownish gray upperparts and black cap on **breeding adult**. Similar to Sooty Tern, but slimmer; wings more pointed; underwings and tail edges more extensively white; tail grayer. Bridled's white forehead patch extends behind the eye, while Sooty's stops at the eye. **Juvenile** has pale mottling above.

RANGE: Nests in the West Indies and locally off Florida Keys. Regular in summer well offshore in the Gulf of Mexico, and in the Gulf Stream to North Carolina; rarely to New Jersey; after tropical storms, casual to New England.

SOOTY TERN *Onychoprion fuscatus* | L 16" (41 cm) WS 32" (81 cm)

Blackish above, white below; white forehead. Lacks white collar of Bridled Tern. Tail is deeply forked, edged with white. **Juvenile** is sooty brown overall, with whitish stippling on back; pale lower belly and undertail coverts; pale wing linings. Typical **call** is a high, nasal *wacky-wack*, given year-round.

RANGE: Large breeding colony on Dry Tortugas, Florida; also nests on islands off Texas and Louisiana. Regular in summer to North Carolina. Casual in southern coastal California. Tropical storms can carry birds inland to Great Lakes and north to Maritime Provinces.

BLACK NODDY *Anous minutus* | L 13½" (34 cm) WS 30" (76 cm)

Compared to Brown Noddies, Black Noddy is smaller, with shorter legs; bill is thinner and disproportionately longer; overall color is slightly blacker. In **immatures**, white area on head is very sharply defined.

RANGE: Tropical species, rare and now irregular visitor among Brown Noddies on Dry Tortugas (mostly immatures). Casual on Texas Gulf Coast.

BROWN NODDY *Anous stolidus* | L 15½" (39 cm) WS 32" (81 cm)

Overall dark gray-brown with whitish gray cap, blending at back; **immature** shows only a whitish line on forehead. Unlike other terns, noddies have long, wedge-shaped tail with only a small notch at tip, and no seasonal variation in appearance. Usually silent; a crowlike *karrk* **call** is heard mostly around the breeding colonies.

RANGE: Nests in a colony on Dry Tortugas, Florida. Casual to Texas coast and off Outer Banks, North Carolina.

LARGE-BILLED TERN *Phaetusa simplex* | L 14½" (37 cm) WS 36" (92 cm)

Mantle and short tail dark gray; white below with white forehead and black cap; legs and stout bill yellow. In flight, shows striking Sabine's Gull-like pattern (page 214).

RANGE: A South American freshwater species. Accidental; recorded in late spring and summer from Illinois, Ohio, and New Jersey; additionally from Cuba and Bermuda.

BLACK SKIMMER *Rynchops niger* | L 18" (46 cm) WS 44" (112 cm)

No other bird has a lower mandible longer than the upper. A long-winged coastal bird, it furrows the shallows with its red, black-tipped bill. Black above and white below; red legs and bill shape are distinctive. Female is distinctly smaller than the male. **Juvenile** is mottled dingy brown above. **Winter adults** show a white collar.

RANGE: Casual inland. West Coast population is increasing.

BRIDLED
TERN
melanoptera

breeding
adult

juvenile

SOOTY TERN
fuscata

juvenile

breeding
adult

BLACK NODDY
americanus

adult

immature

BROWN
NODDY
stolidus

adults

immature

LARGE-BILLED
TERN

breeding
adult

juvenile

BLACK
SKIMMER
niger

winter adults

breeding
adult

SKUAS and JAEGERS (Family Stercorariidae)

Formerly placed with the Gulls, Terns, and Skimmers, recent molecular evidence indicates that they are most closely related to Alcidae and belong in their own family. Predatory and piratic seabirds, skuas are broader winged than jaegers. **Species:** *7 World, 5 N.A.*

GREAT SKUA

Stercorarius skua | L 22" (56 cm) WS 54" (137 cm)

Large, heavy, and barrel-chested; wings broader and more rounded than jaegers (next page); tail shorter and broader. Shows a distinctly hunchbacked appearance in flight and a conspicuous white bar at base of primaries; bill is heavier than in jaegers. Great Skua is distinguished from South Polar Skua by overall reddish or ginger brown color and heavy streaking on back, wing coverts, and much of underparts; sometimes shows dark brown cap. **Juvenile** and immature show less streaking, especially on underparts. A small number of juvenile dark morphs have much less rufous streaking; resemble juvenile South Polar Skuas. Strong, powerful fliers, skuas pursue gulls and other seabirds and rob them of their prey.

RANGE: Uncommon; breeds in Iceland and northern Europe; winters in North Atlantic. Seen well offshore from Nov. to Apr.; rare in summer off Canadian coast.

SOUTH POLAR SKUA

Stercorarius maccormicki | L 21" (53 cm) WS 52" (132 cm)

Large, heavy, and barrel-chested; wings broader and more rounded than jaegers (next page); tail shorter and broader. Like Great Skua, shows a distinctly hunchbacked appearance in flight, a bold white bar at base of primaries, and a heavier bill than in jaegers. In all ages, South Polar Skua shows a uniform mantle coloring and lacks the reddish tones and streaking seen on upperparts of Great Skua. In **light-morph** birds, contrastingly pale gray nape is distinctive; light morph also shows grayish head and underparts. **Dark morph** is uniformly blackish brown across mantle, with golden hackles on nape; distinguished from subadult Pomarine Jaeger by larger size, broader and more rounded wings, more distinct white wing bar. **Juveniles** and immatures of both color morphs are darker than light-morph adults, ranging from dark brown to dark gray. In the field, birds under two years of age are generally indistinguishable from juveniles; birds over two years old are generally indistinguishable from full adults. Silent while in our region.

RANGE: South Polar Skua winters (our summer) in the North Atlantic and North Pacific, usually from May to early Nov. Most numerous in spring and fall off the West Coast, in spring off the East Coast; casual off the south coast of Alaska; accidental in North Dakota; a record from Tennessee after Hurricane Katrina likely this species. Very rarely seen from shore. Difficulty of identification makes range information somewhat speculative for both skua species. Several records (photos) of birds off mid-Atlantic coast could pertain to Brown Skua (*S. antarcticus*), of Southern Hemisphere, or possibly hybrids between that species and South Polar Skua.

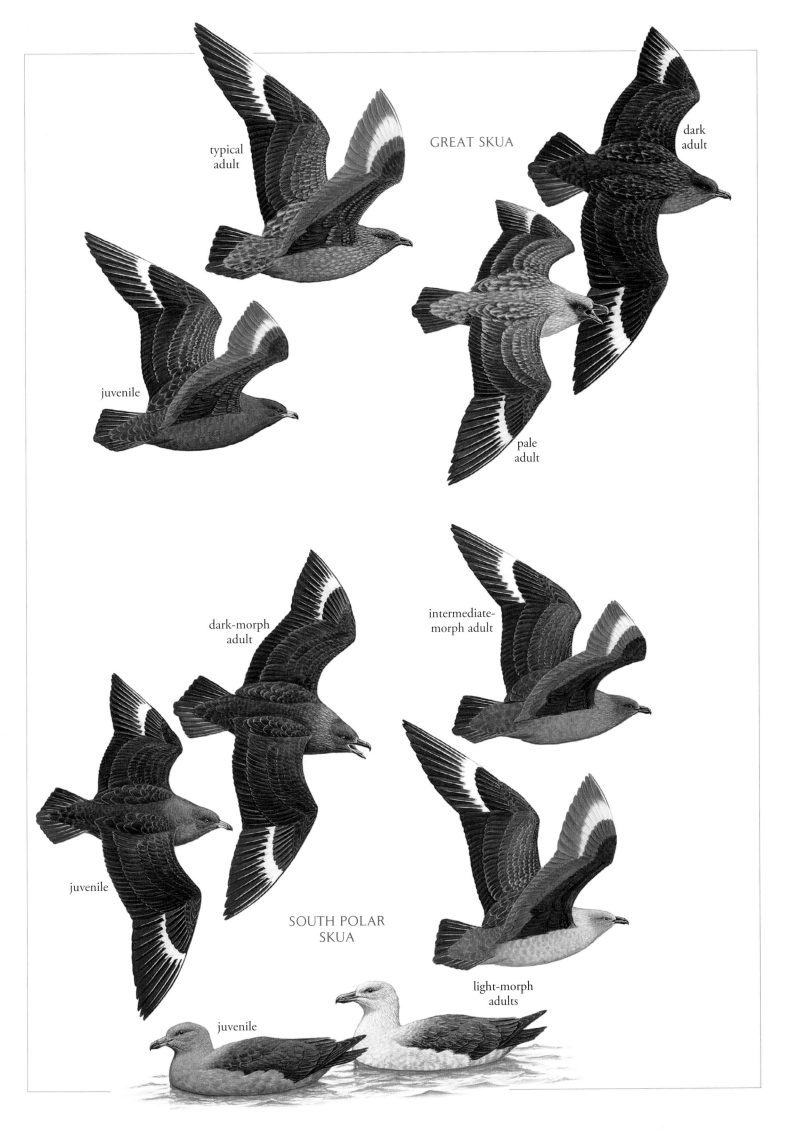

typical adult

juvenile

GREAT SKUA

dark adult

pale adult

dark-morph adult

intermediate-morph adult

juvenile

SOUTH POLAR SKUA

juvenile

light-morph adults

JAEGERS

Arctic breeders. Wings are longer, slimmer, and more powerful than on skuas. Adult plumage and long central tail feathers take three or four years to develop. Complex and variable plumages make identification extremely difficult. Most molts occur after the fall migration.

POMARINE JAEGER
Stercorarius pomarinus | L 21" (53 cm) WS 48" (122 cm)

Body bulkier, bicolored bill longer and thicker, wingbeats slower than Parasitic Jaeger. Most birds show a distinctive second pale underwing patch at base of primaries (fainter or lacking in Parasitic). **Adult**'s tail streamers, twisted at ends, form dark blobs when seen from side; length is variable, averages longer in male. Note extensive helmet on sides of head. Compare **dark-morph adults** and subadults with South Polar Skuas (preceding page). Some grayish brown **juveniles** are dark, some pale, but none shows the foxy red tones of most juvenile Parasitics; underwing is paler than body; pale, barred uppertail covert forms a contrasting patch above. All jaeger are mostly silent away from breeding grounds.
RANGE: Seen less often from shore than Parasitic. Common off West Coast in spring migration and especially late in fall; often migrates in flocks. Uncommon in winter; casual inland away from Great Lakes.

PARASITIC JAEGER
Stercorarius parasiticus | L 19" (48 cm) WS 42" (107 cm)

Smaller size, more slender body, faster wingbeats than Pomarine Jaeger; also smaller head, thinner bill, pointed tail streamers. **Adult** lacks helmeted effect of Pomarine; is paler near bill, as **juveniles** sometimes are also. Juvenile is highly variable; shows rufous tips on primaries; distinctive rusty tones particularly evident on **light morphs**; tail coverts have fainter, wavier bars than Pomarine.
RANGE: Fairly common; the jaeger species most often seen from shore in migration. Casual fall migrant inland; more regular on Great Lakes and Salton Sea.

LONG-TAILED JAEGER
Stercorarius longicaudus | L 22" (56 cm) WS 40" (102 cm)

Most lightly built jaeger, with round chest, flat belly, narrow wings, and disproportionately long tail in all ages; bill rather short and thick. Flight is more graceful, ternlike. Note distinctive contrast between grayish mantle and darker flight feathers; usually has only two to three white primary shafts; no pale underwing patch except on **juvenile. Adult** has well-defined black cap, no breast band as in most jaegers, and usually very long, pointed central tail streamers. Juvenile's central tail feathers have round, often white-edged tips; bill is half dark, half gray; is grayer overall than Parasitic except for dark morph; fringing above whitish, never rusty. **Light-morph juveniles** show distinctive white belly and strong, even, black barring on upper- and undertail coverts; palest birds may have very pale gray heads. **Dark-morph juveniles** often lack barring on uppertail coverts.
RANGE: Common in dry, upland-tundra breeding area; migrating birds uncommon to fairly common well off West Coast (fall); rare closer to shore and off East Coast (mainly in fall); rare inland (mainly in fall), and casual off Gulf Coast.

POMARINE JAEGER

dark-morph
breeding adult

light-morph
breeding adult

light-morph
1st summer

juvenile

PARASITIC JAEGER

dark-morph
breeding adult

light-morph
breeding adult

light-morph
juveniles

light-morph
1st summer

LONG-TAILED JAEGER

light-morph
juvenile

breeding
adult

1st summer

dark-morph
juvenile

AUKS, MURRES, AND PUFFINS  (FAMILY ALCIDAE)

These black-and-white "penguins of the north" have set-back legs that give them an upright stance on land. In flight, wingbeats are rapid and shallow. **Species:** *24 World, 22 N.A.*

DOVEKIE
Alle alle | L 8¼" (21 cm)

Small and plump with short neck, stubby bill. **Breeding adult** is black above, white below; black upper breast contrasts sharply with white underparts; dark wing linings. Usually swims tilted forward in the water. In **winter** plumage, throat, chin, and lower face are white, with white curving around behind eye.

RANGE: Abundant on breeding grounds. Winters irregularly in North Atlantic south to off coast of North Carolina; rarely to Florida, but occasionally in large numbers; casual inland.

COMMON MURRE
Uria aalge | L 17½" (45 cm)

Large, with a long, slender, pointed bill. Upperparts dark sooty gray, head brownish; underparts white. Some Atlantic birds have a "**bridle**," a white eye ring and spur. In **winter** plumage, a dark stripe extends from eye across white cheek. **Juvenile** has shorter bill; distinguished from Thick-billed Murre by white facial stripe, paler upperparts, mottled flanks, and thinner bill.

RANGE: Numerous off East and West Coasts. Nests in dense colonies on rocky cliffs. Chick accompanies adult at sea and can be mistaken for Xantus's Murrelets (page 236).

THICK-BILLED MURRE
Uria lomvia | L 18" (46 cm)

Stocky, with a thick, fairly short bill, arched at tip to form a blunt hook. Upperparts and throat of **adult** are darker than Common Murre; white of underparts usually rises to a sharp point on the foreneck. Most birds show a distinct white line on cutting edge of upper mandible; in Pacific birds, bill is slightly longer and thinner than in Atlantic birds. In immature and **winter adult**, face and neck are more extensively dark than in Common. First-summer bird is browner above than adult; otherwise similar to winter bird.

RANGE: Nests in colonies on rocky cliffs. Common on breeding grounds. On East Coast, much more regular in winter south of Canada than Common; casual to mid-Atlantic states; recorded to Florida. Casual on West Coast south of breeding range, where most records are from the Monterey area in California.

RAZORBILL
Alca torda | L 17" (43 cm)

A chunky bird, with a big head and thick neck; black above, white below. Rather long, pointed tail; heavy head; and massive, arching bill distinguish Razorbill from murres. Swimming birds often hold tail cocked up. A white band crosses the bill; in **breeding** plumage, a white line runs from bill to eye. **Immature** lacks white band; bill is smaller but still distinctively shaped.

RANGE: Nests on rocky cliffs and among boulders. Winters in large numbers on the Grand Banks off Newfoundland. A few winter well offshore as far south as North Carolina, casually to Florida; irregular along the coast south of Long Island.

breeding adult

winter

DOVEKIE

winter

breeding adult

bridled breeding adult

breeding adult

breeding adult

COMMON MURRE

winter

juvenile

THICK-BILLED MURRE

Atlantic breeding adults *lomvia*

Pacific winter *arra*

immature

winter adult

RAZORBILL

breeding adult

breeding adult

BLACK GUILLEMOT
Cepphus grylle | L 13" (33 cm)

Breeding adult black overall, with large white patch on upperwing. **Winter adult** white; upperparts heavily mottled with black except on nape; wing patch less distinct. **Juvenile** is sooty above; sides and wing patches mottled. First-summer birds are patchily black-and-white; wing patches mottled. In all plumages, white axillaries and wing linings distinguish Black from Pigeon Guillemot. In East Coast race *arcticus*, juveniles and winter birds are darker than in high Arctic race, *mandtii*.
RANGE: Fairly common in the East; usually seen close to shore in breeding season.

PIGEON GUILLEMOT
Cepphus columba | L 13½" (34 cm)

Breeding and **winter adults'** plumage similar to smaller Black Guillemot, but note black bar on white upperwing patch; bar may be obscured in swimming bird. **Juvenile** is dusky above; crown and nape darker; wing patch marked with black edgings; breast and sides mottled gray. Compare with juvenile Marbled Murrelet. First-winter Pigeon Guillemot resembles winter adult but is darker. In all plumages, dusky axillaries and wing linings distinguish Pigeon from Black Guillemot.
RANGE: Fairly common; seen near shore in breeding season. Winter range not well-known.

LONG-BILLED MURRELET
Brachyramphus perdix | L 11½" (29 cm)

In **winter**, lacks conspicuous white collar of similar Marbled Murrelet; shows small pale oval patches on sides of nape; in **breeding** plumage upperparts are less rufous, throat paler. Formerly considered a race of Marbled.
RANGE: Found in coastal northeast Asia. Casual throughout North America; most records are in fall of winter-plumaged birds.

MARBLED MURRELET
Brachyramphus marmoratus | T | L 10" (25 cm)

Bill longer than in Kittlitz's Murrelet. Tail all-dark, but white on overlapping uppertail coverts. **Breeding adult** dark above, heavily mottled below. In **winter** plumage, white on scapulars distinguishes Marbled from other murrelets except Long-billed and Kittlitz's with shorter bill and breast band; white on face of Marbled is variable but less than Kittlitz's. **Juvenile** is like winter adult but mottled below; by first winter, underparts are mostly white. Highly vocal; **call** is a series of loud, high *keer* notes. All murrelets have more pointed wings and faster flight than auklets.
RANGE: Often seen close to land, frequenting fjords, deep bays, lagoons. Nests inland, usually in trees. Fairly common in breeding range; rare in southern California.

KITTLITZ'S MURRELET
Brachyramphus brevirostris | L 9½" (24 cm)

Bill shorter than in Marbled Murrelet. Outer tail feathers white. **Breeding adult's** buffy grayish brown upperparts are heavily patterned; throat, breast, and flanks mottled; belly white. In **winter**, note extensive white on face, making eye conspicuous; nearly complete breast band; and white edges on secondaries. **Juvenile** distinguished from Marbled by shorter bill, paler face, white outer tail feathers.
RANGE: Fairly common but local. Accidental to British Columbia and southern California.

BLACK
GUILLEMOT
arcticus

winter
adult

juvenile

winter adult

breeding
adult

winter
adult

juvenile

PIGEON
GUILLEMOT

winter
adult

breeding
adult

LONG-BILLED
MURRELET

breeding adult

winter

winter

juvenile

MARBLED
MURRELET

winter

breeding
adult

breeding
adult

winter

winter

KITTLITZ'S
MURRELET

juvenile

breeding
adult

winter

breeding
adult

XANTUS'S MURRELET
Synthliboramphus hypoleucus | L 9¾" (25 cm)

Slate black above, white below. The more southerly breeding race, nominate *hypoleucus*, has much more white in face than southern California-breeding *scrippsi*. Both races distinguished from Craveri's Murrelet by lack of partial dark collar; slightly shorter, stouter bill; lack of black under the bill; and white wing linings, visible when birds rise to flap wings before taking off. **Call**, a piping whistle or series of whistles, is heard year-round.

RANGE: Uncommon to fairly common; both subspecies move north after breeding as far as Washington; *scrippsi* casually to southern British Columbia; *hypoleucus* is generally rare in North American waters. Usually seen a few miles offshore; nests in colonies on rocky islands, ledges, and sometimes in dense vegetation.

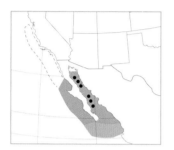

CRAVERI'S MURRELET
Synthliboramphus craveri | L 8½" (22 cm)

Slate black above, white below. Distinguished from Xantus's Murrelet by variably dusky gray wing linings; dark partial collar extending onto breast; slightly slimmer, longer bill; and black color of face extending under the bill. In good light, upperparts have a brownish tinge. **Call**, very different from *scrippsi* Xantus's, is a cicada-like rattle, rising to a reedy trilling when agitated.

RANGE: Breeds on rocky islands off Baja California. Regular late-summer and fall post-breeding visitor to coast of southern and central California. Rarely seen from shore. Usually seen a few or many miles offshore.

ANCIENT MURRELET
Synthliboramphus antiquus | L 10" (25 cm)

Black crown and nape contrast with gray back. White streaks on head and nape of **breeding adult** give it an "ancient" look. Note also black chin and throat, yellowish bill. **Winter adult**'s bib is smaller and flecked with white, streaks on head less distinct. **Immature** lacks head streaks; throat is mostly white; distinguished from winter Marbled Murrelet (preceding page) by heavier, paler bill and by sharp contrast between head and back. In flight, Ancient Murrelet holds its head higher than other murrelets; dark stripe on body at base of wing contrasts with white underparts, white wing linings. **Call** is a short, emphatic *chirrup*.

RANGE: Uncommon to common; breeds primarily on the Aleutians and other Alaska islands; winters to central California, rarely to southern California. Casual inland throughout North America.

CASSIN'S AUKLET
Ptychoramphus aleuticus | L 9" (23 cm)

Small, plump, dark gray bird; wings more rounded than in murrelets; bill short and stout, with pale spot at base of lower mandible; pale eyes. Upperparts are dark gray, shading to paler gray below, with whitish belly. Prominent white crescent above eye. Juvenile is paler overall; throat whitish; has darker eye and black bill. **Call**, heard only on the breeding grounds, is a weak croaking.

RANGE: Common; nests in colonies on islands and on isolated coastal cliffs and headlands. Highly pelagic; usually seen farther from shore than murrelets.

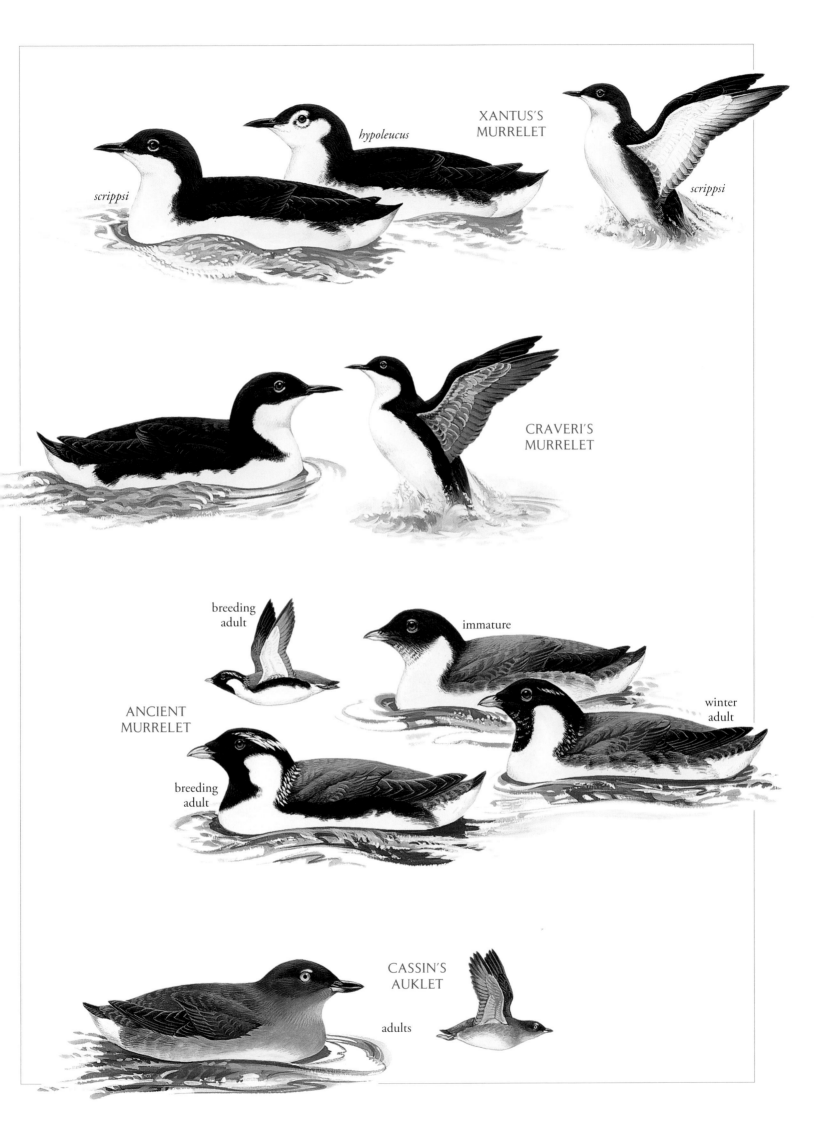

XANTUS'S
MURRELET

hypoleucus

scrippsi

scrippsi

CRAVERI'S
MURRELET

breeding
adult

immature

ANCIENT
MURRELET

winter
adult

breeding
adult

CASSIN'S
AUKLET

adults

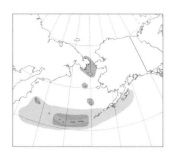

LEAST AUKLET
Aethia pusilla | L 6¼" (16 cm)

Small and chubby, with short neck; dark above, with variably white-tipped scapulars, secondaries and greater coverts; forehead and lores streaked with white bristly feathers. Stubby, knobbed bill is dark red, with pale tip. In **breeding** plumage, acquired by Jan., a streak of white plumes extends back from behind eye; underparts are variable: heavily mottled with gray to nearly all-white. In **winter** plumage, underparts are entirely white. **Juvenile** resembles winter adult.

RANGE: Abundant and gregarious, Least Auklets are found in immense flocks. Nest on boulder-strewn beaches and islands. Often seen far from shore. Accidental in Northwest Territories, Canada, and coastal California. Winters throughout Aleutians.

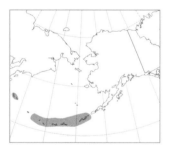

PARAKEET AUKLET
Aethia psittacula | L 10" (25 cm)

In **breeding** plumage, acquired by late Jan., broad upturned bill is orange-red; white plume extends back from behind the eye; dark slate upperparts and throat contrast sharply with white underparts; sides are mottled gray. In **winter** plumage, bill becomes duskier; underparts, including throat, are entirely white. Compare especially with larger Rhinoceros Auklet (next page). **Juvenile** resembles winter adult. Silent except on breeding grounds, when **call** is a musical trill, rising in pitch.

RANGE: Fairly common on breeding grounds; nests in scattered pairs on rocky shores and sea cliffs. Found in pairs or small flocks in winter, well out to sea. Winters irregularly as far south as California. Like other auklets, wings are rounded and wingbeats are fluttery in comparison to those of murrelets.

WHISKERED AUKLET
Aethia pygmaea | L 7¾" (20 cm)

Overall color like Crested Auklet, but note Whiskered is paler on belly and undertail coverts. Three white plumes splay from each side of face; thin crest curls forward. In **breeding** plumage, bill is deep red with white tip. In **winter**, bill is dusky, plumes and crest less conspicuous. **Juvenile** is paler below; bill smaller; lacks crest; less striking head pattern. First-summer bird may lack crest and show reduced plumes.

RANGE: Fairly common but local; nests in Aleutians from Baby Islands off Unalaska Island west to Buldir; and on islands off Russian Far East. Often seen feeding in riptides.

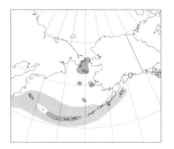

CRESTED AUKLET
Aethia cristatella | L 9" (23 cm)

Sooty black overall; prominent quail-like crest curves forward from forehead; narrow white plume trails from behind yellow eye. **Breeding adult**'s bill is enlarged by bright orange plates. In **winter**, bill is smaller and browner; crest and plume reduced. **Juvenile** has short crest, faint plume; bill smaller. Compare with smaller Whiskered Auklet. **First-summer** bird has more evident plume back from eye, but bill is still small. Common and gregarious; often seen in large flocks that may include Parakeet and Least Auklets. In flight, note more pointed wing tips, and all-dark underparts.

RANGE: Nests in crevices of sea cliffs and rocky shores. Accidental south to Baja California.

LEAST AUKLET

breeding adults

light

dark

breeding adults

winter adults

juvenile

PARAKEET AUKLET

breeding adults

breeding adult

dark

winter adults

juvenile

WHISKERED AUKLET

breeding adult

winter adult

breeding adults

juvenile

winter adult

winter adult

CRESTED AUKLET

1st summer

breeding adult

breeding adult

winter adult

juvenile

winter adult

winter adult

RHINOCEROS AUKLET
Cerorhinca monocerata | L 15" (38 cm)

Large, heavy-billed auklet with large head and short, thick neck. Blackish brown above; paler on sides, neck, and throat. In flight, whitish on belly blends into dark breast; compare with extensively white underparts of similar Parakeet Auklet (preceding page). In **breeding** plumage, acquired by Feb., Rhinoceros Auklet has distinct white plumes and a pale yellow "horn" at base of orange bill. **Winter adult** lacks horn; plumes are less distinct, bill paler. Juvenile and **immature** lack horn and plumes; bill is dusky, eyes darker. Compare with much smaller Cassin's Auklet (page 236).
RANGE: Rhinoceros Auklets are common along most of the West Coast in fall and winter; often seen in large numbers close inshore.

ATLANTIC PUFFIN
Fratercula arctica | L 12½" (32 cm)

The only East Coast puffin. **Breeding adult** identified by massive, brightly colored bill; pale face and underparts contrast with dark upperparts. **Winter adult** has smaller, duller bill, dusky face. In **juvenile** and first-winter birds, face is even duskier, bill much paler and smaller. Full adult bill takes about five years to develop. In flight, distinguished from murres and Razorbill (page 232) by red-orange legs, rounded wings, grayish wing linings, absence of white trailing edge on wing.
RANGE: Locally common in breeding season; winters at sea, rarely south to North Carolina.

HORNED PUFFIN
Fratercula corniculata | L 15" (38 cm)

A stocky North Pacific species with thick neck, large head, massive bill; underparts are white in all plumages. **Breeding adult**'s face is white, bill brightly colored. Dark, fleshy "horn" extending up from eye is visible only at close range. **Winter adult**'s bill is smaller, duller; face is dusky. Bill of **juvenile** and first-winter birds smaller and duskier than adult; full adult bill takes several years to develop. In flight, bright orange legs are conspicuous; wings are rounded; wing linings grayish; wings lack white trailing edge.
RANGE: Locally common; winters well out to sea. Rare and irregular. Irregular straggler off the West Coast to southern California, mainly in late spring.

TUFTED PUFFIN
Fratercula cirrhata | L 16" (41 cm)

Stocky, with thick neck, large head, massive bill. Underparts are dark in adults. **Breeding adult**'s face is white, bill brightly colored; pale yellow head tufts droop over back of neck. **Winter adult** has smaller, duller bill; face is gray, tufts shorter or absent. **Juvenile** has smaller, dusky bill; dark eye; white or dark underparts. First-winter bird looks like juvenile until spring molt. As in other puffins, full adult bill and plumage take several years to develop. Red-orange feet are conspicuous in flight; wings are rounded; wing linings grayish; wings lack white trailing edge.
RANGE: Common to abundant in northern breeding range; uncommon to rare off California. Winters far out at sea.

RHINOCEROS
AUKLET

winter
adult

immature

winter
adult

breeding
adult

breeding
adult

ATLANTIC
PUFFIN

breeding
adult

juvenile

winter
adult

juvenile

HORNED
PUFFIN

winter adult

breeding
adult

WH

TUFTED
PUFFIN

juvenile

juvenile

winter
adult

breeding
adult

OWLS (Families Tytonidae and Strigidae)

Distinctive birds of prey, divided by structural differences into two families, Barn Owls and Typical Owls.
All have immobile eyes in large heads. Fluffy plumage makes their flight nearly soundless. Many species hunt
at night and roost during the day. To find owls, search the ground for regurgitated pellets of fur and bone
below a nest or roost. Listen for flocks of small songbirds noisily mobbing a roosting owl.
Species: *Tytonidae 15 World, 1 N.A.; Strigidae 180 World, 21 N.A.*

BARN OWL
Tyto alba | L 16" (41 cm)

A pale owl with dark eyes in a heart-shaped face. Rusty brown above; underparts vary from white to cinnamon. Darkest birds are always **females**, palest birds **males**. Compare with Snowy Owl (next page). Typical **call** is a raspy, hissing screech.

RANGE: Rare to uncommon in parts of western range, uncommon and declining in the East. Barn Owl roosts and nests in dark cavities in city and farm buildings, cliffs, and trees.

SHORT-EARED OWL
Asio flammeus | L 15" (38 cm)

Tawny; boldly streaked on breast; belly paler, more lightly streaked. Ear tufts are barely visible. In flight, long wings show buffy patch above, black "wrist" mark below; these markings are usually more prominent than in Long-eared Owl, which also has less distinct but more individual bars on primaries. Flight is wavering, wingbeats erratic. Typical **call**, heard in breeding season and sometimes in winter, is a raspy, high barking.

RANGE: Fairly common. A bird of open country, marshes, tundra, and weedy fields; nests on the ground. During the day it roosts on the ground or on open, low perches: short poles, muskrat houses, and duck blinds. Somewhat gregarious in winter; groups may gather where prey is abundant. Short-eared Owls sighted on Florida Keys and Dry Tortugas are thought to have originated from West Indian populations, likely *domingensis* from Cuba.

LONG-EARED OWL
Asio otus | L 15" (38 cm)

A slender owl with long, close-set ear tufts. Boldly streaked and barred on breast and belly. Wings generally have a less prominent buffy patch with more dark barring and a smaller black "wrist" mark than Short-eared Owl; facial disk is rusty. Lives in thick woods; hunts at night over open fields, marshes. By day it roosts in a tree, close to the trunk. Generally silent except in breeding season. Common **call** is one or more long *hooo* notes.

RANGE: Uncommon. More gregarious in winter; flocks may roost together.

GREAT HORNED OWL
Bubo virginianus | L 22" (56 cm)

Size, bulky shape, and white throat separate this owl from the Long-eared Owl; ear tufts distinguish it from other large species. Chiefly nocturnal. Takes prey as large as skunks and grouse. **Call** is a series of three to eight loud, deep hoots; the second and third hoots are often short and rapid.

RANGE: Common; habitats vary from forest to city to open desert. Widespread interior race, *subarcticus,* is paler. Nests in trees, caves, or on the ground.

BARN OWL
pratincola

SHORT-EARED OWL
flammeus

LONG-EARED
OWL

subarcticus

GREAT
HORNED OWL

HUMMINGBIRDS (Family Trochilidae)

These birds hover at flowers to sip nectar with needlelike bills. Often identified by twittery calls.
*Males' throat feathers (gorget) look black in poor light. **Species:** 331 World, 23 N.A.*

GREEN VIOLETEAR *Colibri thalassinus* | L 4¾" (12 cm)

Green overall with dark subterminal tail band; bill slightly downcurved. **Adult male** has blue-violet patches on face and breast. Female and **immature** slightly duller, immature grayer on belly. **Song** is a repeated *tsip-tsup*.
RANGE: Tropical species. Most records in summer in the Hill Country in Texas, where nearly annual; casual in eastern North America, accidental elsewhere.

GREEN-BREASTED MANGO *Anthracothorax prevostii* | L 4¾" (12 cm)

Has curved bill and purplish color in outer tail feathers. **Adult male** is deep green overall. **Female** shows broad blackish green stripe on underparts bordered in white. **Immature** is similar to female but shows cinnamon border to sides of underparts.
RANGE: Found from eastern Mexico to northern South America. Casual in fall and winter to southern Texas.

BUFF-BELLIED HUMMINGBIRD *Amazilia yucatanensis* | L 4¼" (11 cm)

Green overall with buff belly and broad, mostly chestnut tail; bill pinkish red with black tip. **Calls** are shrill and squeaky.
RANGE: Mexican species, fairly common in lower Rio Grande Valley. Rare in winter along Gulf Coast; casual to Florida.

BERYLLINE HUMMINGBIRD *Amazilia beryllina* | L 4¼" (11 cm)

Green above and below, with chestnut wings, rump, and tail. Base of lower mandible red. **Male**'s lower belly is chestnut, **female**'s grayish.
RANGE: Very rare summer visitor from Mexico to mountains of southeastern Arizona; has bred there. Casual to southwestern New Mexico and west Texas.

BAHAMA WOODSTAR *Calliphlox evelynae* | L 3¾" (10 cm)

Adult male has broad white collar, light purple throat, long, forked tail; mixed olive and rich buff below. In **female** note tail projection past primary tips, cinnamon tips to outer tail feathers, slightly curved bill. **Calls** suggestive of Anna's Hummingbird.
RANGE: Bahamian endemic. Four records of five birds from southeast Florida.

VIOLET-CROWNED HUMMINGBIRD *Amazilia violiceps* | L 4½" (11 cm)

Crown violet; underparts entirely white; upperparts bronze green; tail greenish. Long bill is mostly red. **Call** is a loud chattering; male's **song** is a series of sibilant ts notes.
RANGE: Uncommon. Casual in California and west Texas.

LUCIFER HUMMINGBIRD *Calothorax lucifer* | L 3½" (9 cm)

Bill downcurved. **Adult male** has green crown, purple throat, long tail. **Female** is rich buff below; broadening pale stripe behind eye; outer tail feathers reddish at base.
RANGE: Uncommon; rare in southeastern Arizona and southwestern New Mexico.

GREEN
VIOLETEAR
thalassinus

adult ♂

immature

GREEN-BREASTED
MANGO
prevostii

adult ♂

immature

adult ♀

BUFF-BELLIED
HUMMINGBIRD
chalconota

BERYLLINE
HUMMINGBIRD
viola

♂

♀

♂

adult ♂

immature ♂

♀

BAHAMA
WOODSTAR
evelynae

♀

adult ♂

LUCIFER
HUMMINGBIRD

adult ♂

♀

VIOLET-CROWNED
HUMMINGBIRD
ellioti

♀

immature ♂

BROAD-BILLED HUMMINGBIRD *Cynanthus latirostris* | L 4" (10 cm)

Adult male is dark green above and below, with white undertail coverts, blue gorget, and mostly red bill. Broad, forked tail is blackish blue. **Adult female** is duller above, gray below; often shows a narrow white eye stripe; has a square-tipped tail. Juveniles resemble female; by late summer, male begins to show blue and green flecks on throat, green on sides; dark, forked tail helps distinguish it from White-eared Hummingbird. Chattering *je-dit* **call** is similar to Ruby-crowned Kinglet. Male's display call is a whining *zing*.
RANGE: Common in desert canyons and low mountain woodlands. Very rare in southern California and Texas during fall and winter; casual in East.

WHITE-EARED HUMMINGBIRD *Hylocharis leucotis* | L 3¾" (10 cm)

Bill shorter than in Broad-billed Hummingbird; broad white stripe extends back from eye; has black ear patch, square tail. **Adult male** has dark purple crown and chin, emerald green gorget; display **call** is a repeated, silvery *tink tink tink*. Chattering calls are loud and metallic.
RANGE: Summer visitor to mountains of Southwest: rare in southeastern Arizona; very rare in southwest New Mexico and west Texas.

BLUE-THROATED HUMMINGBIRD *Lampornis clemenciae* | L 5" (13 cm)

Adult male's throat is blue, **female**'s gray. Broad white eye stripe and faint white malar stripe border dark ear patch. Tail has broad white tips on outer feathers; compare female Magnificent Hummingbird. Male's **call** is a loud, high, repeated *seep*.
RANGE: Fairly common; found in mountain canyons, especially near streams. Casual north of mapped range.

XANTUS'S HUMMINGBIRD *Hylocharis xantusii* | L 3½" (9 cm)

Plumages and **calls** similar to White-eared Hummingbird, but note buff on underparts and rufous in tail; **male** has black forehead and ear patches.
RANGE: Endemic to southern Baja California, Mexico. Accidental vagrant to southern California and southwestern British Columbia.

MAGNIFICENT HUMMINGBIRD *Eugenes fulgens* | L 5¼" (13 cm)

Adult male is green above, with purple crown; metallic green throat; breast and upper belly black and green; lower belly dull brown. Tail is dark green and deeply notched. **Female** is duller, lacks purple crown; squarish tail has small, grayish white tips on outer feathers; compare with female Blue-throated Hummingbird, which has stronger eyebrow, shorter bill. Main **call** is a sharp *chip*.
RANGE: Fairly common in high mountain meadows and canyons. Casual away from mapped range.

PLAIN-CAPPED STARTHROAT *Heliomaster constantii* | L 5" (13 cm)

Broad white malar stripe, white eye stripe, and white patches on sides of rump are conspicuous. Throat shows variable amount of red; note also very long bill. **Call** is a sharp *chip*.
RANGE: Casual stray in summer and fall from Mexico to arid foothills and deserts of southeastern Arizona.

BROAD-BILLED
HUMMINGBIRD
magicus

♀

immature ♂

♂

adult ♂

adult ♀

WHITE-EARED
HUMMINGBIRD
borealis

adult ♂

♂

BLUE-THROATED
HUMMINGBIRD
bessophilus

♀

♂

♂

XANTUS'S
HUMMINGBIRD

♀

♀

MAGNIFICENT
HUMMINGBIRD
fulgens

adult ♂

PLAIN-CAPPED
STARTHROAT
pinicola

RUBY-THROATED HUMMINGBIRD
Archilochus colubris | L 3¾" (10 cm)

The only hummingbird regularly seen throughout most of the East. Metallic green above. **Adult male** has a brilliant red throat and black chin; underparts are whitish; sides and flanks dusky green; tail forked. **Female's** throat is whitish; underparts grayish white, with buffy wash on sides; tail is similar to female Black-chinned Hummingbird. Immatures resemble adult female. Some **immature males** begin to show red spotting on throat by early fall. As with all hummingbirds, adult males migrate earlier than females and immatures. *Archilochus* hummingbirds seen in the Southeast in winter are more likely to be Black-chinned, but females and immatures of these two species are almost indistinguishable. Ruby-throated generally has a greener crown, shorter bill; note darker face and more pointed shape of darker primaries, especially outermost. **Calls** very similar.

RANGE: Ruby-throated Hummingbirds are fairly common in parts of range; found in gardens and woodland edges. Casual west to California.

BLACK-CHINNED HUMMINGBIRD
Archilochus alexandri | L 3¾" (10 cm)

Metallic green above. In good light, **adult male** shows violet band at lower border of black throat. Underparts whitish; sides and flanks dusky green. **Female's** throat can be all-white or show faint dusky or greenish streaks. Immatures resemble adult female; **immature male** may begin to show violet on lower throat in the fall. **Call** is a soft *tchew*; chase note combines high squeals and *tchew* notes. Twitches tail more than Ruby-throated while feeding.

RANGE: Common in lowlands and low mountains. A few Black-chinned Hummingbirds winter in Southeast.

COSTA'S HUMMINGBIRD
Calypte costae | L 3½" (9 cm)

Adult male has deep violet crown and gorget extending far down sides of neck. **Female** is generally grayer above, whiter below, than female Black-chinned Hummingbird; note also tail differences. Best distinguished by voice. **Call** is a high, metallic *tink*, often given in a series. Male's call is a loud *zing*.

RANGE: Fairly common in desert washes, dry chaparral. Casual north to south coastal Alaska and east to Texas.

ANNA'S HUMMINGBIRD
Calypte anna | L 4" (10 cm)

Adult male's head and throat are deep rose red, the color extending a short distance onto sides of neck. **Female's** throat usually shows red flecks, often forming a patch of color. In both sexes, underparts are grayish, washed with a varying amount of green. Bill is disproportionately short. Immatures resemble female; **immature male** usually shows some red on crown. **Juveniles** lack red on throat; compare with smaller female Black-chinned and Costa's Hummingbirds. Common **call** note, a sharp *chick*; chase call is a rapid dry rattling. Male's **song** is a jumble of high squeaks and raspy notes.

RANGE: Abundant in coastal lowlands and mountains; also in deserts, especially in winter. Casual north to south coastal Alaska and in eastern North America.

RUBY-THROATED
HUMMINGBIRD

immature ♂

adult ♂

Ruby-
throated

Black-
chinned

adult ♀ wings

immature ♂

♀

adult ♂

immature ♂

♀

BLACK-CHINNED
HUMMINGBIRD

adult ♂

immature ♂

COSTA'S
HUMMINGBIRD

♀

ANNA'S
HUMMINGBIRD

Black-chinned ♀

Costa's ♀

Anna's ♀

juvenile

adult ♀

immature ♂

adult ♂

BROAD-TAILED HUMMINGBIRD *Selasphorus platycercus* | L 4" (10 cm)

Except during winter molt, all *Selasphorus* **adult males**' wingbeats produce a loud whistle, harsh and trilling in Broad-tailed. Male has rose-red throat. **Female** has blended buff on underparts; similar to female Calliope, but note tail tip extends well past primaries. **Calls** include a metallic *chip*, often given in a short series.
RANGE: Common; summers in mountains. Casual fall and winter to Gulf Coast.

CALLIOPE HUMMINGBIRD *Stellula calliope* | L 3¼" (8 cm)

Very short bill and tail; primary tips extend well past end of tail. Carmine streaks on **adult male**'s throat form a V-shaped gorget. **Female**'s underparts similar to female Broad-tailed. Immature male like female but some show some red on throat by late summer. Relatively silent.
RANGE: Common; summers in mountains. Very rare in Southeast in fall and winter.

RUFOUS HUMMINGBIRD *Selasphorus rufus* | L 3¾" (10 cm)

Tail mainly rufous. **Adult male** has rufous back, sometimes marked with green, very rarely entirely green; orange-red gorget. **Immatures** resemble **adult female**; immature male may show reddish brown back by winter before acquiring full gorget. Buzzy wing whistle and all **calls** identical to Allen's: Calls include a sibilant *chip*, often given in a series; chase note, *zeee-chuppity-chup*.
RANGE: Common fall migrant through Rockies. Rare in fall over much of the East.

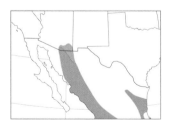

ALLEN'S HUMMINGBIRD *Selasphorus sasin* | L 3¾" (10 cm)

Wing whistle and **calls** are identical to Rufous. **Adult male** has full, orange-red gorget; usually distinguishable from Rufous by solid green back. Adult female and immatures inseparable in the field from female Rufous, though tail feathers are comparatively narrower for each sex and age class.
RANGE: Allen's migrates earlier in fall and spring than Rufous. Casual vagrant to eastern U.S.

TROGONS (FAMILY TROCHILIDAE)

Colorful tropical birds with short, broad bills. **Species: 39 World, 2 N.A.**

ELEGANT TROGON *Trogon elegans* | L 12½" (32 cm)

Yellow bill, white breast band; undertail delicately barred. **Male** is bright green above and has bright red belly. **Female** and juvenile are browner and duller, with a broad, white teardrop below and behind eye. **Song** is a series of croaking *co-ah* notes.
RANGE: Found in streamside woodlands, mostly at altitudes from 4,000 to 6,000 feet. Casual in southernmost Texas.

EARED QUETZAL *Euptilotis neoxenus* | L 14" (36 cm)

Larger and with thicker body than Elegant Trogon; bill is black or gray; lacks white breast band and barring on undertail. **Calls** include a loud upslurred squeal ending in a *chuck* note; also a loud, hard cackling. Male's **song** is a long, quavering series of whistled notes that increase in volume.
RANGE: Casual in mountain streamside woodlands of southeastern Arizona. Wary.

BROAD-TAILED
HUMMINGBIRD

adult ♂

♀

CALLIOPE
HUMMINGBIRD

adult ♂

♀

adult ♂

♀

RUFOUS
HUMMINGBIRD

immature ♂

green-flecked ♂

ALLEN'S
HUMMINGBIRD
sasin

adult ♂

Broad-tailed ♀   Rufous ♀   Allen's ♀   Calliope ♀

ELEGANT
TROGAN
canescens

♀

♂

EARED
QUETZAL

♀

♂

♂   ♀

KINGFISHERS (Family Alcedinidae)

Stocky and short-legged, with a large head, a large bill, and, in two North American species, a ragged crest. Look for kingfishers near woodland streams and ponds and in coastal areas. They hover over water or watch from low perches and then plunge headfirst to catch a fish. With strong bills and feet, they dig nest burrows in stream banks. **Species:** *91 World, 3 N.A.*

BELTED KINGFISHER
Megaceryle alcyon | L 13" (33 cm)

The only kingfisher in most of North America. Both **male** and **female** have slate blue breast band; white belly and undertail coverts. Female has rust belly band and flanks; may be confused with female Ringed Kingfisher, note white belly and smaller size. Juvenile resembles adult but has rust spotting in breast band. **Call** is a loud, dry rattle.

RANGE: Common and conspicuous along rivers and brooks, ponds and lakes, and estuaries. Generally solitary. Rare in winter north into summer range.

RINGED KINGFISHER
Megaceryle torquatus | L 16" (41 cm)

Larger than Belted Kingfisher. Rufous underwing coverts of **female** distinctive in flight; white in **male**. Male is rust below with white undertail coverts. Female has slate blue breast, narrow white band, rust belly and undertail coverts. Juveniles resemble adult female, but juvenile male's breast is largely rust. **Calls** include a harsh rattle, lower and slower than in Belted Kingfisher, and single *chack* notes, given chiefly in flight.

RANGE: Resident in lower Rio Grande Valley; casual elsewhere in southern Texas. Generally frequents large rivers and ponds, perches on higher branches than Belted.

GREEN KINGFISHER
Chloroceryle americana | L 8¾" (22 cm)

Smallest of our kingfishers; crest inconspicuous. **Male** is green above, with white collar; white below, with dark green spotting. **Female** has a band of green spots. Juvenile resembles adult female. One **call**, a faint but sharp *tick tick*, often ends in a short rattle; another, a squeaky *cheep*, is given in flight. Flight is direct and very fast; white outer tail feathers conspicuous in flight.

RANGE: Uncommon and often hard to see. Often perches on low, sheltered branches. Resident of lower Rio Grande Valley; rarer on Edwards Plateau. Casual wanderer along the Texas coast; also to Big Bend region. In southeastern Arizona, a few pairs are resident along San Pedro River; also occasionally seen on Santa Cruz River near Nogales.

BELTED
KINGFISHER

RINGED
KINGFISHER
torquatus

GREEN
KINGFISHER

WOODPECKERS AND ALLIES (FAMILY PICIDAE)

Strong claws, short legs, and stiff tail feathers enable woodpeckers to climb tree trunks. Sharp bill is used to chisel out insect food and nest holes and to drum a territorial signal. **Species:** *210 World, 25 N.A.*

RED-HEADED WOODPECKER
Melanerpes erythrocephalus | L 9¼" (24 cm)

Entire head, neck, and throat are bright red in **adults**, contrasting with blue-black back and snowy white underparts. **Juvenile** is brownish; acquires red head during gradual winter molt. Distinctive white inner wing patches and white rump are visible in all ages in perched and flying birds. In breeding season utters a loud *queark*, similar to Red-bellied Woodpecker (next page) but harsher and sharper; uncommon to fairly common **call** is a guttural rattle.
RANGE: Now vagrant in the Northeast, chiefly in fall; no longer breeds there, due in part to habitat loss and competition with European Starlings for nest holes. Rare to Maritimes; casual to California and Arizona. Inhabits a variety of open and densely wooded habitats; often seen catching flies.

ACORN WOODPECKER
Melanerpes formicivorus | L 9" (23 cm)

Black chin, yellowish throat, white cheeks and forehead, red cap. **Female** has smaller bill than **male**, less red on crown. In flight, white rump and small white patches on outer wings are conspicuous. Most frequent **call**, *waka*, usually repeated several times.
RANGE: Common in oak woods or pine forests where oak trees are abundant. Sociable; generally found in small, noisy colonies. Eats chiefly acorns and other nuts in winter, insects in summer. In the fall, is sometimes seen drilling small holes in a tree trunk and pounding a nut into each hole for a winter food supply. Colonies maintain and protect the same "granary tree" year after year.

WHITE-HEADED WOODPECKER
Picoides albolarvatus | L 9¼" (24 cm)

Head and throat white. **Male** has a red patch on back of head. Body is black except for white wing patches. Juveniles have a variable patch of pale red on crown. **Calls** include a sharp *pee-dink* or *pee-dee-dink*.
RANGE: Nests in coniferous mountain forests, especially ponderosa and sugar pine; feeds primarily on seeds from their cones; also pries away loose bark in search of insects and larvae. Casual at lower altitudes in winter. Fairly common over most of range; rare and local in the North.

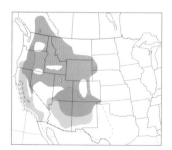

LEWIS'S WOODPECKER
Melanerpes lewis | L 10¾" (27 cm)

Greenish black head and back; gray collar and breast; dark red face, pinkish belly. In flight, darkness, large size, and slow, steady wingbeats give it a crowlike appearance. **Juvenile** lacks collar and red face; belly may be only faintly pink; acquires more adultlike plumage from late fall through winter.
RANGE: Uncommon to fairly common in open woodlands of interior; rare on coast. Often gregarious; fall and winter movements unpredictable. Largely eliminated in coastal Northwest; accidental in East. Main food, insects, mostly caught in the air; also eats fruit and nuts. Stores acorns, which it first shells, in tree bark crevices. Generally silent.

RED-HEADED
WOODPECKER

juvenile

adults

ACORN
WOODPECKER

♂

♀

♂

WHITE-HEADED
WOODPECKER

♂

♂

♀

juvenile

LEWIS'S
WOODPECKER

adults

GOLDEN-FRONTED WOODPECKER
Melanerpes aurifrons | L 9¾" (25 cm)

Black-and-white barred back, white rump, usually an all-black tail; golden orange nape, paler in **females**; yellow feathering above bill. **Male** has a small red cap. Yellow tinge on belly not easily seen. Juvenile has streaked breast, brownish crown. In flight, all plumages show white wing patches, white rump as in Red-bellied and Gila Woodpeckers; unlike them, Golden-fronted shows black, not barred, tail. **Calls**, a rolling *churr-churr* and cackling *kek-kek*, are slightly louder and raspier than in Red-bellied.
RANGE: Fairly common in dry woodlands, pecan groves, and mesquite brushlands.

RED-BELLIED WOODPECKER
Melanerpes carolinus | L 9¼" (24 cm)

Black-and-white barred back; white uppertail coverts; barred central tail feathers. Crown and nape are red in **males**. **Females** have red nape only; small reddish patch or tinge on belly. **Call**, a rolling *churr* or *chiv-chiv*, is slightly softer than that of Golden-fronted.
RANGE: Red-bellied Woodpeckers are common in open woodlands, suburbs, parks. Breeding range extending northward. Rare to Maine and Maritimes; casual to New Mexico.

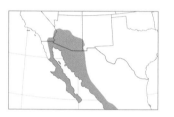

GILA WOODPECKER
Melanerpes uropygialis | L 9¼" (24 cm)

Black-and-white barred back and rump; central tail feathers barred. **Male** has a small red cap. **Calls**, a rolling *churr* and a loud, sharp, high-pitched *yip*, often given in a series.
RANGE: Inhabits towns, scrub desert, cactus country, and streamside woods.

NORTHERN FLICKER
Colaptes auratus | L 12½" (32 cm)

Two distinct groups occur: "**Yellow-shafted Flicker**" in the East and far north, and "**Red-shafted Flicker**"in the West. These flickers have brown, barred back; spotted underparts, with black crescent bib. White rump is conspicuous in flight; no white wing patches. Intergrades are regularly seen in the Great Plains. "Yellow-shafted Flicker" has yellow wing lining and undertail color, gray crown, and tan face with red crescent on nape. "Red-shafted Flicker" has brown crown and gray face, with no red crescent. "Yellow-shafted" male has a black moustachial stripe (red stripe in "Red-shafted" male); **females** lack these stripes. **Call** heard on the breeding ground is a long, loud series of *wick-er* notes; a single, loud *klee-yer* is given year-round.
RANGE: Common in open woodlands and suburban areas. "Yellow-shafted" is rare in the West in fall and winter.

GILDED FLICKER
Colaptes chrysoides | L 11½" (29 cm)

Gilded Flicker's head pattern more like "Red-shafted" Northern, but underwings and base of tail yellow; crown more cinnamon. Note also smaller size; larger black chest patch; paler back with narrower black bars; more crescent-shaped markings below. **Female** lacks red moustachial stripe. **Calls** are like Northern's.
RANGE: Inhabits low desert woodlands; favors saguaro. Hybrids with "Red-shafted" are noted in cottonwoods at middle elevations in southern Arizona and along the lower Colorado River.

GOLDEN-FRONTED
WOODPECKER
aurifrons

♂

♀

RED-BELLIED
WOODPECKER

♀

♂

♂

GILA
WOODPECKER
uropygialis

♂

♀

"Yellow-
shafted" ♂

NORTHERN
FLICKER

"Red-
shafted" ♂

"Yellow-
shafted" ♀

♂

GILDED
FLICKER

Gilded
Flicker ♀

"Red-
shafted" ♀

SAPSUCKERS

These woodpeckers drill evenly spaced rows of holes in trees and then visit these "wells" for sap and the insects it attracts. Red-breasted and Red-naped Sapsuckers were formerly considered subspecies of Yellow-bellied.

WILLIAMSON'S SAPSUCKER
Sphyrapicus thyroideus | L 9" (23 cm)

Male has black back, white rump, and large white wing patch; black head with narrow white stripes, bright red chin and throat. Breast is black, belly yellow; flanks are barred with black and white. **Female**'s head is brown; back, wings, and sides barred with dark brown and white; rump white; lacks white wing patch and red chin; breast has large dark patch; belly variably yellow. Juveniles resemble adults but are duller; attain adultlike plumage by Nov. Juvenile male has white throat; juvenile female lacks black breast patch. Gives a harsh, shrill raptorlike **call**.
RANGE: Fairly common in dry, piney forests of the western mountains; moves south or to lower elevations in winter. Accidental in eastern North America.

RED-BREASTED SAPSUCKER
Sphyrapicus ruber | L 8½" (22 cm)

This and next two species formerly considered one species. All show white wing patches and give a querulous descending mewing **call**. Red head, nape, and breast; large white wing patch; white rump. Back is black, lightly spotted with yellow in northern subspecies, *ruber*; more heavily marked with white in southern *daggetti*. Belly is yellow in *ruber*; *daggetti* has paler belly and duller head with longer white moustachial stripe. In both races, briefly held juvenal plumage is brownish, showing little or no red. Red-breasted frequently hybridizes with Red-naped Sapsucker.
RANGE: Common in coniferous or mixed forests in coastal ranges, usually at lower elevations and in moister forests than Williamson's Sapsucker. Most migrate south or move to lower elevations in winter.

YELLOW-BELLIED SAPSUCKER
Sphyrapicus varius | L 8½" (22 cm)

Red forecrown on black-and-white head; chin and throat red in **male**, white in **female**. Back is blackish, with white rump and large white wing patch. Underparts yellowish, paler in female. **Juvenile** retains largely brownish plumage until late in the winter. Hybridizes with Red-naped Sapsucker.
RANGE: Common in deciduous forests. Highly migratory; rare to very rare in the West during fall and winter.

RED-NAPED SAPSUCKER
Sphyrapicus nuchalis | L 8½" (22 cm)

Very similar to Yellow-bellied, but has variable red patch on back of head; spotting on back more clearly organized into two rows. On **male**, extensive red on throat penetrates the surrounding black "frame"; on **female**, throat is partly red to almost entirely red on some birds. Juvenile is brownish overall; resembles adult by first fall except for lack of black chest.
RANGE: Common in deciduous forests.

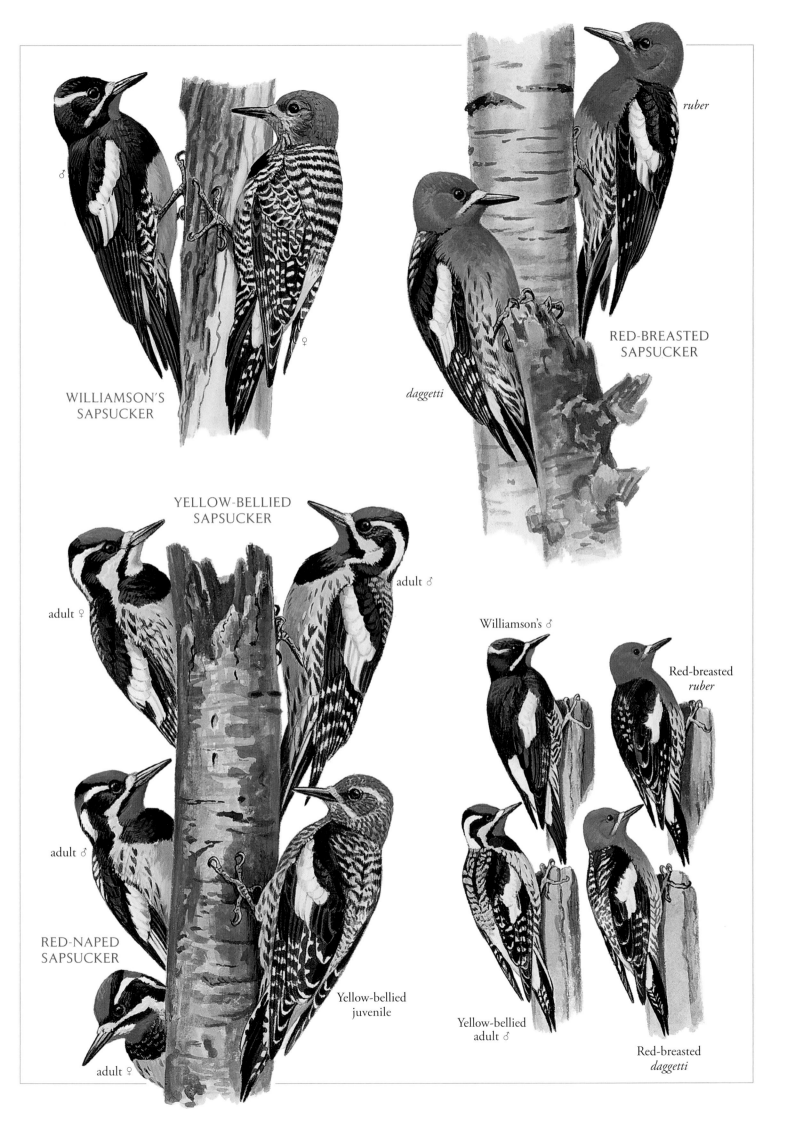

WILLIAMSON'S
SAPSUCKER

♂

♀

RED-BREASTED
SAPSUCKER

ruber

daggetti

YELLOW-BELLIED
SAPSUCKER

adult ♀

adult ♂

adult ♂

RED-NAPED
SAPSUCKER

adult ♀

Yellow-bellied
juvenile

Williamson's ♂

Red-breasted
ruber

Yellow-bellied
adult ♂

Red-breasted
daggetti

LADDER-BACKED WOODPECKER
Picoides scalaris | L 7¼" (18 cm)

Black-and-white barred back, spotted sides; face and underparts slightly buffy or grayish; face marked with black lines. **Male** has red crown. In California, Ladder-backed may be confused with Nuttall's Woodpecker. Ladder-backed shows less black on face and is buffy-tinged rather than white below; white barring on back is more pronounced and extends to nape; white outer tail feathers are evenly barred rather than spotted; nasal tufts buffy. **Call** is a crisp *pik*, very similar to Downy Woodpecker's call and different from Nuttall's rattled *prrrt*; also gives a descending whinny. May hybridize with Nuttall's where ranges overlap. RANGE: Common in dry brushlands, mesquite and cactus country; also towns and rural areas. Feeds on beetle larvae; also eats cactus fruits and forages on the ground for insects.

RED-COCKADED WOODPECKER
Picoides borealis | **E** | L 8½" (22 cm)

Black-and-white barred back, black cap, and large white cheek patch identify this woodpecker; red tufts, seldom visible, on the **male's** head. Similar Hairy and Downy Woodpeckers (next page) have solid white down backs. Distinctive **calls,** a raspy *sripp* and high-pitched *tsick*, are much more buzzy; recall a loud Brown-headed Nuthatch. RANGE: Red-cockaded inhabits open, mature pine or pine-oak woodlands. Bores nest hole only in a large living pine afflicted with heartwood disease, then drills small holes around the nest opening. Pine pitch oozing down the trunk from these holes may repel predators; also makes the tree a distinctive signpost. Endangered: Populations continue to decline. Almost eliminated from Virginia, and has disappeared from Kentucky and Tennessee in the last decade. Accidental to Illinois and Ohio.

ARIZONA WOODPECKER
Picoides arizonae | L 7½" (19 cm)

Solid brown back distinguishes this species from all other woodpeckers. **Female** lacks red patch on back of head. **Call** is a sharp *peek*, similar to call of Hairy Woodpecker but hoarser. RANGE: Uncommon resident in foothills and mountains; generally found in oak or pine-oak forests or canyons.

NUTTALL'S WOODPECKER
Picoides nuttallii | L 7½" (19 cm)

Closely resembles Ladder-backed Woodpecker. Nuttall's shows more black on face; white bars on back are narrower; more extensive solid black just below the nape. White outer tail feathers are sparsely spotted rather than barred; nasal tufts white. **Call** is a low, rattled *prrrt*, much lower than Ladder-backed's *pik*; also a series of loud, spaced, descending notes. RANGE: Nuttall's prefers less arid habitat than Ladder-backed; usually seen in chaparral mixed with scrub oak and in wooded canyons and streamside trees.

GREAT SPOTTED WOODPECKER
Dendrocopos major | L 9" (25 cm)

Large size with large white cheek and scapular patch and red undertail coverts. Female lacks red nape patch. **Call,** a sharp *kick*. RANGE: Widespread in Eurasia. Casual to western Aleutians and Pribilofs; accidental in winter north of Anchorage.

LADDER-BACKED
WOODPECKER

RED-COCKADED
WOODPECKER

ARIZONA
WOODPECKER

GREAT SPOTTED
WOODPECKER

NUTTALL'S
WOODPECKER

Ladder-backed ♂   Nuttall's ♂   Arizona ♂   Red-cockaded ♂

DOWNY WOODPECKER

Picoides pubescens | L 6¾" (17 cm)

White back generally identifies both Downy and similar Hairy Woodpecker. Downy is much smaller, with a smaller bill; outer tail feathers generally have faint dark bars or spots. Birds in the Pacific Northwest have pale gray-brown back and underparts. Rocky Mountain birds have less white spotting on wings. Downy's **call**, *pik*, and whinny are softer and higher-pitched than Hairy Woodpecker's.

RANGE: Common; active, and somewhat unwary; often seen in suburbs, parklands, and orchards, as well as in forests. A familiar visitor to feeders.

HAIRY WOODPECKER

Picoides villosus | L 9¼" (24 cm)

White back generally identifies both Hairy and similar Downy Woodpecker. Hairy is much larger, with a larger bill; outer tail feathers are entirely white. Birds in the Pacific Northwest have pale gray-brown back and underparts. Rocky Mountain birds have less white spotting on wings. **Juveniles**, particularly in the Maritime Provinces, have some barring on back and flanks; sides may be streaked. Juveniles on the Queen Charlotte Islands have heavily barred outer tail feathers. In young males, the forehead is spotted with white; crown streaked with red or orange. Hairy's **calls** include a loud, sharp *peek* and a slurred whinny.

RANGE: Fairly common; inhabits both open and dense forests. Uncommon to rare in the South and in Florida.

AMERICAN THREE-TOED WOODPECKER

Picoides dorsalis | L 8¾" (22 cm)

Black-and-white barring down center of back distinguishes most races of American Three-toed Woodpecker from similar Black-backed. Both have heavily barred sides. **Male**'s yellow cap is usually more extensive in Three-toed but less solid. Density of barring on back is intermediate in northwestern race, *fasciatus*. In Rocky Mountain race, nominate *dorsalis*, back is almost entirely white. Back is much darker in eastern race, *bacatus*; thinner, white submoustachial stripe helps distinguish it from Black-backed.

RANGE: American Three-toed is found in coniferous forests, especially in burned-over areas. Scarcer than Black-backed in the East. Accidental south to Rhode Island in winter; east to southwest Kansas (in summer!). New English and scientific names reflect a recent split: The eight Old World subspecies are now considered their own species, *P. tridactylus.*

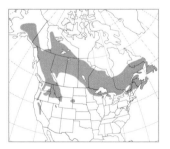

BLACK-BACKED WOODPECKER

Picoides arcticus | L 9½" (24 cm)

Solid black back, heavily barred sides. **Male** has a solid yellow cap. Compare especially with eastern race of Three-toed Woodpecker, *bacatus*, which has a darker back than other Three-toeds. Black-backed Woodpecker is larger; has longer, stouter bill; and lacks white streak behind the eye. **Call** note is a single, sharp *pik*, lower-pitched than call of Three-toed.

RANGE: Black-backed inhabits coniferous forests; often found in burned-over areas. Forages on dead conifers, flaking away large patches of loose bark rather than drilling into it, in search of beetle larvae and insects. Casual south of mapped range in the East, fewer records in recent decades.

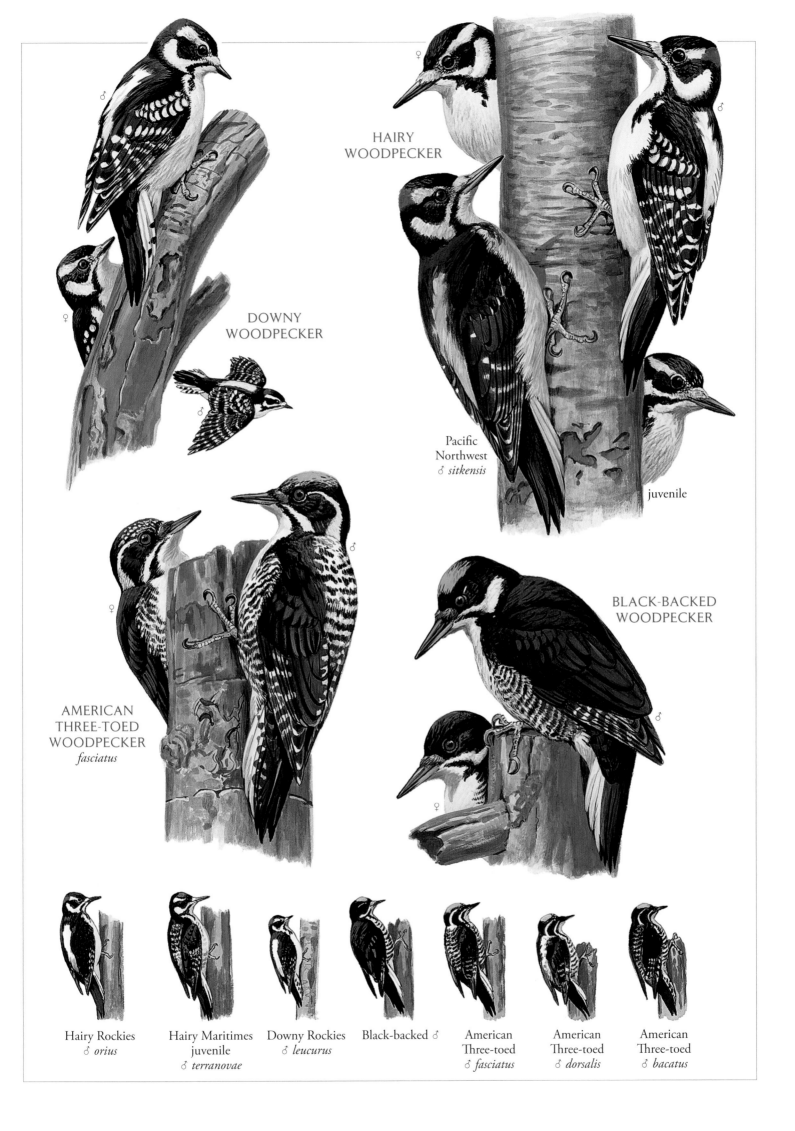

DOWNY
WOODPECKER

HAIRY
WOODPECKER

Pacific
Northwest
♂ sitkensis

juvenile

BLACK-BACKED
WOODPECKER

AMERICAN
THREE-TOED
WOODPECKER
fasciatus

Hairy Rockies
♂ *orius*

Hairy Maritimes
juvenile
♂ *terranovae*

Downy Rockies
♂ *leucurus*

Black-backed ♂

American
Three-toed
♂ *fasciatus*

American
Three-toed
♂ *dorsalis*

American
Three-toed
♂ *bacatus*

IVORY-BILLED WOODPECKER
Campephilus principalis | **E** | L 19½" (50 cm)

Note Ivory-billed Woodpecker's black chin, striking ivory bill, and the extensive white wing patches and scapular lines, visible in perched birds; tail longer than Pileated, especially visible in flight. **Females** have black rather than red crests. Compare also the black-and-white wing patterns of Ivory-billed and Pileated Woodpeckers in flight; in Ivory-billed from below, the white leading edge and the white secondaries are divided by a black bar; viewed from above, the white secondaries are diagnostic; flight of Ivory-billed is more direct and swift. Distinctive **call** note sounds like a toy trumpet, a high-pitched, nasal *yank,* given singly or in short series, like a loud version of the call of eastern White-breasted Nuthatch, but beware of Blue Jays giving similar calls; double-rap drum is characteristic of genus *Campephilus.* Our largest woodpecker, Ivory-billed required large tracts of old-growth river forest; dead and dying trees supplied nesting sites and food: the larvae of wood-boring beetles. Destruction of habitat at the end of the 19th century and early in the 20th century led to the disappearance of this never-common species.

RANGE: Formerly found north to the Ohio River, near its confluence with the Mississippi. Perhaps extinct; last definite records from U.S. were in 1944 in the Singer Tract, near Tallulah in northeastern Louisiana. Possibly valid sightings recorded into the 1950s in Florida; from 1948, perhaps into the 1980s, in Cuba (subspecies *bairdii*), but now probably extinct there. Unconfirmed sightings over the last 50 years come from eastern Texas, Louisiana, Georgia, and Florida. In Apr. 2005 came the much publicized announcement that the species had been rediscovered more than a year earlier in the Big Woods of the White River-Cache River system of eastern Arkansas. Documentation was provided in the form of sound recordings and videotape. However, intense searching subsequently has yet to produce more documentation, and some authorities now question the conclusions reached from the original evidence. The search continues.

PILEATED WOODPECKER
Dryocopus pileatus | L 16½" (42 cm)

Perched bird is almost entirely black on back and wings, lacking Ivory-billed Woodpecker's large white wing patches. White chin and dark bill also distinguish Pileated Woodpecker, along with smaller size. Compare also the wing patterns of the two species in flight; also note Pileated's deep, slow, crowlike wingbeats. This is the largest woodpecker now generally seen in North America. **Female**'s red cap is less extensive than in **male**. Juvenale plumage, held briefly, resembles adult but is duller and browner overall. Generally shy. **Call** is a loud *wuck* note or series of notes, given all year, often in flight; similar call of Northern Flicker is given only in the breeding season.

RANGE: Prefers dense, mature forest. In woodlots and parklands as well as deep woods, listen for its loud, resonant, territorial drumming, given by both sexes but less frequently by females; look for the long rectangular or oval holes it excavates. Carpenter ants in fallen trees and stumps are its major food. Common in Southeast; uncommon and local elsewhere, but increasing in East.

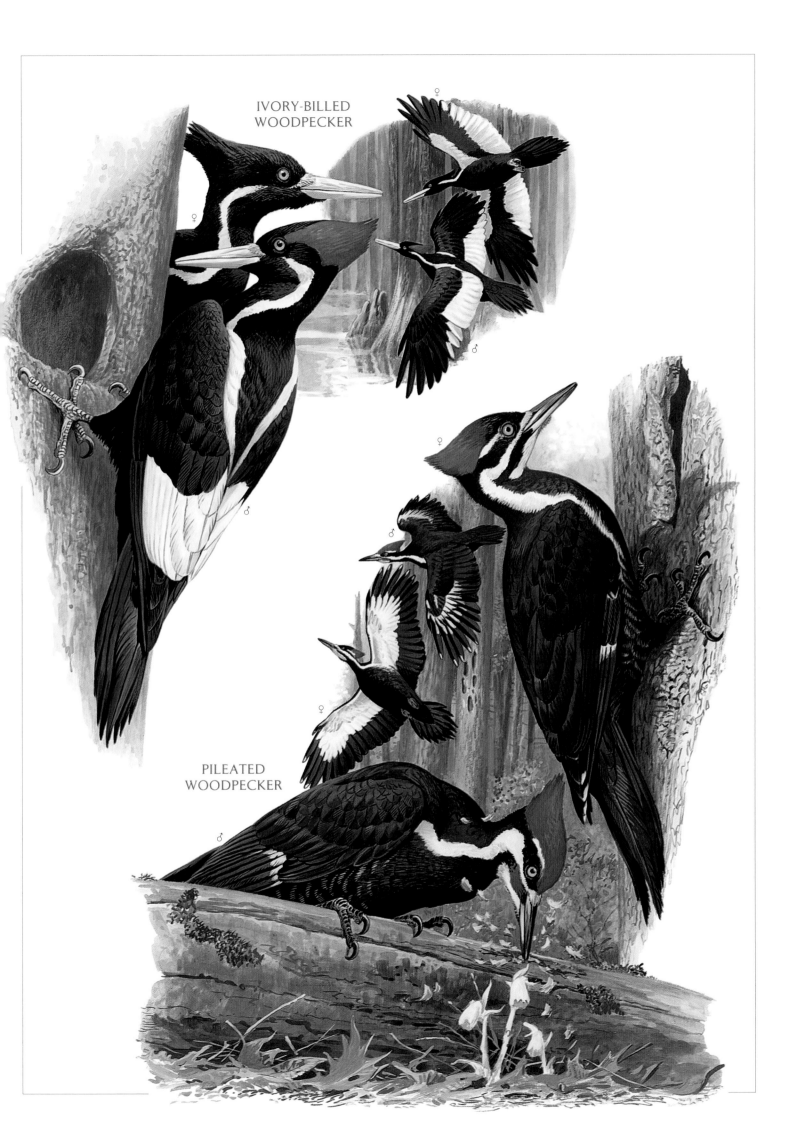

IVORY-BILLED
WOODPECKER

PILEATED
WOODPECKER

TYRANT FLYCATCHERS (FAMILY TYRANNIDAE)

A typical flycatcher darts out from a fixed perch to catch insects. Most have a large head, bristly "whiskers," and a broad-based, flat bill. **Species:** *400 World, 44 N.A.*

GREATER PEWEE

Contopus pertinax | L 8" (20 cm)

Note longer tail, more slender crest (usually visible), and more uniformly colored underparts than Olive-sided. Worn summer birds are overall grayer than freshly molted winter ones. Unlike *Empidonax* flycatchers (next page), most *Contopus* (pewees) do not wag their tails. **Song**, a whistled *ho-say ma-re-ah*; **call** is a repeated *pip*.

RANGE: Fairly common in mountain pine-oak woodlands. Very rare in winter in southern Arizona, southern and central California; casual in south and west Texas.

OLIVE-SIDED FLYCATCHER

Contopus cooperi | L 7½" (19 cm)

Large, with rather short tail. Brownish olive above; white tufts on sides of rump distinctive but often not visible. Throat, center of breast, and belly dull white. Sides and flanks brownish olive and streaked. Bill is mostly black; center and sometimes base of lower mandible dull orange. Distinctive **song**, a clear *quick-three-beers*, the second note higher; typical **call**, a repeated *pip*.

RANGE: Fairly common in coniferous forests and bogs. Casual in winter on coastal slope of southern California.

EASTERN WOOD-PEWEE

Contopus virens | L 6¼" (16 cm)

Plumage generally dark grayish olive above, with dull white throat, darker breast; underparts whitish or pale yellow. Bill of **adult** has black upper mandible, dull orange lower mandible. **Juvenile** and immature may have all-dark bill. Distinctive **song** is a clear, slow, plaintive *pee-a-wee*, the second note lower; this phrase often alternates with a downslurred *pee-yer*. **Calls** include a loud *chip* and clear, whistled, rising *pweee* notes; often given together, *chip pweee*.

RANGE: Common in woodland habitats. Casual in the West. No winter records in U.S.

WESTERN WOOD-PEWEE

Contopus sordidulus | L 6¼" (16 cm)

Plumage variable; slightly darker and less greenish than Eastern Wood-Pewee; base of lower mandible usually shows some yellow-orange. Identification very difficult; best done by range and voice. **Calls** include a harsh, slightly descending *peeer* and clear whistles suggestive of Eastern's *pee-yer*. **Song**, heard chiefly on breeding grounds, has three-note *tswee-tee-teet* phrases mixed with the *peeer* note.

RANGE: Common in open woodlands. Casual in East. No winter records in U.S.

CUBAN PEWEE

Contopus caribaeus | L 6" (15 cm)

Short primary projection makes species look like an *Empidonax*, but Cuban does minimal tail flicking. Note expansion of prominent white partial eye ring behind eye; dull wing bars; faint "vest." **Call**, a clear, steady *dee-dee-dee*, also a soft *dep* note.

RANGE: West Indian species; accidental in south Florida.

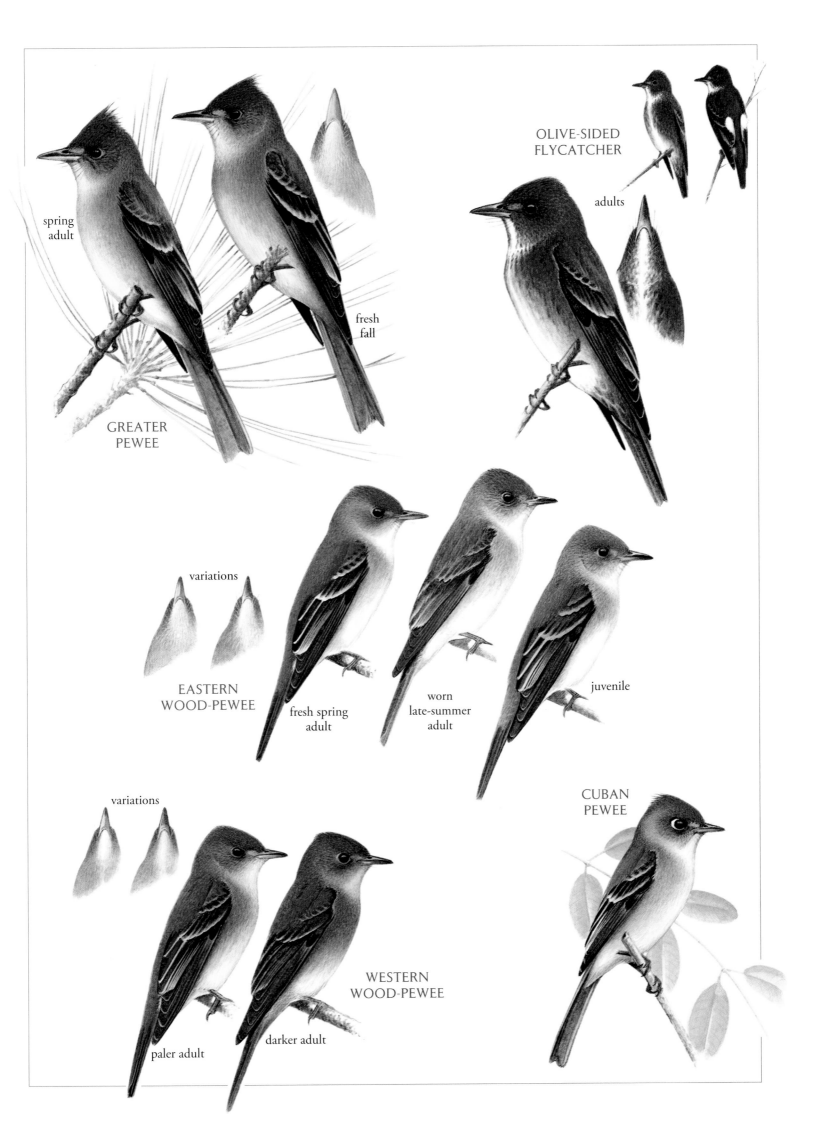

spring
adult

fresh
fall

GREATER
PEWEE

OLIVE-SIDED
FLYCATCHER

adults

variations

EASTERN
WOOD-PEWEE

fresh spring
adult

worn
late-summer
adult

juvenile

CUBAN
PEWEE

variations

WESTERN
WOOD-PEWEE

paler adult

darker adult

EMPIDONAX FLYCATCHERS

All empids are drab, with pale eye rings and wing bars. From spring to summer, plumages grow duller from wear. Some species molt before fall migration, acquiring fresh plumage in late summer. Identification depends on voice, habitat, behavior, and subtle differences in size, bill shape, primary projection, and tail length. Most flip their tails up.

ACADIAN FLYCATCHER
Empidonax virescens | L 5¾" (15 cm)

Olive above, with yellow eye ring, two buffy or whitish wing bars; very long primary projection. Long, broad-based bill, with mostly yellowish lower mandible. Most birds show pale grayish throat, pale olive wash across upper breast, white lower breast, and yellow belly and undertail coverts. Molts before migration; **fall** birds have buffy wing bars. Juvenile is brownish olive above, edged with buff. **Call** is a soft *peace*, extended in **song** to an emphatic *pee-tsup*, accented on first syllable. On breeding grounds, also gives a flickerlike *ti ti ti ti ti*.
RANGE: Found in woodlands and swamps. Range is expanding in the Northeast.

YELLOW-BELLIED FLYCATCHER
Empidonax flaviventris | L 5½" (14 cm)

Has rather short tail and big head. Olive above, yellow below. Broad yellow eye ring. Lower mandible entirely pale orange. Shows a more extensive olive wash across breast than Acadian Flycatcher; lacks pale area between olive and yellow belly. Also, throat is yellow, rather than whitish; bill smaller. Molts after migration; **worn fall** migrants slightly grayer above, duller below. **Song** is a liquid *je-bunk*; also a plaintive, rising *per-wee*. **Call**, a sharp, whistled *chiu* that sounds somewhat like Acadian.
RANGE: Found in bogs, swamps, and damp coniferous woods.

ALDER FLYCATCHER
Empidonax alnorum | L 5¾" (15 cm)

Very similar to Willow Flycatcher, but bill is slightly shorter, eye ring usually more prominent, back greener. Distinguished from eastern race of Willow by darker head; from western races by well defined tertial edges, bolder wing bars, long primary projection. Best identified by voice. **Call**, a loud *pip*, similar to Hammond's but louder. Distinctive **song**, a falling, wheezy *weeb-ew*. On breeding grounds, also gives a descending *wheer*.
RANGE: Common in brushy habitats near bogs, birch and alder thickets.

WILLOW FLYCATCHER
Empidonax traillii | L 5¾" (15 cm)

Lacks prominent eye ring. Color ranges from pale gray head and greenish back of nominate eastern race to darker-headed, browner *brewsteri* in the Northwest. Great Basin race, *adastus* (not illustrated), is paler than *brewsteri*; endangered southwestern *extimus* (**E**) even paler. Western races have duller wing bars, blended tertial edges, and shorter primary projection. Distinguished from pewees (preceding page) by shorter wings and upward flicks of tail. **Call** is a liquid *wit*. **Songs**, a sneezy *fitz-bew*; on breeding grounds, also a rising *brreet*; often sings in spring migration.
RANGE: Found in brushy habitats in wet areas; also in pastures, mountain meadows. Very rare migrant in Southeast.

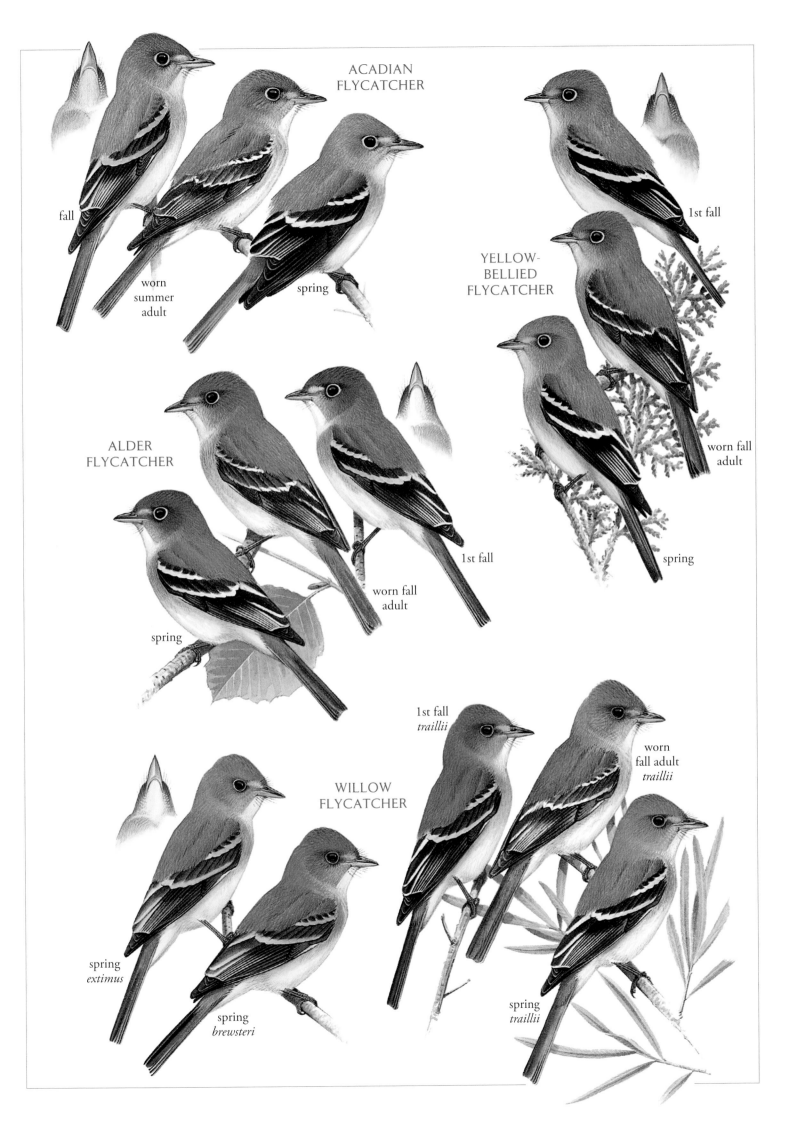

ACADIAN
FLYCATCHER

fall

worn
summer
adult

spring

1st fall

YELLOW-
BELLIED
FLYCATCHER

worn fall
adult

spring

ALDER
FLYCATCHER

spring

worn fall
adult

1st fall

1st fall
traillii

worn
fall adult
traillii

WILLOW
FLYCATCHER

spring
extimus

spring
brewsteri

spring
traillii

LEAST FLYCATCHER
Empidonax minimus | L 5¼" (13 cm)

Smallest eastern empid. Large-headed; bold white eye ring; rather short primary projection. Throat whitish; breast washed with gray; belly and undertail coverts pale yellow. Underparts are usually paler than similar Hammond's Flycatcher. Bill short, triangular; lower mandible mostly pale. Molt occurs after fall migration. **First-fall** has buffier wing bars. **Song**, a dry *che-bek* accented on the second syllable, is usually delivered in a rapid series; **call**, a sharp *whit*, is sometimes also given in a series.
RANGE: Inhabits deciduous woods, orchards, and parks. Common in East, rare migrant through most of West.

HAMMOND'S FLYCATCHER
Empidonax hammondii | L 5½" (14 cm)

A small empid, with a fairly large head and short tail. White eye ring, usually expanded in a "teardrop" at rear. Grayish head and throat; grayish olive back; gray or olive wash on breast and sides; belly tinged with pale yellow. Molt occurs before migration; **fall** birds are much brighter olive above and on sides of breast, yellower below. Medium-length, slightly notched tail is edged with gray. Bill is slightly shorter, thinner, and usually somewhat darker than in similar Dusky Flycatcher; primary projection is longer. **Call** note is a sharp *peek*. **Song** resembles Dusky's but is hoarser and lower-pitched, especially on the second note.
RANGE: Nests chiefly in coniferous forests. Most Hammond's Flycatchers migrate earlier in spring and later in fall than Dusky. Casual in the East.

GRAY FLYCATCHER
Empidonax wrightii | L 6" (15 cm)

Gray above, with a slight olive tinge in fresh fall plumage; whitish below, belly washed with pale yellow by late fall. Head is disproportionately small and rounded; white eye ring inconspicuous on pale gray face. Long bill; on most birds, the lower mandible is mostly pinkish orange at the base and sharply divided from dark tip; on a few, entire lower mandible is pinkish orange. Short primary projection. Long tail, with thin whitish outer edge. Perched bird dips its tail down slowly, like a phoebe. Juvenile is brownish gray above, with pale, buffy wing bars; underparts are tinged brownish buff. **Song** is a vigorous *chi-wip* or *chi-bit*, followed by a liquid *whilp*, trailing off in a gurgle. **Call**, a loud *wit*.
RANGE: Fairly common in dry habitat of Great Basin, in pine or piñon-juniper. Regular migrant on California coast. Accidental in the East.

DUSKY FLYCATCHER
Empidonax oberholseri | L 5¾" (15 cm)

Grayish olive above; yellowish below, with whitish throat, pale olive wash on upper breast. White eye ring. Bill partly dark, orange at base of lower mandible blending into dark tip. Bill and tail slightly longer than Hammond's Flycatcher. Short primary projection. Molt occurs after fall migration; fresh late-fall birds are quite yellow below. **Calls** include a *wit* note, softer than Gray Flycatcher; a mournful *deehic*, heard on breeding grounds. **Song** has several phrases: A clear *sillit*; an upslurred *ggrrreep*; another high *sillit*, often omitted; and a clear, high *pweet*.
RANGE: Breeds in open woodlands and brush of mountainsides.

LEAST
FLYCATCHER

1st fall

worn fall
adult

spring

HAMMOND'S
FLYCATCHER

fall

spring

GRAY
FLYCATCHER

winter

spring

DUSKY
FLYCATCHER

1st fall

worn fall
adult

winter

spring

PACIFIC-SLOPE FLYCATCHER
Empidonax difficilis | L 5½" (14 cm)

Formerly considered same species as Cordilleran; known together as Western Flycatcher. Brownish green above; yellowish below with brownish tinge on breast. Broad pale eye ring, broken above, expanded behind eye; lower mandible entirely orange. Tail longer, wing tip slightly shorter than Yellow-bellied; wings and back slightly browner; less contrast in wing bars and tertial edges. Pacific-slope molts after arrival on winter grounds, so migrating **fall adults** appear more worn than spring birds. **First-fall** birds duller; wing bars buffy; variably whitish below, compare with Least Flycatcher (preceding page). Channel Islands race *insulicola* is slightly duller. **Call** is a sharp *seet*; male gives upslurred *psee-yeet* note. **Song**, a complex series of notes, including call notes.

RANGE: Common in moist woodlands, coniferous forests, and shady canyons. Winters in lowlands of western Mexico. Common migrant through Southwest lowlands east to southeastern Arizona. Accidental in eastern North America.

CORDILLERAN FLYCATCHER
Empidonax occidentalis | L 5¾" (15 cm)

Formerly considered same species as Pacific-slope Flycatcher. Nearly identical but slightly larger, darker, and greener above; more olive and yellow below. Separable in field only by male's **call**, a two-note *pit peet*; some populations in western portion of breeding range give more intermediate notes. *Seet* note seems sharper in Cordilleran than in Pacific-slope Flycatcher. **Song** is very similar to Pacific-slope's.

RANGE: Breeds in coniferous forests and canyons in mountains of the West. Rare in lowlands in migration, even within breeding range. Casual on the Great Plains. Winters in mountains of Mexico.

BUFF-BREASTED FLYCATCHER
Empidonax fulvifrons | L 5" (13 cm)

Smallest and most distinctive *Empidonax* flycatcher. Brownish above; breast cinnamon-buff, paler on worn summer birds. Whitish eye ring; pale wing bars; small bill, with lower mandible entirely pale orange. Molts before migration. **Call** note is a soft *pwit*. Typical **song**, a quick *chicky-whew* or *chee-lick*.

RANGE: Small colonies nest in dry woodlands of canyon floors. Very local in Huachuca and Chiricahua Mountains, Arizona; recently in Davis Mountains, Texas; now casual in New Mexico, but recorded recently in Peloncillo Mountains. Formerly more widespread as a breeder.

NORTHERN BEARDLESS-TYRANNULET
Camptostoma imberbe | L 4½" (11 cm)

Grayish olive above and on breast; dull white or pale yellow below. Indistinct whitish eyebrow; small, slightly curved bill. Crown is darker than nape in many birds and often raised in a bushy crest. Distinguished from similar Ruby-crowned Kinglet (page 346) by buffy wing bars and lack of bold eye ring. Small and easily overlooked; most easily located by voice. **Song** on breeding grounds is a descending series of loud, clear peer notes; **call**, an innocuous, whistled *pee-yerp*.

RANGE: Rather uncommon in U.S. Often found near streams in sycamore, mesquite, and cottonwood groves.

PACIFIC-SLOPE
FLYCATCHER

pale
1st fall

1st fall

worn fall
adult

spring

spring

CORDILLERAN
FLYCATCHER

spring

Yellow-bellied Flycatcher
(for comparison)

worn
summer
adult

BUFF-BREASTED
FLYCATCHER

fresh

worn
adult

NORTHERN
BEARDLESS-
TYRANNULET

fresh

TUFTED FLYCATCHER
Mitrephanes phaeocercus | L 5" (13 cm)

Distinctive small, crested flycatcher with cinnamon underparts and face; brownish olive above with faint cinnamon wing bars. Behavior suggests pewees. **Call**, a whistled *tchurree-tchurree*. Sometimes given singly; also a soft *peek* like Hammond's Flycatcher.
RANGE: Widespread tropical species; partially migratory at northern end of range (northern Mexico). Accidental in winter and early spring to west Texas and western Arizona.

EASTERN PHOEBE
Sayornis phoebe | L 7" (18 cm)

Brownish gray above, darkest on head, wings, and tail. Underparts mostly white with pale olive wash on sides and breast; **fresh fall** birds are washed with yellow below. Molts before migration. All phoebes are distinguished from pewees (page 294) by their habit of pumping down and spreading their tails; Eastern Phoebe also by all-dark bill and lack of distinct wing bars. Also compare lack of eye rings and wing bars with *Empidonax* flycatchers (preceding pages). Distinctive **song**, a harsh, emphatic *fee-be*, accented on first syllable. Typical **call** note is a sharp *chip*.
RANGE: Common in woodlands, farmlands, suburbs; often nests under bridges, in eaves and rafters. Rare late fall migrant and winter visitor in the Southwest and on the West Coast.

BLACK PHOEBE
Sayornis nigricans | L 6¾" (17 cm)

Black head, upperparts, and breast; white belly and undertail coverts. **Juvenal** plumage, held briefly, is browner, with two cinnamon wing bars, cinnamon rump. Four-syllable **song**, a rising *pee-wee* followed by a descending *pee-wee*. **Calls** include a loud *tseee* and a sharper *tsip*, slightly more plaintive than Eastern Phoebe's call.
RANGE: Common near water; casual to Oklahoma.

SAY'S PHOEBE
Sayornis saya | L 7½" (19 cm)

Grayish brown above, darkest on head, wings, and tail; breast and throat pale grayish brown; belly and undertail coverts tawny. **Song** is a fast *pit-tse-ar*, often given in fluttering flight. Typical **call**, a plaintive, whistled *pee-ee*, slightly downslurred.
RANGE: Fairly common in dry, open areas, canyons, cliffs; perches on bushes, boulders, fences. Highly migratory; casual in eastern North America.

VERMILION FLYCATCHER
Pyrocephalus rubinus | L 6" (15 cm)

Adult male strikingly red and brown. **Adult female** grayish brown above, with blackish tail; throat and breast white, with dusky streaking; belly and undertail coverts are peach; note also whitish eyebrow and forehead. **Juvenile** resembles adult female but is spotted rather than streaked below; belly white, often with yellowish tinge. **Immature male** begins to resemble adult by midwinter. Male in breeding season sings during fluttery display flight. **Song**, a soft, tinkling *pit-a-see pit-a-see*; also sings while perched. Typical **call** note is a sharp, thin *pseep*. Frequently pumps and spreads its tail down.
RANGE: Fairly common and approachable; found along streamsides, and near small wooded ponds. Rare winter visitor to coastal southern California and the Gulf Coast. Casual elsewhere in eastern North America, primarily in fall.

TUFTED
FLYCATCHER

EASTERN
PHOEBE

worn summer
adult

fresh fall

BLACK
PHOEBE

juvenile

SAY'S PHOEBE

VERMILION
FLYCATCHER

immature ♀

immature ♂

adult
♀

adult ♂

juvenile

BROWN-CRESTED FLYCATCHER *Myiarchus tyrannulus* | L 8¾" (22 cm)

Brownish olive above; as in all *Myiarchus* flycatchers, shows a bushy crest, rufous in primaries; bill longer, thicker, broader than Ash-throated Flycatcher. Throat and breast are pale gray; belly slightly paler yellow than in Great Crested Flycatcher. Tail feathers show reddish on outer two-thirds of inner webs. Texas race, *cooperi*, is smaller than southwestern *magister*. **Song** is a clear musical whistle, a rolling *whit-will-do*. **Call** is a sharp *whit*.

RANGE: Fairly common in saguaro desert, river groves, lower mountain woodlands. Casual in Louisiana and Florida.

GREAT CRESTED FLYCATCHER *Myiarchus crinitus* | L 8½" (21 cm)

Dark olive above. Gray throat and breast; bright lemon yellow belly and undertail coverts. Note broad, sharply contrasting edge to inner tertial. Outer tail feathers show entirely reddish inner webs. Distinctive **call**, a loud whistled *wheep*.

RANGE: Common in a wide variety of open woods; feeds high in the canopy. Very rare on California coast during fall migration.

NUTTING'S FLYCATCHER *Myiarchus nuttingi* | L 7¼" (18 cm)

Similar to Ash-throated Flycatcher but belly yellower; slightly more olive above; rufous primary edges blend to yellow-cinnamon secondary edges. Dark on outer webs of outer tail feathers does not extend across tip as in Ash-throated; orange, not flesh-colored, mouth lining. **Call**, a rather sharp *wheep*, different from Ash-throated.

RANGE: Tropical species; found from northwest Mexico to Costa Rica; three certain winter records from southeastern Arizona and coastal southern California.

ASH-THROATED FLYCATCHER *Myiarchus cinerascens* | L 7¾" (19 cm)

Grayish brown above; throat and breast pale gray; underparts paler than in Brown-crested. Tail shows rufous on inner webs with dark tips. As in all *Myiarchus* flycatchers, brief **juvenal** plumage shows mostly reddish tail. Distinctive **call**, heard year-round, is a rough *prrrt*. **Song**, heard on breeding grounds, is a series of burry *ka-brick* notes.

RANGE: Common in a wide variety of habitats. Very rare fall and winter visitor to East.

LA SAGRA'S FLYCATCHER *Myiarchus sagrae* | L 7¾" (19 cm)

Grayish brown upperparts, mainly white underparts suggestive of Ash-throated Flycatcher, but bill longer; inner tertial edge stronger; rufous on outer tail less extensive. Distinctive **call**, a rather high-pitched *wink*, is often doubled.

RANGE: West Indian species from Bahamas, Cuba, and Caymans. Casual visitor, mainly in winter and spring, to south Florida; accidental to Alabama.

DUSKY-CAPPED FLYCATCHER *Myiarchus tuberculifer* | L 6¾" (17 cm)

Smaller, bill larger, belly and undertail coverts usually brighter yellow than in Ash-throated Flycatcher; tail shows less rufous. Secondaries have rufous edges, unlike other *Myiarchus*. **Call** is a mournful, descending *peeur*.

RANGE: Fairly common in wooded mountain ranges. Casual in west Texas. Very rare in late fall and winter in southern and central California.

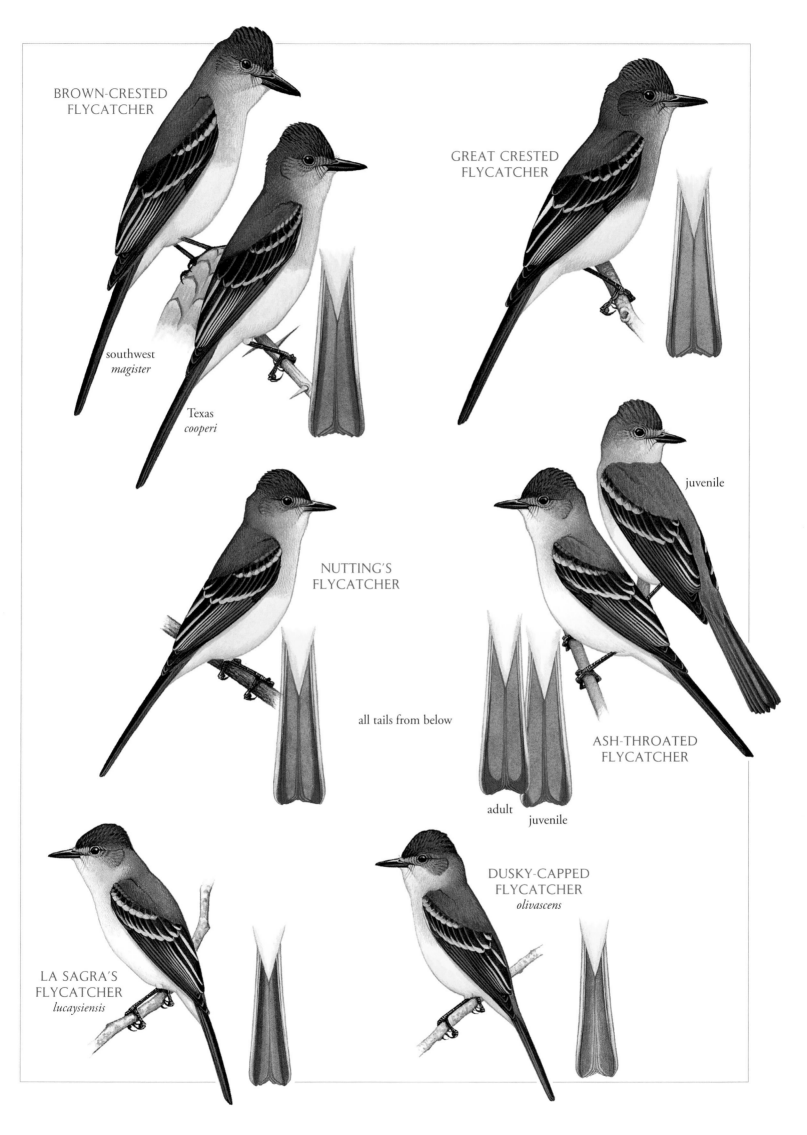

BROWN-CRESTED
FLYCATCHER

GREAT CRESTED
FLYCATCHER

southwest
magister

Texas
cooperi

NUTTING'S
FLYCATCHER

juvenile

all tails from below

ASH-THROATED
FLYCATCHER

adult

juvenile

LA SAGRA'S
FLYCATCHER
lucaysiensis

DUSKY-CAPPED
FLYCATCHER
olivascens

CASSIN'S KINGBIRD
Tyrannus vociferans | L 9" (23 cm)

Dark brown tail; narrow buffy tips and lack of white edges on outer tail feathers help distinguish this species from Western Kingbird. Bill is much shorter than in Tropical and Couch's Kingbirds. Upperparts darker gray than in Western, washed with olive on back; paler wings contrast with darker back. White chin contrasts with dark gray head and breast. Belly dull yellow. **Juvenile** is duller, slightly browner above, with bold buffy edges on wing coverts; paler below. Most common **call**, given year-round, is a short, loud *chi-bew*, accented on second syllable.

RANGE: Fairly common in varied habitats; usually prefers denser foliage and hillier country than does Western Kingbird.

WESTERN KINGBIRD
Tyrannus verticalis | L 8¾" (22 cm)

Black tail, with white edges on outer feathers. Bill much shorter than in Tropical and Couch's Kingbirds. Upperparts ashy gray, paler than in Cassin's Kingbird, tinged with olive on back; dark wings contrast with paler back. Throat and breast pale gray; belly bright lemon yellow. **Juvenile** has slightly more olive on back and buffy edges on wing coverts, brownish tinge on breast, paler yellow belly. Common **call** is a sharp *whit*.

RANGE: Common in dry, open country; perches on fences, telephone lines. Regular straggler in fall and early winter along the East Coast from the Maritime Provinces south; winters in small numbers in central and southern Florida.

COUCH'S KINGBIRD
Tyrannus couchii | L 9¼" (24 cm)

Almost identical to Tropical Kingbird, with thicker, broader-based bill; back slightly greener and less gray; at close range, tips of individual primaries are evenly spaced on adults. Distinguished from Western and Cassin's Kingbirds by larger bill, darker ear patch, and slightly notched brown tail. Juvenile is duller overall, with buffy edges on wing coverts. Distinctive **calls**, a shrill, rolling *breeeer*; and a more common *kip*, similar to call of Western Kingbird, given singly or in a series.

RANGE: Common in the lower Rio Grande Valley in summer; uncommon in winter. Found in groves and shrubs. Casual on Gulf Coast in fall and winter. Accidental to New Mexico and southern California.

TROPICAL KINGBIRD
Tyrannus melancholicus | L 9¼" (24 cm)

Almost identical to Couch's Kingbird. Bill is thinner and longer; back slightly grayer, less green; at close range, tips of individual primaries are unevenly staggered on adults. Distinguished from Western and Cassin's Kingbirds by larger bill, darker ear patch, and slightly notched brown tail. Distinctive **call** is a rapid, twittering *pip-pip-pip-pip*.

RANGE: Uncommon and local in southeastern Arizona and Rio Grande Valley, Texas; found in lowlands near water; often nests in cottonwoods. Rare but regular during fall and winter along the West Coast to British Columbia, casually to southeast Alaska.

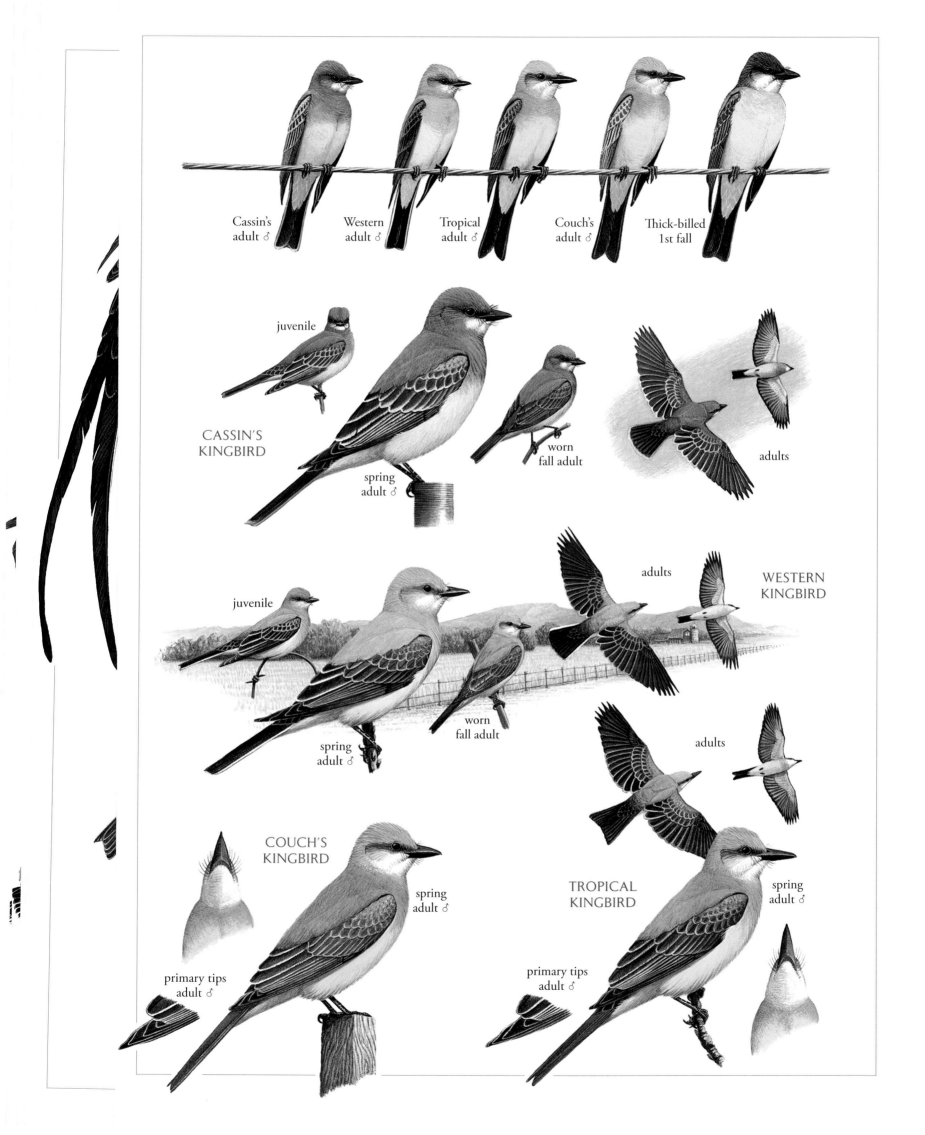

Cassin's
adult ♂

Western
adult ♂

Tropical
adult ♂

Couch's
adult ♂

Thick-billed
1st fall

juvenile

CASSIN'S
KINGBIRD

spring
adult ♂

worn
fall adult

adults

juvenile

adults

WESTERN
KINGBIRD

spring
adult ♂

worn
fall adult

adults

COUCH'S
KINGBIRD

spring
adult ♂

primary tips
adult ♂

TROPICAL
KINGBIRD

spring
adult ♂

primary tips
adult ♂

WESTERN SCRUB-JAY
Aphelocoma californica | L 11" (28 cm)

Long tail; blue above; variable bluish band on chest. Coastal races, including nominate *californica*, deeper blue above; contrasting brown patch; distinct white eyebrow and blue breast band; undertail coverts geographically variable; may or may not be bluish. Interior races (possibly a separate species) range from duller, slender-billed *nevadae* of Great Basin to bluer, stouter-billed *texana* (neither shown); slender-billed *woodhouseii* is intermediate in color. **Calls** include raspy *shreep*, often in a short series. Western, Island, and Florida were formerly considered one species, Scrub Jay.
RANGE: Tame and widespread; found in urban areas. Interior races rather shy; inhabit lower mountain woodland. All U.S. subspecies hold individual territories.

ISLAND SCRUB-JAY
Aphelocoma insularis | L 12" (30 cm)

Larger and with much larger bill than Western; darker blue above; always shows rich blue undertail coverts. **Calls** very similar to coastal Western Scrub-Jay.
RANGE: Restricted to Santa Cruz Island, California, where it is the only scrub-jay. Birds hold individual territory; takes several years for young birds to acquire territory and breed.

FLORIDA SCRUB-JAY
Aphelocoma coerulescens | **T** | L 11" (28 cm)

Distinguished from other scrub-jays by whitish forehead and eyebrow; shorter, broader bill; paler back; distinct collar; indistinct streaking below; disproportionately longer tail. Varied **calls** include raspy, hoarse notes. Has cooperative breeding system: Fledged young remain on territory and help rear nestlings.
RANGE: Restricted to Florida scrub region where population has declined some 90 percent in 20th century due to habitat destruction. Optimum habitat is transitional, produced by fire: consists of scrub, mainly oak, about ten feet high with small openings.

MEXICAN JAY
Aphelocoma ultramarina | L 11½" (29 cm)

Blue above, with slight grayish cast on back, brownish patch on center of back. Lacks crest. Distinguished from scrub-jays by absence of white throat and white eyebrow and by chunkier shape; flight is more direct. Texas race, *couchii*, has richer blue head. Arizona **juvenile** *arizonae* retains pale bill past post-juvenal molt. **Calls** include a loud, ringing *week*, given singly or in a series.
RANGE: Common in montane pine-oak canyons of the Southwest, where it greatly outnumbers scrub-jays. Has cooperative breeding system similar to Florida Scrub-Jay.

PINYON JAY
Gymnorhinus cyanocephalus | L 10½" (27 cm)

Blue overall; blue throat streaked with white; bill long and spiky; tail short. Immature is duller. Flight is direct, with rapid wingbeats, unlike scrub-jays' undulating flight. Typical flight **call** is a high-pitched, piercing *mew*, audible over long distances. Also gives a rolling series of *queh* notes.
RANGE: Generally seen in large flocks, often numbering in the hundreds; nests in loose colonies. Common in piñon-juniper woodlands of interior mountains and high plateaus; also yellow pine woodlands. Casual to Plains states and coastal California.

WESTERN
SCRUB-JAY

interior
woodhouseii

interior
juvenile
woodhouseii

coastal
californica

ISLAND
SCRUB-JAY

FLORIDA
SCRUB-JAY

MEXICAN
JAY

adult

Texas
couchii

Arizona
arizonae

PINYON JAY

juvenile
arizonae

BROWN JAY
Cyanocorax morio | L 16½" (42 cm)

Very large jay with long, broad tail. Dark, sooty brown overall except for pale belly. **Adult** has black bill. **Juvenile** has yellow bill and eye ring, turning black by second winter; in transition, many have blotchy yellow-and-black bills. A noisy species; its harsh scream is similar to the **call** of Red-shouldered Hawk. Another call sounds like a hiccup.
RANGE: Tropical species; range barely extends to Texas, where it is resident but rare in woodlands and mesquite along the Rio Grande in vicinity of Falcon Dam.

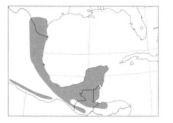

GREEN JAY
Cyanocorax yncas | L 10½" (27 cm)

Green and blue plumage blends with dappled sun and shade in woodland habitat. Somewhat inquisitive. Gregarious and noisy; most common **call** is a series of raspy *cheh-cheh-cheh* notes.
RANGE: Tropical species; range extends to southern Texas. Resident and locally common in brushy areas and streamside growth of the lower Rio Grande Valley.

BLACK-BILLED MAGPIE
Pica hudsonia | L 19" (48 cm)

Readily identified as a magpie by black and white markings and unusually long tail with iridescent green highlights. White wing patches flash in flight. Black bill and range distinguish this species from look-alike Yellow-billed Magpie. Gregarious and noisy; typical **calls** include a whining *mag* and a series of loud, harsh *chuck* notes. Calls and many behavioral traits resulted in North American Black-billed Magpie being split from Old World populations of magpie, whose calls are faster and lower-pitched.
RANGE: Common inhabitant of open woodlands and thickets in rangelands and foothills, especially along watercourses. Black-billed and Yellow-billed ranges almost overlap, and Black-billed Magpies casually stray south and east of normal range in winter. Birds seen casually throughout the East may be escaped cage birds.

YELLOW-BILLED MAGPIE
Pica nuttalli | L 16½" (42 cm)

Similar to Black-billed Magpie, but never occurs in Black-billed's normal range. Distinguished by its yellow bill and by a yellow patch of bare skin around the eye; extent of yellow variable, sometimes fully encircles eye; may be related to state of molt rather than individual variation, or may be a combination of both. **Calls** are similar to Black-billed; both species roost and feed in flocks, usually nest in loose colonies, but Yellow-billed Magpie's behavior is more colonial than Black-billed's.
RANGE: Prefers oaks, especially more open oak grassland, also orchards and parks. Common resident of rangelands and foothills of central and northern Central Valley, California, and coastal valleys south to Santa Barbara County. Not prone to wandering, but casual north almost to Oregon. Some authorities believe that Yellow-billed Magpie is more closely related to the North American Black-billed than either is to the Old World races of magpies.

juvenile

adult

BROWN JAY
palliatus

GREEN JAY
glaucescens

BLACK-BILLED
MAGPIE

YELLOW-BILLED
MAGPIE

EURASIAN JACKDAW *Corvus monedula* | L 13" (33 cm)

Small, black overall, with gray nape and face, pale grayish eyes. Lively and inquisitive. **Calls** include a metallic *kow* and a softer *jack* note.

RANGE: Arrived in Northeast in early 1980s, most perhaps ship-assisted. Found from Atlantic Canada to Pennsylvania. The last report was in Apr. 1999 from Newfoundland.

TAMAULIPAS CROW *Corvus imparatus* | L 14½" (37 cm)

Smaller, glossier than American Crow. **Call**, a low, froglike croak.

RANGE: First appeared in U.S. in late 1960s near Brownsville, Texas, where it has nested. Present each winter since in varying numbers, sharply declining over last two decades.

AMERICAN CROW *Corvus brachyrhynchos* | L 17½" (45 cm)

Our largest crow. Long, heavy bill is noticeably smaller than in ravens. Fan-shaped tail distinguishes all crows from ravens in flight. Adult American Crow is readily identified by familiar *caw* call, but juvenile's higher-pitched, nasal *cah* begging call resembles the call of the similar Fish Crow.

RANGE: Generally common throughout most of its range in a wide variety of habitats.

NORTHWESTERN CROW *Corvus caurinus* | L 16" (41 cm)

Not shown. Nearly identical to American Crow but slightly smaller. **Call** is somewhat hoarser and lower than that of American. Best clue is range. Considered by some to be a subspecies of American Crow. Common scanenger along the shore.

RANGE: Inhabits northwestern coastal areas and islands.

FISH CROW *Corvus ossifragus* | L 15½" (39 cm)

Smaller than American Crow with smaller bill and feet; wings more pointed; wingbeats faster. Best distinguished by voice: **Call**, a high, nasal *uh uh*, the second note lower; also low, short *car* notes.

RANGE: Favors tidewater marshes and low valleys along eastern river systems; less frequent inland, except along rivers. Often seen in winter in flocks with American Crows.

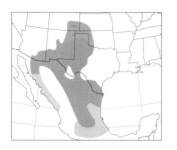

CHIHUAHUAN RAVEN *Corvus cryptoleucus* | L 19½" (50 cm)

Heavier bill and wedge-shaped tail distinguish both raven species from crows. Distinguished from Common Raven by shorter wings and shorter, less wedge-shaped tail; bristles extend farther out on shorter, thicker-appearing bill. Neck feathers white rather than grayish at base, but usually obscured. Frequent **call**, a drawn-out croak, usually slightly higher-pitched than call of Common Raven.

RANGE: Common in desert areas and scrubby grasslands.

COMMON RAVEN *Corvus corax* | L 24" (61 cm)

Large, with long, heavy bill and long, wedge-shaped tail. Most common **call** is a low, drawn-out croak. Larger than Chihuahuan Raven; note thicker, shaggier throat feathers and nasal bristles that do not extend as far out on larger bill.

RANGE: Found in a variety of habitats, including mountains, deserts, and coastal areas. Numerous in western and northern part of range; uncommon and local, but slowly spreading, in Appalachians.

EURASIAN
JACKDAW

TAMAULIPAS
CROW

AMERICAN
CROW

FISH
CROW

CHIHUAHUAN
RAVEN

COMMON
RAVEN

LARKS (Family Alaudidae)

Ground dwellers of open fields, larks are slender-billed seed- and insect-eaters.
*They seldom alight on trees or bushes. On the ground, they walk rather than hop. **Species:** 92 World, 2 N.A.*

SKY LARK
Alauda arvensis | L 7¼" (18 cm)

Plain brown bird with slender bill; slight crest is raised when bird is agitated. Upperparts heavily streaked; buffy white underparts streaked on breast and throat. Dark eye prominent. All juveniles have a scaly brown mantle. In flight, Sky Larks show a conspicuous white trailing edge on the inner wing and white edges on tail. **Song** is a continuous outpouring of trills and warblings, delivered in high hovering or circling song flight. **Call** is a liquid *chirrup* with buzzy overtones.

RANGE: Nominate *arvensis*, a widespread European race introduced to Vancouver Island in the early 1900s, is resident there on open slopes and fields. The population on San Juan Islands is now extirpated. Highly migratory Asian subspecies *pekinensis*, rare on western Aleutians and Pribilofs and casual on St. Lawrence Island, is darker and more heavily streaked above; accidental in winter in Washington and northern California.

HORNED LARK
Eremophila alpestris | L 6¾–7¾" (17–20 cm)

Head pattern distinctive in all subspecies: black "horns"; white or yellowish face and throat with broad black stripe under eye; black bib. **Female** duller overall than **male**, horns less prominent. Conspicuous in flight is the mostly black tail with white outer feathers, brown central feathers. Brief **juvenal** plumage has whitish markings above, streaks below; can be confused with Sprague's Pipit (page 370). Three widespread races are found in the East: "**Prairie Horned Lark**," *praticola*, which breeds in southern Canada and the eastern U.S., is pale, with white eyebrows and throat. "**Northern Horned Lark**," *alpestris*, is much darker, with a yellow throat. The central Arctic coast race, *hoyti*, is pale like *praticola*, but larger; *giraudi* from western Gulf Coast is quite yellow below. Western subspecies vary widely in overall color; selected extremes are shown here: *enthymia* (Plains, not illustrated), the Southwest deserts' *ammophila*, and large *arcticola* (northwest Canada, Alaska) are very pale; *sierrae* (northeast California) and *strigata* (coastal Northwest) are yellower on head and chest, *strigata* is also more streaked underneath; *rubea* (Central Valley, California) is redder dorsally; *insularis* (Channel Islands, California) is streaked below; *flava* from Palearctic (casual fall vagrant to Alaska) has a yellow throat and supercilium. Horned Lark's **calls** include a high *tsee-ee* or *tsee-titi*. **Song** is a weak twittering, delivered from the ground or in flight.

RANGE: Widespread and common, Horned Lark prefers dirt fields, gravel ridges, and shores. The flocks in winter in the eastern U.S. are mainly *alpestris*. Some subspecies are highly migratory, others are largely resident.

SKY LARK
arvensis

fresh

pekinensis

worn

juvenile

HORNED
LARK

flava

Lapland Longspur
(for comparison)

hoyti

winter
♂

"Northern Horned
Lark" *alpestris*
♂

♀

arcticola

"Prairie Horned
Lark" *praticola*

juvenile
alpestris

sierrae

rubea

strigata

giraudi

ammophila

SWALLOWS (Family Hirundinidae)

Slender bodies with long, pointed wings resemble swifts, but "wrist" angle is sharper and farther from the body; flight is more fluid. Adept aerialists, swallows dart to catch flying insects. Flocks perch in long rows on branches and wires. **Species:** *84 World, 15 N.A.*

TREE SWALLOW
Tachycineta bicolor | L 5¾" (15 cm)

Dark, glossy greenish blue above, greener in fall plumage; white below. White cheek patch does not extend above eye as in Violet-green Swallow. **Juvenile** is gray-brown above; usually has more diffuse breast band than Bank Swallow (next page). **First-spring female** shows varying amount of adult color on crown and back.
RANGE: Common in wooded habitat near water, and where dead trees provide nest holes. Also nests in fence posts, barn eaves, nest boxes. Migrates in huge flocks; goes north earlier in spring and lingers farther north in fall than other swallows.

BAHAMA SWALLOW
Tachycineta cyaneoviridis | L 5¾" (15 cm)

Greenish above. Deeply forked tail and white underwing coverts separate this species from similar Tree Swallow. Immatures have shorter tail fork, dusky wash on breast and wing linings.
RANGE: Endemic Bahamian species. Breeds in northern Bahamas and vicinity; casual visitor to the Florida Keys, and nearby mainland but unrecorded for over a decade.

VIOLET-GREEN SWALLOW
Tachycineta thalassina | L 5¼" (13 cm)

White on cheek extends above eye; white flank patches extend onto sides of rump; compare with larger Tree Swallow. May also be confused with White-throated Swift (page 270). Female is duller above than **male**. **Juvenile** is gray-brown above; white except on rump may be mottled or grayish.
RANGE: Common in a variety of woodland habitats. Nests in hollow trees or rock crevices, often forming loose colonies. Casual in the East.

PURPLE MARTIN
Progne subis | L 8" (20 cm)

Male is dark, glossy purplish blue. **Female** and juvenile are gray below. **First-spring males** have some purple below. In flight, male resembles European Starling (page 366); but note forked tail, longer wings, and typical swallow flight, short glides alternating with flapping.
RANGE: Locally common where suitable nest sites are available. Declining over much of North America, especially in Pacific states. Very early spring migrant in south, arriving on Gulf Coast in February; winters in South America.

COMMON HOUSE-MARTIN
Delichon urbicum | L 5" (13 cm)

Deep, glossy blue above; mostly white below with white rump; underwing coverts pale smoky gray. Female slightly grayer below; juvenile duller. Soars for long periods. **Call**, a rough scratchy *prrit*, somewhat similar to Rough-winged Swallow. One Alaska specimen of eastern race *lagopoda* has more extensive white on rump.
RANGE: Old World species. Casual mainly in spring in western Alaska; one record for Saint-Pierre, off Newfoundland.

juvenile

adult

1st spring ♀

fall adult

adult

spring adult

TREE
SWALLOW

♂

♂

♂

adults

BAHAMA
SWALLOW

adults

juvenile

♂

VIOLET-GREEN
SWALLOW

adult ♂

COMMON
HOUSE-MARTIN
lagopoda

PURPLE
MARTIN

♀

adult ♂

1st spring ♂

♀

adult ♂

BANK SWALLOW
Riparia riparia | L 4¾" (12 cm)

Our smallest swallow. Distinct brownish gray breast band, often extending in a line down center of breast. Throat is white; white curves around rear border of ear patch. **Juvenile** has thin buffy wing bars; compare with juvenile Northern Rough-winged Swallow and juvenile Tree Swallow (preceding page). Locally common throughout most of range. Unlike Northern Rough-winged, wingbeats are shallow and rapid; also paler rump contrasts with wings. RANGE: Nests in large colonies, excavating nest burrows in steep riverbank cliffs, gravel pits, and highway cuts. Winters chiefly in South America; often migrates in large flocks.

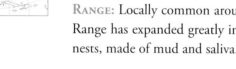

CLIFF SWALLOW
Petrochelidon pyrrhonota | L 5½" (14 cm)

Squarish tail and buffy rump distinguish this swallow from all others except Cave Swallow. Most Cliff Swallows have dark chestnut and blackish throat, pale forehead. A primarily southwestern race, *melanogaster*, has cinnamon forehead like Cave Swallow, but throat is dark chestnut. All **juveniles** are much duller and grayer than adults; throat is paler, forehead darker. RANGE: Locally common around bridges, rural settlements, and in open country on cliffs. Range has expanded greatly in last two decades. Nests in colonies, building gourd-shaped nests, made of mud and saliva.

NORTHERN ROUGH-WINGED SWALLOW
Stelgidopteryx serripennis | L 5" (13 cm)

Brown above, whitish below, with gray-brown wash on chin, throat, and upper breast. Lacks Bank Swallow's distinct breast band; wings are longer, wingbeats deeper and slower. **Juvenile** has cinnamon wing bars. RANGE: Nests in single pairs in riverbanks, cliffs, culverts, and under bridges. Migrates singly or in small flocks.

BARN SWALLOW
Hirundo rustica | L 6¾" (17 cm)

Long, deeply forked tail. Throat is reddish brown; underparts usually cinnamon or buffy. Two Eurasian, white-bellied races have occurred in western and northern Alaska: *rustica*, which has a solid dark breast band, and *gutturalis*, with incomplete breast band, which also has been found on Queen Charlotte Islands. In all **juveniles**, tail is shorter but still noticeably forked; underparts pale. RANGE: Common; generally nests on or inside farm buildings, under bridges, and inside culverts, in pairs or small colonies.

CAVE SWALLOW
Petrochelidon fulva | L 5½" (14 cm)

Squarish tail; distinguished from Cliff by buffy throat color extending through auriculars and around nape setting off dark cap; rump averages a richer color; cinnamon forehead; juvenile much paler; compare with southwestern subspecies of Cliff also with cinnamon forehead. RANGE: Mexican and West Indian species: West Indies race, *fulva*, is smaller than *pelodoma* and has more buff below and darker rump. Mexican *pelodoma* is widespread in the southwest; West Indies *fulva* is a local breeder in south Florida. Nests in colonies in limestone caves, sinkholes, culverts, and under bridges, sometimes with Barn and Cliff Swallows. Very rare on East Coast, mainly in fall; casual to southern Arizona, southeastern California.

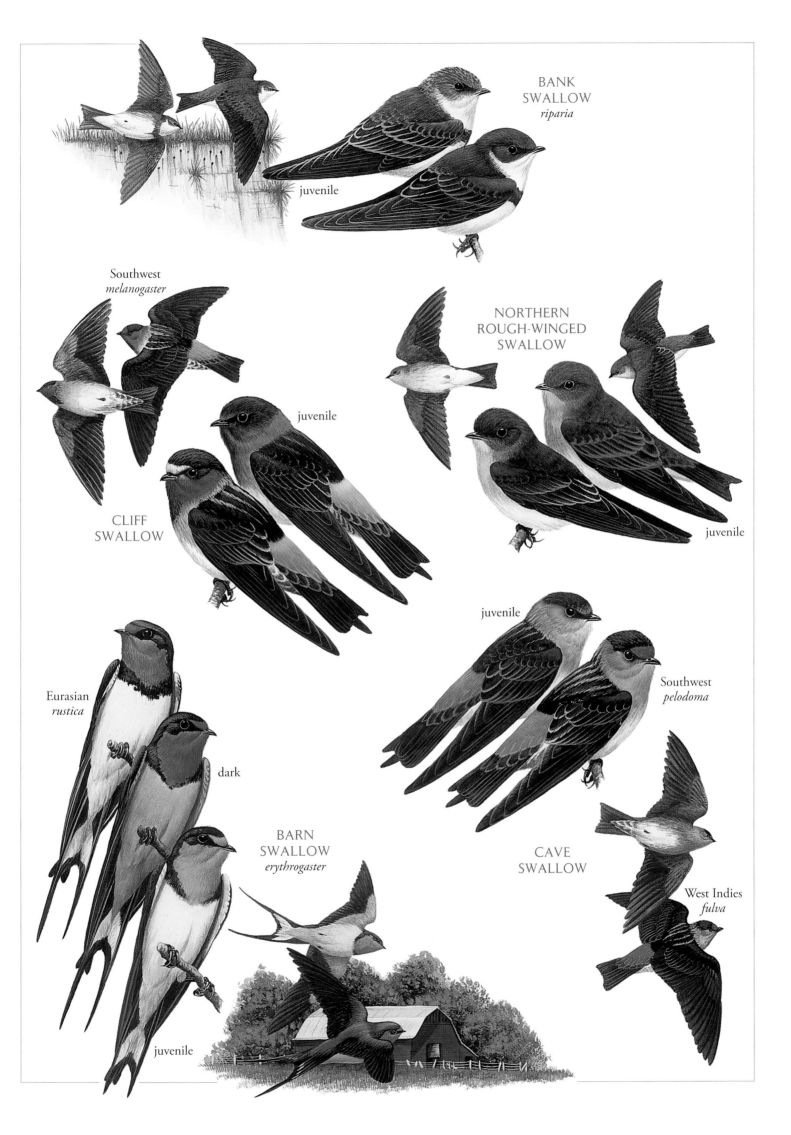

BANK
SWALLOW
riparia

juvenile

Southwest
melanogaster

NORTHERN
ROUGH-WINGED
SWALLOW

juvenile

juvenile

CLIFF
SWALLOW

juvenile

juvenile

Southwest
pelodoma

Eurasian
rustica

dark

BARN
SWALLOW
erythrogaster

CAVE
SWALLOW

West Indies
fulva

juvenile

BABBLERS (Family Timaliidae)

Species: 273 World, 1 N.A.

WRENTIT
Chamaea fasciata | L 6½" (17 cm)

Color varies from reddish brown in northern populations to grayer in southern birds. Note distinct cream-colored eye and lightly streaked buffy breast; long, rounded tail usually cocked. Wrentits are usually heard before they are seen. Male's loud **song**, sung year-round, begins with a series of accelerating notes and runs into a descending trill: *pit-pit-pit-tr-r-r-r*. Female's song lacks trill.

RANGE: Common in chaparral and coniferous brushland.

CHICKADEES AND TITMICE (Family Paridae)

Species: 54 World, 12 N.A.

BRIDLED TITMOUSE *Baeolophus wollweberi* | L 5¼" (13 cm)

Note distinct crest, black-and-white facial pattern, black throat. Most common **call** is a rapid, high-pitched variation of *chick-a-dee-dee*, similar to Juniper Titmouse.

RANGE: Stands of oak, juniper, and sycamore in southern Arizona, New Mexico mountains.

OAK TITMOUSE *Baeolophus inornatus* | L 5" (13 cm)

Grayish brown with a short crest. Northern race *inornatus* is slightly smaller, paler, and smaller billed than *affabilis* (shown here) from southwestern California and northern Baja; birds from Little San Bernadino Mountains are paler, grayer than *affabilis*. **Song** is variable, a repeated series of syllables made up of whistled, alternating, high and low notes. **Call** is a hoarse *tschick-a-dee*.

RANGE: Common in warm, dry oak woodland.

JUNIPER TITMOUSE *Baeolophus ridgwayi* | L 5¼" (13 cm)

Like Oak Titmouse; range overlaps on northern California Modoc plateau. Larger, paler, grayer than Oak. **Song**, a rolling series of syllables, rapid, with uniform pitch. **Call**, a hoarse *tschick-a-dee* similar to Bridled. Overall, chattering call notes are more clipped and delivered much more rapidly than Oak's call.

RANGE: Uncommon to fairly common in juniper or piñon-juniper woodland.

TUFTED TITMOUSE *Baeolophus bicolor* | L 6¼" (16 cm)

Note gray crest and distinct blackish forehead. **Juvenile** has brownish forehead and pale crest. In overlap zone in Texas, hybrids show variable brown foreheads, dark gray crests. Active, noisy; typical song is loud, whistled *peter peter peter*; but less vocal than Black-crested; calls softer and less nasal.

RANGE: Deciduous woodlands, parks, and suburbs.

BLACK-CRESTED TITMOUSE *Baeolophus atricristatus* | L 5¾" (15 cm)

Resplit from Tufted. **Adult** has black crest, pale forehead. **Juvenile** crown darker than upperparts; forehead dirty white. **Calls** louder, sharper than Tufted.

RANGE: Wooded areas.

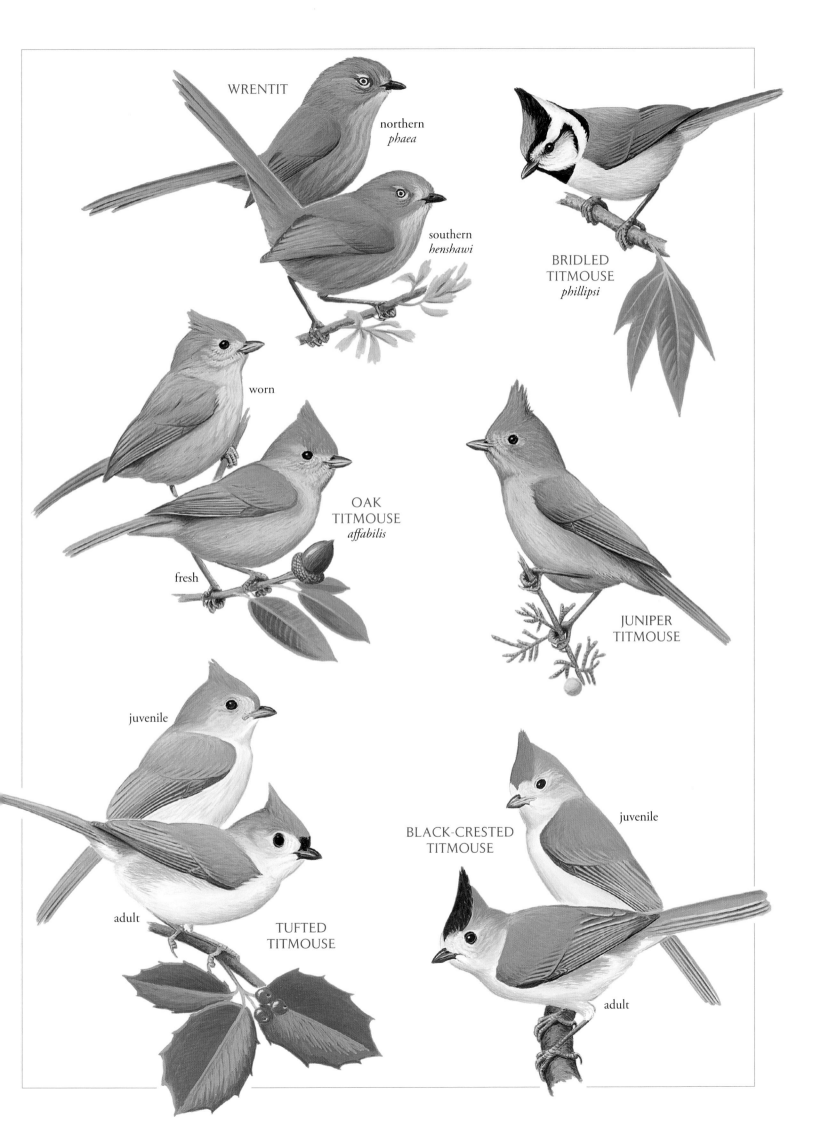

WRENTIT

northern
phaea

southern
henshawi

BRIDLED
TITMOUSE
phillipsi

worn

OAK
TITMOUSE
affabilis

fresh

JUNIPER
TITMOUSE

juvenile

adult

TUFTED
TITMOUSE

BLACK-CRESTED
TITMOUSE

juvenile

adult

BLACK-CAPPED CHICKADEE
Poecile atricapillus | L 5¼" (13 cm)

Black cap and bib; cheeks more extensively and purer white than similar Carolina Chicka-dee. Note that Black-capped Chickadee's greater wing coverts and secondaries are broadly edged in white; tertials more boldly edged, with darker centers than Carolina's; flanks more olive, lower edge of black bib a bit more ragged. These differences are obscured in **worn summer** birds. Plumage is geographically variable: *occidentalis* from Pacific Northwest is darker; *nevadensis* from Great Basin is palest subspecies. Best distinction is voice. Black-capped's **call** is a lower, slower *chick-a-dee-dee-dee* than Carolina's; typical **song**, a clear, whistled *fee-bee* or *fee-bee-ee*, the first note higher in pitch; vocalizations show some geographic variation.

RANGE: Common in open woodlands, clearings, and suburbs. Usually forages in thickets and low branches of trees. The usual ranges of Black-capped and Carolina barely overlap, but periodic fall irruptions push Black-capped's range south of mapped range. Where breeding ranges overlap, the two species hybridize. In the Appalachians, Black-capped inhabits higher elevations.

CAROLINA CHICKADEE
Poecile carolinensis | L 4¾" (12 cm)

Very similar to Black-capped Chickadee: black cap and bib, white cheeks. Note that Caro-lina lacks broad white edgings on greater wing coverts; lower edge of black bib is usually neater, has less olive on flanks than Black-capped. Westernmost race, *atricapilloides,* is grayer than nominate. Best distinction for separating species is voice. Carolina's **call** is a higher, faster version of *chick-a-dee-dee-dee* than Black-capped; typical **song** is a four-note whistle, *fee-bee fee-bay,* the last note lowest in pitch.

RANGE: Common in open deciduous forests, woodland clearings and edges, suburban areas, and urban parks. Feeds in trees and thickets; seldom descends to ground. At northern edge, its range in some winters is invaded by Black-capped. In the Appalachians, Carolina prefers valleys and foothills.

MEXICAN CHICKADEE
Poecile sclateri | L 5" (13 cm)

The only breeding chickadee in its range. Extensive black bib is distinctive, along with dark gray flanks. Lacks white eyebrow of Mountain Chickadee. **Song** is a warbled whistle; **call** note, a husky buzz.

RANGE: A Mexican species, fairly common resident in coniferous and pine-oak forests; found in U.S. only in Chiricahua Mountains of southeastern Arizona and Animas and Pel-oncillo Mountains of southwestern New Mexico.

MOUNTAIN CHICKADEE
Poecile gambeli | L 5¼" (13 cm)

White eyebrow and pale gray sides distinguish this species from other chickadees; lack of crest separates it from the Bridled Titmouse (preceding page). Birds of Rocky Mountain nominate race *gambeli* are tinged with buff on back, sides, flanks, and have broader white eyebrow than *baileyae*. **Call** is a hoarse *chick-adee-adee-adee*; typical **song**, a three- or four-note descending whistle, *fee-bee-bay* or *fee-bee fee-bee*.

RANGE: Common resident in coniferous and mixed woodlands. Some descend to lower elevations in winter.

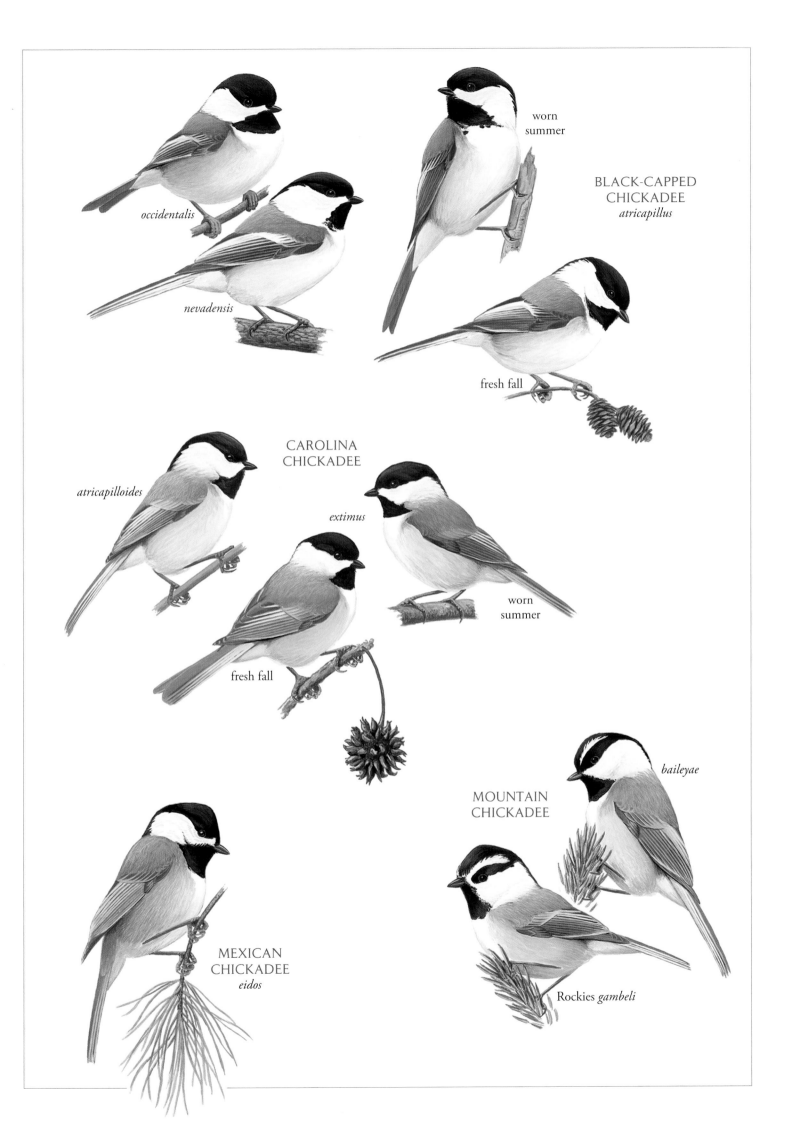

occidentalis

nevadensis

worn
summer

BLACK-CAPPED
CHICKADEE
atricapillus

fresh fall

CAROLINA
CHICKADEE

atricapilloides

extimus

worn
summer

fresh fall

MOUNTAIN
CHICKADEE

baileyae

MEXICAN
CHICKADEE
eidos

Rockies *gambeli*

CHESTNUT-BACKED CHICKADEE *Poecile rufescens* | L 4¾" (12 cm)

Sooty brown cap, white cheeks, black bib; chestnut back and rump. Over most of its range, this species has bright chestnut sides and flanks; *barlowi*, on central California coast south of Golden Gate Bridge, shows almost no chestnut below. **Call** is a hoarse, rapid *tseek-a-dee-dee*.

RANGE: Found in coniferous forests, deciduous woodlands. Usually feeds high in trees.

BOREAL CHICKADEE *Poecile hudsonicus* | L 5½" (14 cm)

Grayish brown on crown and back, with pinkish brown flanks. Note that rear portion of cheeks is heavily washed with gray. **Call** is a nasal *tseek-a-day-day*.

RANGE: Fairly common in coniferous forests. In some winters, small numbers wander hundreds of miles south of normal eastern range.

GRAY-HEADED CHICKADEE *Poecile cinctus* | L 5½" (14 cm)

Gray-brown above, whitish below, with white cheek patch, black bib, buffy sides and flanks. Distinguished from Boreal Chickadee by more extensively white cheeks, longer tail, paler flanks, and pale edges on wing coverts; also by **call**, a series of peevish *dee deer* notes.

RANGE: Rare; found in willows and spruces edging tundra. Still known as the Siberian Tit in the Old World.

PENDULINE TITS AND VERDINS (FAMILY REMIZIDAE)

Small, spritely birds with finely pointed bills. They inhabit arid scrub country, feed in brush chickadee-style, and build spherical nests. **Species: 10 World, 1 N.A.**

VERDIN *Auriparus flaviceps* | L 4½" (11 cm)

Adult has dull gray plumage, chestnut shoulder patches, yellow head and throat. **Juvenile** is brown-gray overall; shorter tail helps separate it from Bushtit. Compare also with Lucy's Warbler (page 378). **Song** is a plaintive three-note whistle, the second note higher. **Calls** include rapid *chip* notes.

RANGE: Common in mesquite and other dense thorny shrubs of the southwestern desert.

LONG-TAILED TITS AND BUSHTITS (FAMILY AEGITHALIDAE)

A longer tail distinguishes these tiny birds from other chickadee-like species. Except during nesting, usually feed in large, busy, twittering flocks. Nest is an elaborate hanging structure. **Species: 11 World, 1 N.A.**

BUSHTIT *Psaltriparus minimus* | L 4½" (11 cm)

Gray above, paler below; female has pale eyes, male's eyes are dark. Coastal birds have brown crown; interior birds show brown ear patch and gray cap, and have sharper, slower, twittering **calls**. **Juvenile male** and some adult males in the Southwest have a black mask, formerly considered a separate species, the "**Black-eared Bushtit**."

RANGE: Common in a wide variety of woodlands.

rufescens

CHESTNUT-BACKED
CHICKADEE

coastal
central California
barlowi

juvenile

worn
summer

fresh fall

BOREAL
CHICKADEE
hudsonicus

GRAY-HEADED
CHICKADEE
lathami

interior ♂
plumbeus

BUSHTIT

VERDIN

juvenile

interior ♀
plumbeus

"Black-eared Bushtit"
juvenile ♂ *plumbeus*

coastal ♂
minimus

CREEPERS (Family Certhiidae)

With curved bills, these little tree-climbers dig insects and larvae from bark. Stiff tail feathers serve as props.
Species: *8 World, 1 N.A.*

BROWN CREEPER *Certhia americana* | L 5¼" (13 cm)

Camouflaged by streaked brown plumage, Creepers spiral upward from base of a tree, then fly to a lower place on another tree. **Call** is a soft, sibilant *see*; **song**, a high-pitched, variable *see see see titi see*. Fairly common but hard to spot.
Range: Nests in coniferous, mixed, or swampy forests. Generally solitary, but sometimes seen in winter flocks of titmice and nuthatches.

NUTHATCHES (Family Sittidae)

These short-tailed acrobats climb up, down, and around tree trunks and branches.
Species: *25 World, 4 N.A.*

WHITE-BREASTED NUTHATCH *Sitta carolinensis* | L 5¾" (15 cm)

Black cap tops all-white face and breast; extent of rust below is variable. **Females** in the Northeast have gray crowns more consistently than in the South. Western birds have longer, thinner bills. Typical **song** is a rapid series of nasal whistles on one pitch. **Call** is usually a low-pitched, repeated, nasal *yank* in eastern nominate race; high-pitched in West Coast *aculeata*; and higher-pitched and given in a rapid series by Great Basin *tenuissima* and Rockies *nelsoni*.
Range: Common; found in leafy trees in the East, oaks and conifers in the West.

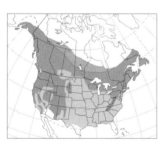

RED-BREASTED NUTHATCH *Sitta canadensis* | L 4½" (11 cm)

Black cap and eye line, white eyebrow, rust underparts; **female** and juveniles have duller head, paler underparts. High-pitched, nasal **call** sounds like a toy tin horn.
Range: Resident in northern and subalpine conifers; gleans small branches and outer twigs. Irruptive migrant; numbers and winter range vary yearly. In the East, resident range is expanding southward.

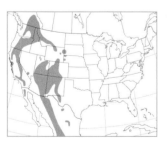

PYGMY NUTHATCH *Sitta pygmaea* | L 4¼" (11 cm)

Gray-brown cap; creamy buff underparts. Pale nape spot visible at close range. Dark eye line bordering cap, most distinct in interior populations. Typical **calls**, a high, rapid *peep peep* and a piping *wee-bee*; grouped in three or more notes in coastal nominate race, *pygmaea*.
Range: Favors yellow-pine forest, except for birds in coastal California pines. Roams in loose flocks.

BROWN-HEADED NUTHATCH *Sitta pusilla* | L 4½" (11 cm)

Brown cap; dull buff underparts. Pale nape spot visible at close range. Narrow dark eye line borders cap. **Call** is a repeated double note like the squeak of a rubber duck. Feeding flocks also give twittering, chirping, and talky *bit bit bit* calls.
Range: Fairly common; found in pine woodlands. Accidental north to Wisconsin, Ohio, and New Jersey.

BROWN
CREEPER

eastern
carolinensis

WHITE-BREASTED
NUTHATCH

Great Basin
♂ *tenuissima*

RED-BREASTED
NUTHATCH

PYGMY
NUTHATCH

BROWN-HEADED
NUTHATCH

WRENS \text{(Family Troglodytidae)}

Found throughout most of North America, wrens are chunky birds with slender, slightly curved bills. Tails are often uptilted. Loud song and vigorous territorial defense belie the small size of most species.
***Species:** 76 World, 9 N.A.*

HOUSE WREN
Troglodytes aedon | L 4¾" (12 cm)

Brown above with faint eyebrow. Separated from Winter Wren by longer tail, less prominent barring on belly, and larger overall size. Western *parkmanii* breeds east to Ontario; grayer above, paler below. **Juvenile** shows a bright rufous rump and darker buff below. Birds from mountains of southeastern Arizona, formerly known as "**Brown-throated Wren**," have a slightly buffier throat and breast and a bolder eyebrow. Exuberant **song** is a cascade of bubbling whistled notes.
RANGE: Common in shrubs, farms, gardens, parks. Winters rarely north into summer range.

WINTER WREN
Troglodytes troglodytes | L 4" (10 cm)

Stubby tail; dark barring on belly. Widespread eastern subspecies *hiemalis* from the north breeds west to northeastern British Columbia; similar *pullus* (not shown) breeds in the Appalachians. Western subspecies *pacificus* is richer buff on throat and breast, darker on back. Races from Bering Sea islands and Aleutians are larger, paler, and longer billed. Eastern subspecies give a *kelp-kelp* **call** like Song Sparrow; all western races a *timp-timp* call like Wilson's Warbler. **Song**, a rapid series of melodious trills, is much faster in western birds.
RANGE: Rather secretive, nests in dense brush, especially along stream banks, in moist coniferous woods; in winter may be found in any type of woodland.

CAROLINA WREN
Thryothorus ludovicianus | L 5½" (14 cm)

Deep rusty brown above, warm buff below; white throat and prominent white eye stripe. Vivacious, melodious **song**, a loud, clear *teakettle tea-kettle teakettle* or *cheery cheery cheery*. Sings any time of day or year.
RANGE: Common in the concealing underbrush of moist woodlands and swamps, wooded suburbs, and gardens. Nonmigratory, but after mild winters resident populations expand north of mapped range. After harsh winters, range limits retract. Casual to Colorado, New Mexico, and Arizona.

BEWICK'S WREN
Thryomanes bewickii | L 5¼" (13 cm)

Long, sideways-flitting tail, edged with white spots; long white eyebrow. Subspecies differ mainly in dorsal color: Eastern *bewickii*, is reddish brown above; south Texas *cryptus* (not shown) duller, but still tinged red. Widespread *eremophilus* of the western interior is the grayest; western coastal races grow browner and darker as one travels north. Northwest *calaphonus* (not shown) is dark, richly colored, with a rufous cast. **Song** variable, a high, thin buzz and warble, similar to Song Sparrow. Calls include a flat, hollow *jip*.
RANGE: Found in brushland, hedgerows, stream edges, open woods, and clear-cuts in the East. Sharply declining east of the Rockies, especially east of the Mississippi.

\text{---} 342 \text{---}

"Brown-throated Wren"
southeast Arizona

juvenile

western
parkmanii

HOUSE
WREN

eastern
aedon

Aleutians

WINTER
WREN

western
pacificus

CAROLINA
WREN

eastern
hiemalis

western interior
eremophilus

eastern
bewickii

BEWICK'S WREN

CACTUS WREN *Campylorhynchus brunneicapillus* | L 8½" (22 cm)

Large; dark crown, streaked back, heavily barred wings and tail, broad white eyebrow. Breast is densely spotted with black; threatened Californian subspecies, *sandiegense*, is less densely spotted. **Song**, heard all year, is a low-pitched, harsh, rapid *cha cha cha cha cha*.

RANGE: Common in cactus country and arid hillsides and valleys. Bulky nests are tucked into the protective spines of cholla cactus or thorny bushes.

ROCK WREN *Salpinctes obsoletus* | L 6" (15 cm)

Dull gray-brown above with contrasting cinnamon rump, buffy tail tips, broad blackish tail band. Breast finely streaked. Frequently bobs its body, especially when alarmed. **Song** is a variable mix of buzzes and trills; **call**, a buzzy *tick-ear*.

RANGE: Fairly common in arid and semiarid habitats, sunny talus slopes, scrublands, and dry washes. Casual in fall and winter to the East.

CANYON WREN *Catherpes mexicanus* | L 5¾" (15 cm)

White throat and breast, chestnut belly. Long bill aids in extracting insects from deep crevices. Loud, silvery **song**, a decelerating, descending series of liquid *tee* and *tew* notes. Typical **call** is a sharp *jeet*.

RANGE: Common in canyons and cliffs, often near water; may also build its cup nest in stone buildings and chimneys.

MARSH WREN *Cistothorus palustris* | L 5" (13 cm)

Much plumage variation in eastern and western races. Where ranges overlap on Great Plains, eastern birds darker, more richly colored, with black-and-white speckled neck; western birds duller, with brownish smudges on neck. **Songs** more liquid in the East; harsher and much more variable in the West. Alarm **call**, a sharp *tsuk*, often doubled.

RANGE: Common in reedy marshes and cattail swamps. Football-shaped nest attached to reeds above water.

SEDGE WREN *Cistothorus platensis* | L 4½" (11 cm)

Crown and back streaked; eyebrow whitish and indistinct; underparts largely buff. **Song** begins with a few single notes followed by a weak staccato trill or chatter; **call** note, a rich *chip*, often doubled. Globular nest similar to that of Marsh Wren.

RANGE: Found in wet meadows or sedge marshes. Generally common but local; uncommon to rare in the East. Rare and local in winter to New Mexico. Casual to California in late fall.

DIPPERS (FAMILY CINCLIDAE)

Aquatic birds that wade and even swim underwater in clear, rushing mountain streams to feed.
Species: 5 World, 1 N.A.

AMERICAN DIPPER *Cinclus mexicanus* | L 7½" (19 cm)

Adult sooty gray; dark bill; tail and wings short. **Juvenile** has paler, mottled underparts and pale bill. **Song**, loud, musical, wrenlike.

RANGE: Found along mountain streams. Descends to lower elevations in winter; casual vagrant well outside mapped range.

CACTUS
WREN

ROCK
WREN

CANYON
WREN

SEDGE
WREN

MARSH
WREN

AMERICAN
DIPPER

juvenile

KINGLETS (Family Regulidae)

Small, active birds that often hover to feed. **Species:** *5 World, 2 N.A.*

GOLDEN-CROWNED KINGLET *Regulus satrapa* | L 4" (10 cm)

Orange crown patch of **male** is bordered in yellow and black; **female**'s crown is yellow. Head pattern and paler underparts are unlike Ruby-crowned Kinglet. **Call** is a series of high, thin *tsee* notes. **Song**, almost inaudibly high, is a series of *tsee* notes accelerating into a trill. **RANGE:** Common in coniferous woodlands.

RUBY-CROWNED KINGLET *Regulus calendula* | L 4¼" (11 cm)

Male's red crown patch seldom visible; dusky underparts. Compare carefully with Golden-crowned Kinglet. Active; flicks wings rapidly. **Calls** include a scolding *je-ditt*. **Song**, several high, thin *tsee* notes followed by descending *tew* notes, ends with warbled three-note phrases. **RANGE:** Common in woodlands and thickets.

OLD WORLD WARBLERS
AND GNATCATCHERS (Family Sylviidae)

Old World Warblers are a large, diverse group. Sexes differ little, except in genus Sylvia. *Gnatcatchers are found only in the New World.* **Species:** *279 World, 12 N.A.*

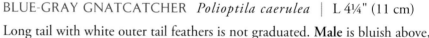

BLUE-GRAY GNATCATCHER *Polioptila caerulea* | L 4¼" (11 cm)

Long tail with white outer tail feathers is not graduated. **Male** is bluish above, in **breeding** plumage has black line on sides of crown. **Female** is grayer. **Call** is a querulous *pwee*. **RANGE:** Active; favors woodlands, thickets, and chaparral.

BLACK-CAPPED GNATCATCHER *Polioptila nigriceps* | L 4¼" (11 cm)

Separated from Blue-gray and Black-tailed Gnatcatchers by more graduated white outer tail feathers and longer bill; **breeding male**'s black cap extends below eye. **Female** and winter male best identified by tail shape and pattern and by voice. **Calls** like California Gnatcatcher or Bewick's Wren.
RANGE: West Mexican species, very rare, in southeastern Arizona.

BLACK-TAILED GNATCATCHER *Polioptila melanura* | L 4" (10 cm)

White terminal spots on graduated tail feathers; short bill. **Breeding male** has glossy black cap, contrasting with eye ring. **Female** washed with brown. **Calls** include rasping *cheeh* and hissing *ssheh*; **song** is a rapid series of *jee* notes.
RANGE: Desert resident; partial to washes.

CALIFORNIA GNATCATCHER *Polioptila californica* | **T** | L 4¼" (11 cm)

Similar to Black-tailed, but is darker with less white in outer tail feathers, less distinct eye ring. **Call**, a rising and falling, kitten-like *zeeer*; **song**, a series of *jzer* or *zew* notes.
RANGE: Local resident in sage scrub of southwest California. Dark northern nominate race now threatened, due to habitat destruction.

GOLDEN-
CROWNED
KINGLET
satrapa

RUBY-CROWNED
KINGLET
calendula

BLUE-GRAY
GNATCATCHER
caerulea

breeding ♂

BLACK-CAPPED
GNATCATCHER

breeding ♂

BLACK-TAILED
GNATCATCHER

breeding ♂

breeding ♂

CALIFORNIA
GNATCATCHER
californica

LANCEOLATED WARBLER
Locustella lanceolata | L 4½" (11 cm)

Resembles Middendorff's Grasshopper-Warbler but smaller; less broadly streaked above, including crown and rump, but streaks extend to feather tips; clear brown fringe on tertials. Breast, undertail coverts, and flanks are streaked. Highly secretive, keeps well concealed. Walks and runs; flicks its wings. Distinctive **call**, a metallic *rink-tink-tink*, delivered infrequently; also an explosive *pwit* and excited *chack* when disturbed. **Song**, a thin, insectlike reeling sound, like a fishing line makes.

RANGE: Mainly Asian species. Many occurred in spring and summer of 1984 on Aleutian island of Attu. Accidental in fall in California.

MIDDENDORFF'S GRASSHOPPER-WARBLER
Locustella ochotensis | L 6" (15 cm)

Big, chunky warbler with whitish-tipped, wedge-shaped tail; hefty bill. Indistinct dark markings above; yellowish buff below with a faintly streaked breast, rustier above. By late spring, underparts are mostly whitish and lack streaking. Like all *Locustella* warblers, Middendorff's is very secretive.

RANGE: East Asian species, casual migrant on westernmost Aleutians in fall; three spring to summer records. Recorded on St. Lawrence and Nunivak Islands and Pribilofs.

DUSKY WARBLER
Phylloscopus fuscatus | L 5½" (14 cm)

Dusky brown, not greenish, upperparts and lack of wing bar distinguish this species from Arctic Warbler; tail slightly rounded; bill shorter and thinner than Arctic. Usually has dark, slender legs; long distinct eyebrow; dull white in front of eye; broad, dark brown eye line; faint whitish eye ring. Underparts creamy white, with buffy brown wash on flanks and undertail coverts. **Calls** include a hard, sharp *tschick*, not unlike *chip* note of Lincoln's Sparrow. Constantly flicks wings.

RANGE: Asian species, casual on islands off western Alaska, and in fall off south coastal Alaska and in California.

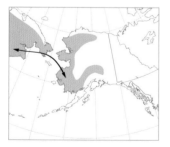

ARCTIC WARBLER
Phylloscopus borealis | L 5" (13 cm)

Long, yellowish white eyebrow, often curving upward behind eye; straw-colored legs and feet. Broad, dark eye line, mottled ear patches. Has square tail, olive upperparts, and pale wing bar on tips of greater coverts; faint second wing bar. Olive wash on sides and flanks; long primary projection. Stout bill is thicker, straighter than Orange-crowned Warbler (page 376); lacks streaking below. Compare also to accidental Willow Warbler (page 476) and Wood and Yellow-browed Warblers (page 477). Larger specimens taken in the western Aleutians with larger bills are either nominate race *borealis*, with paler underparts, or yellower *xanthrodyas* (not shown). Alaskan race, *kennicotti*, is smaller, with smaller bill than of either the Asian races. Arctic Warbler's **song** is a long, loud series of toneless buzzy notes. **Calls** include a buzzy *dzik*.

RANGE: Fairly common in western and central Alaska; four fall records for coastal California; nests on grassy tundra or in willow thickets.

adult

LANCEOLATED
WARBLER

fall

MIDDENDORFF'S
GRASSHOPPER-WARBLER

spring

DUSKY
WARBLER

fall
kennicotti

ARCTIC
WARBLER

spring
kennicotti

borealis

OLD WORLD FLYCATCHERS (Family Muscicapidae)

Short-legged birds that perch upright and obtain insects primarily through fly-catching.
May flick wings or tail. Species of genus Ficedula *nest in cavities; genus* Muscicapa *build exposed nests.*
Not related to New World tyrant flycatchers. **Species:** *275 World, 7 N.A.*

NARCISSUS FLYCATCHER
Ficedula narcissina | L 5¼" (13 cm)

Adult male overall black and yellow-orange; most orange on eyebrow and throat. Has yellow rump; white patch on inner secondary coverts. **First-spring male** similar, but duller. **Female** drab; brownish olive above, with green on rump; contrasting reddish-tinged uppertail coverts and tail; whitish throat; brownish mottling on breast. First-fall male similar to female.
RANGE: East Asian species; two spring records of males on Attu in the western Aleutians.

DARK-SIDED FLYCATCHER
Muscicapa sibirica | L 5¼" (13 cm)

Dark grayish brown upperparts and wash on sides and flanks; center of breast diffusely streaked. Whitish half collar; brownish supraloral spot; short bill; long primary projection; dark centers on undertail coverts may be concealed. Northern nominate race, *sibirica*, is darker and more diffusely streaked than southern races.
RANGE: Asian species; casual to western Aleutians; four spring records for Pribilofs; one fall record for Bermuda.

TAIGA FLYCATCHER
Ficedula albicilla | L 5¼" (13 cm)

Distinct white oval patches at base of outer tail feathers visible in flight, barely visible on folded tail from below; prominent eye ring. **Breeding male** with reddish throat. **Females** and winter males have whitish throats; grayish wash on breast. All show extensive patch of black on uppertail coverts. Formerly considered a subspecies of Red-breasted Flycatcher (*F. parva*). Perches low; often drops to ground to catch prey, then returns to perch. Frequently flicks tail up while giving rattled *trrt* call; also a metallic *tic* and harsh *ze-it*.
RANGE: Asian species; casual in late spring to western Aleutians; one spring and one fall record on St. Lawrence Island.

GRAY-STREAKED FLYCATCHER
Muscicapa griseisticta | L 6" (15 cm)

Larger and with smaller head than Dark-sided; primary projection longer. Note distinctive, but variable, streaking below; paler supraloral spot; more distinct submoustachial stripe usually shows some markings; undertail coverts white.
RANGE: East Asian species; casual to western Aleutians in late spring and fall; also several Pribilof records.

ASIAN BROWN FLYCATCHER
Muscicapa dauurica | L 5¼" (13 cm)

Grayish brown above; largely whitish below; grayish wash across chest or, rarely, some diffuse streaks. Bill larger than Dark-sided or Gray-streaked, extensively flesh-colored at base of lower mandible; primary projection shorter; supraloral area paler.
RANGE: Asian species; two spring records, for Attu Island, and St. Lawrence Island.

NARCISSUS
FLYCATCHER

adult ♂

1st spring ♂

spring ♀

♀

breeding
adult ♂

♀

1st fall

DARK-SIDED
FLYCATCHER
sibirica

spring

1st fall

TAIGA
FLYCATCHER

spring ♀

spring

GRAY-STREAKED
FLYCATCHER

1st fall

1st fall

spring

ASIAN BROWN
FLYCATCHER
dauurica

THRUSHES (Family Turdidae)

Eloquent songsters of many habitats. With narrow, notched bills, they feed on insects and fruit.
Species: *165 World, 29 N.A.*

SIBERIAN RUBYTHROAT

Luscinia calliope | L 6" (15 cm)

Male has a ruby red throat and broad, white submoustachial stripe. **Females** have white throats, often with some pink on adults and buffy on immatures; compare with smaller Bluethroat, which has rufous tail patches, dark breast band, paler underparts.

RANGE: Asian species; rare spring and fall migrant on western Aleutians, very rare on Pribilofs, casual on St. Lawrence Island.

BLUETHROAT

Luscinia svecica | L 5½" (14 cm)

Colorful throat pattern distinguishes **breeding male** from all other birds. In all plumages, rufous patches at base of tail are conspicuous in flight, which is low off the ground. In **female** and immature, note dark breast band. Bluethroat runs on ground, usually with tail cocked. Generally furtive, but in courtship males sing from high perches and in elaborate display flight. Varied, melodious **song** often begins with a crisp, metallic *ting ting ting*; **call**, *tchak*, often given in a series.

RANGE: Uncommon; nests in tundra thickets near water. Regular migrant on St. Lawrence Island; casual on Pribilofs and western Aleutians.

RED-FLANKED BLUETAIL

Tarsiger cyanurus | L 5½" (14 cm)

Note bluish tail, often flicked down; orangish flanks; colors subdued on female. **Adult male** has bright blue upperparts, but much individual variation; brightest birds may be several years old. Immature males closely resemble **females** until second fall. Rather secretive. **Calls** include a *hueet* and dry *keck-keck*.

RANGE: Primarily Asian species; casual to western Aleutians and Pribilofs, mainly in spring; one late fall record for Farallones, off California.

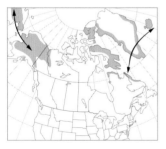

NORTHERN WHEATEAR

Oenanthe oenanthe | L 5¾" (15 cm)

Tail pattern distinctive: white rump, tail with dark central and terminal band. Greenland race, *leucorhea*, averages a little larger, is richer buff below; western birds are whitish, with a buff tinge. **Males** in fall and winter resemble females. Active; bob their tails. **Calls** include *chak* and whistled *wheet*, often combined. **Song**, a scratchy warbling mixed with call notes, often given in flight with tail spread.

RANGE: Prefers open, stony habitats. Uncommon; very rare along Atlantic coast during fall; casual elsewhere.

STONECHAT

Saxicola torquatus | L 5¼" (13 cm)

All records from the eastern *maurus* group of races, known as "Siberian Stonechat." Compact body; pale spot on inner coverts; paler rump. Note pattern of black on head of **adult male** obscured by fresh pale feather tips in **fall**; orange-buff wash on breast; extensive white on sides of neck, belly, and rump. **Female** and **first-fall male** have pale throat; pale buffy rump.

RANGE: Eurasian species; casual from scattered locations in Alaska; accidental in fall from New Brunswick and California. Favors open country.

SIBERIAN
RUBYTHROAT

adult ♂

1st fall ♂

pink-throated
adult ♀

♀

BLUETHROAT
svecica

1st fall ♀

♀

♀

breeding ♂

winter
adult ♂

RED-FLANKED
BLUETAIL
cyanurus

♀

adult ♂

breeding adult ♂
oenanthe

spring ♀
oenanthe

fall adult ♂
oenanthe

NORTHERN
WHEATEAR

♀

1st fall
leucorhoa

breeding adult ♂
leucorhoa

spring ♀

STONECHAT
maurus group

1st fall

spring ♂

fall adult ♂

EASTERN BLUEBIRD
Sialia sialis | L 7" (18 cm)

Chestnut throat, sides of neck, breast, sides and flanks; contrasting white belly, white under-tail coverts. **Male** is uniformly deep blue above; **female** grayer. The subspecies resident in the mountains of southeastern Arizona, *fulva*, is paler overall. All subspecies distinguished from Western Bluebird by chestnut on throat and sides of neck and cinnamon flanks that sharply contrast with the white, not grayish, belly and undertail. **Call** note is a musical, rising *chur-lee*, extended in **song** to *chur chur-lee chur-lee*.
RANGE: Found in open woodlands, farmlands, and orchards. Nests in holes in trees and posts; also in nest boxes. Serious decline in recent decades was due largely to competition with starling and House Sparrow for nesting sites. The provision of specially designed boxes by concerned conservationists has resulted in a promising comeback.

WESTERN BLUEBIRD
Sialia mexicana | L 7" (18 cm)

Male's upperparts and throat are deep purple-blue; breast, sides, and flanks chestnut; belly and undertail coverts grayish. Most birds show some chestnut on shoulders and upper back. **Female** duller, brownish gray above; breast and flanks tinged with chestnut, throat pale gray. **Call** note is a mellow *few*, extended in brief **song** to *few few fawee*.
RANGE: Nests in holes in trees and posts; also in nest boxes. Common in woodlands, farm-lands, orchards; in desert areas during winter, found in mesquite-mistletoe groves.

MOUNTAIN BLUEBIRD
Sialia currucoides | L 7¼" (18 cm)

Male is sky blue above, paler below, with whitish belly and undertail coverts. **Female** is brownish gray overall, with white belly and undertail coverts; white edges on coverts give folded wing a scalloped look. In fresh fall plumage, female's throat and breast are tinged with red-orange; brownish rear flank contrasting with white undertail coverts distinguishes her from female Eastern Bluebird, which has reddish flank. Note also longer, thinner bill and longer primary tip projection of Mountain Bluebird. **Call** is a thin *few*; **song**, a low, warbled *tru-lee*. More often than other bluebirds, hovers above prey, chiefly insects, before dropping to catch them; also catches insects in flight.
RANGE: Nests in tree cavities and buildings. Inhabits open rangelands, meadows, gener-ally at elevations above 5,000 feet; in winter, found primarily in open lowlands and desert. Highly migratory; casual in the East during migration and winter.

TOWNSEND'S SOLITAIRE
Myadestes townsendi | L 8½" (22 cm)

Large and slender; gray overall, with bold white eye ring. Buff wing patches and white outer tail feathers are most conspicuous in flight. **Call** note is a high-pitched *eek*; **song**, heard all year, a loud, complex, melodious warbling. Often seen on a high perch, from which it sometimes fly-catches.
RANGE: Nests on the ground. Fairly common in coniferous forests on high mountain slopes; in winter, also in wooded valleys, canyons, wherever juniper berries are available. Highly migratory; casual in fall and winter in Midwest and Northeast.

juvenile

EASTERN
BLUEBIRD
sialis

♀

♂

southwestern
♂ *fulva*

WESTERN
BLUEBIRD

♀

♂

MOUNTAIN
BLUEBIRD

♀

♂

juvenile

TOWNSEND'S
SOLITAIRE

WOOD THRUSH *Hylocichla mustelina* | L 7¾" (20 cm)

Reddish brown above, brightest on crown and nape; rump and tail brownish olive. White eye ring conspicuous on streaked face. Large dark spots on whitish throat, breast, and sides. Loud, liquid **song** of three- to five-note phrases, each phrase usually ending with a complex trill. **Calls** include a rapid *pit pit pit*.
RANGE: Common in moist deciduous or mixed woods. Casual in the West.

VEERY *Catharus fuscescens* | L 7" (18 cm)

Reddish brown above, white below, with gray flanks, grayish face, incomplete and indistinct gray eye ring. Upperparts duller, breast more spotted in more westerly *salicicola* than in eastern *fuscescens*. **Song** is a descending series of *veer* notes; **call**, a sharp, descending, whistled *veer*.
RANGE: Fairly common; found in dense, moist woodlands and streamside thickets. Western birds limited by suitable riparian habitat. Casual in Southwest.

GRAY-CHEEKED THRUSH *Catharus minimus* | L 7¼" (18 cm)

Gray-brown above, with faint, incomplete eye ring. Dark spots on breast, which is usually less buffy than Swainson's; flanks brownish gray. Breeding *minimus* on Newfoundland can be warmer colored above, more like Bicknell's. Thin, nasal **song** is somewhat like Veery's, but first and last phrases drop, middle one rises; **call**, a sharp *pheu* similar to Veery's, but higher-pitched, not descending.
RANGE: Favors coniferous or mixed woodlands.

BICKNELL'S THRUSH *Catharus bicknelli* | L 6¼" (16 cm)

Identification of Bicknell's Thrush when not singing is very difficult due to variation within Gray-cheeked, of which it formerly was considered a subspecies. Bicknell's is smaller, warmer brown above, especially on tail; lower mandible has more yellow. **Song** usually comes in three parts, the first and last rising.

SWAINSON'S THRUSH *Catharus ustulatus* | L 7" (18 cm)

Brownish above, with buffy lores and bold buffy eye ring; bright buffy breast with dark spots; brownish gray sides and flanks. Pacific coast races such as *ustulatus* are reddish brown above, less distinctly spotted below; distinguished from *salicicola* race of Veery by face pattern, buffy brown sides and flanks, and voice. **Song** is an ascending spiral of varied flutelike whistles; common **call**, a liquid *whit* in Pacific coast races, a sharper *quirk* in others; at night a peeping *queep* is heard.
RANGE: Fairly common; found in moist woods and swamps.

HERMIT THRUSH *Catharus guttatus* | L 6¾" (17 cm)

Complete, often whitish eye ring; reddish tail. Upperparts vary from rich brown to gray-brown. Eastern races such as widespread *faxoni* have buff-brown flanks. Larger, paler western mountain races, such as *auduboni,* and smaller, darker north Pacific coast races, such as *guttatus,* have grayish flanks. **Song** is a serene series of clear, flutelike notes, the similar phrases repeated at different pitches. **Calls** include a deeper *chuck,* often doubled, and a whiny, upslurred *wee.*
RANGE: Fairly common; found in coniferous or mixed woodlands, and thickets.

WOOD THRUSH

VEERY

western
salicicola

eastern
fuscescens

1st fall
aliciae

GRAY-CHEEKED
THRUSH

aliciae

BICKNELL'S
THRUSH

minimus

Pacific coast
ustulatus

juvenile
faxoni

Alaska
incanus

faxoni

HERMIT
THRUSH

SWAINSON'S
THRUSH

swainsoni

auduboni

guttatus

VARIED THRUSH

Ixoreus naevius | L 9½" (24 cm)

Male has grayish blue nape and back, orange eyebrow; underparts orange with black breast band; buffy orange bar on underwing prominent in flight. **Female** distinguished from American Robin (next page) by orange eyebrow and wing bar, dusky breast band, and unmarked throat. **Juvenile** resembles female but has white belly, scalier-looking throat and breast. In a very rare variant morph, all orange color is replaced by white. **Call** is a soft, low *tschook*; **song**, a slow series of variously pitched notes, rapidly trilled.

RANGE: Common in dense, moist woodlands, especially coniferous forests. Generally feeds in trees. Very rare in winter as far east as New England and south to Virginia. Numbers vary from year to year in southern part of mapped winter range.

EYEBROWED THRUSH

Turdus obscurus | L 8½" (22 cm)

Brownish olive above, with distinct white eyebrow. Belly is white, sides pale buffy orange. **Male** has dark gray throat and breast; **female**'s throat is white and streaked; browner head shows little contrast with rest of upperparts. Wing linings pale gray. Flight **call** is a high, piercing, drawn-out *dzee*.

RANGE: Asian species. Regular spring migrant on the Aleutians, casual in fall; casual on St. Lawrence Island and in northern Alaska; accidental California (late spring).

DUSKY THRUSH

Turdus naumanni | L 9½" (24 cm)

White eyebrow conspicuous on blackish head. Upperparts strongly patterned, with rust-colored rump; wings extensively rust, underwing almost entirely rufous. Below, white edgings give a scaly look to dark breast and sides. Note also distinctive white crescent across breast. Female and immatures average duller overall. Several sight records of redder nominate race for western Alaska's islands. **Call** is a series of *shack* notes; also a shrill, wheezy *shrree* similar to European Starling.

RANGE: Asian species. Casual spring migrant on westernmost Aleutians; accidental on St. Lawrence Island, Point Barrow, and in winter south to coastal Washington.

FIELDFARE

Turdus pilaris | L 10" (25 cm)

Large thrush with gray head and rump contrast with purplish brown upper back, blackish tail. Below, dark arrowhead-shaped spots pattern the buffy breast and extend along sides. White wing linings flash in flight. **Song** is a noisy twittering; **call** note is a series of *shack* notes, like Dusky Thrush; also gives a thin *seeh*.

RANGE: Eurasian species. Breeds from Greenland to Russian Far East. Casual vagrant to Alaska and northeastern North America. Accidental to Ontario and Minnesota.

REDWING

Turdus iliacus | L 8¼" (21 cm)

Distinctive whitish to buffy eyebrow; boldly streaked below, with rusty red flanks; rusty red wing linings visible in flight. **Call** is a thin, penetrating *seeeh*, usually heard in flight; also a hard *kuk* note.

RANGE: Eurasian species; casual visitor to Newfoundland, mainly in winter; accidental south to Long Island, New York; once coastal Washington.

VARIED
THRUSH
meruloides

♀

♂

juvenile

EYEBROWED
THRUSH

♀

♂

DUSKY
THRUSH
eunomus

♀

♂

♂ *naumanni*

FIELDFARE

REDWING

immature

immature

AMERICAN ROBIN
Turdus migratorius | L 10" (25 cm)

Gray-brown above, with darker head and tail; bill yellow; underparts brick red; lower belly white. Most western birds paler and duller overall than eastern nominate *migratorius* (shown here), which breeds to western Alaska; in most, tail has white corners, visible in flight. Northwestern race, *caurinus*, is equally dark but lacks white tail spots; breeds north to southeast Alaska. **Juvenile**'s underparts are tinged with cinnamon, heavily spotted with brown. Loud, liquid **song**, is a variable *cheerily cheer-up cheerio*. **Calls** include a rapid *tut tut tut;* a high, thin *ssip* in flight.
RANGE: Common, widespread. Often seen on lawns, head cocked as it searches for earthworms; also eats insects and berries. Nests in shrubs, trees, on sheltered windowsills, and eaves. In winter, found in moist woodlands, swamps, suburbs, and parks. Numbers vary greatly from winter to winter.

WHITE-THROATED ROBIN
Turdus assimilis | L 9½" (24 cm)

Distinct white collar in front; white throat with dark brown streaking; head and upperparts brownish; often shows yellow orbital ring; underparts mostly gray. Compare to Clay-colored Robin's less marked throat; lack of collar; overall tawnier color; more extensively yellow bill. **Call** is a nasal *rreeuh*, often doubled.
RANGE: Tropical species; casual to southernmost Texas in winter.

RUFOUS-BACKED ROBIN
Turdus rufopalliatus | L 9¼" (24 cm)

Distinguished from American Robin by reddish brown back and wing coverts, uniformly gray head with no white around eye, and more extensively streaked throat. **Calls** include a plaintive, drawn-out, whistled *teeeuu*, a clucking series of *chuk* notes, and in flight, a high, thin *ssi*. Somewhat secretive; found in treetops and dense shrubbery.
RANGE: West Mexican species, very rare winter visitor to southern Arizona, casual from southern and southwestern Texas to southern California.

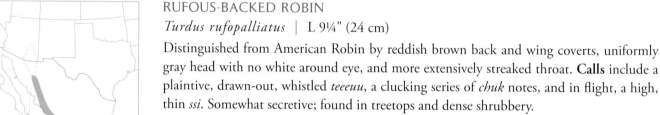

CLAY-COLORED ROBIN
Turdus grayi | L 9" (23 cm)

Brownish olive above; tawny buff below; pale buffy throat is lightly streaked with olive. Lacks white around eye conspicuous in American Robin. **Calls** include a slurred *reeeur-ee*, a clucking note, and, in flight, a high, thin *ssi*; **song** resembles American Robin's but is slower, clearer, much less varied.
RANGE: Species from east Mexico to north Colombia; rare visitor and very rare breeder in southernmost Texas. Rather secretive; forages in dense thickets, streamside brush, and woodlands. Will come to feeders.

AZTEC THRUSH
Ridgwayia pinicola | L 9¼" (24 cm)

Male is blackish brown above, with white patches on wings, white uppertail coverts; tail broadly tipped with white; contrasty underparts. **Female** is browner, more streaked on throat and breast. **Juvenile** is heavily streaked above with creamy white; underparts whitish and heavily scaled with brown. **Calls**, a quavering *wheeerr*, a metallic *wheer*, and a clear *sweee-uh*.
RANGE: Mexican species, rare and irregular to southeast Arizona, mainly late summer; casual to southwestern Texas and the central Texas coast.

juvenile

AMERICAN
ROBIN
migratorius

♀

♂

WHITE-
THROATED
ROBIN
suttoni

RUFOUS-
BACKED ROBIN
rufopalliatus

juvenile

AZTEC
THRUSH

♂

♀

CLAY-COLORED
ROBIN
tamaulipensis

MOCKINGBIRDS AND THRASHERS (FAMILY MIMIDAE)

Notable singers, unequaled in North America for the rich variety and volume of their song.
Some mimic the songs of other species. **Species:** *34 World, 12 N.A.*

GRAY CATBIRD *Dumetella carolinensis* | L 8½" (22 cm)

Plain dark gray with a black cap and a long, black tail, often cocked; undertail coverts chestnut. **Song** is a mixture of melodious, nasal, and squeaky notes interspersed with catlike *mew* notes; some are good mimics. Most readily identified by harsh, downslurred *mew* call; also gives a low *quirt* and a clucking noise.

RANGE: Generally common but rather secretive in thickets.

NORTHERN MOCKINGBIRD *Mimus polyglottos* | L 10" (25 cm)

White outer tail feathers and white wing patches flash in flight and in territorial and displays. **Song** is a mixture of original and imitative phrases, each repeated several times. Often sings at night. Imitates other species' songs and calls. Both sexes sing in fall, claiming feeding territories. **Call** is a loud, sharp *check*.

RANGE: Found in a variety of habitats, including towns. Casual well north of mapped range, as far as Alaska.

BAHAMA MOCKINGBIRD *Mimus gundlachii* | L 11" (28 cm)

Larger and browner than Northern Mockingbird, with streaking on neck and flanks; white only on tail tip. Lacks white patches on wings; flight more direct. **Song** is varied but not known to include imitations; **call** slightly harsher, more downslurred than Northern's.

RANGE: Caribbean species; casual south Florida.

BLUE MOCKINGBIRD *Melanotis caerulescens* | L 10" (25 cm)

Adult deep slaty blue with black mask and red eye. Immature slightly duller, brownish tinge to wings, darker eye.

RANGE: Mexican species which moves altitudinally. Casual in winter to southeast Arizona. Records from coastal southern California, New Mexico, and southern Texas of more questionable origin (though accepted).

BROWN THRASHER *Toxostoma rufum* | L 11½" (29 cm)

Reddish brown above, heavily streaked below. Immature's eyes darker. Compare to Wood Thrush (page 356). **Sings** a series of varied melodious phrases, each phrase usually given only two or three times. Seldom imitates other birds. **Calls** include a sharp *spuck* and a low *churr*.

RANGE: Common in hedgerows and woodland edges. Rare to West and Maritimes; casual to Newfoundland.

LONG-BILLED THRASHER *Toxostoma longirostre* | L 11½" (29 cm)

Closely resembles Brown Thrasher but much grayer above, with longer, more strongly curved bill; also has darker malar stripe, blacker streaking below, shorter primary projection. **Song** similar to Brown Thrasher. Gives *tsuck* **call** like Brown; other calls, a mellow *kleak*, and a loud, whistled *cheeooep*.

RANGE: Inhabits dense bottomland thickets. Very rare in west Texas; casual in New Mexico and Colorado.

GRAY
CATBIRD

NORTHERN
MOCKINGBIRD
polyglottos

juvenile

BAHAMA
MOCKINGBIRD
gundlachii

BLUE
MOCKINGBIRD

adult

BROWN
THRASHER
rufum

LONG-BILLED
THRASHER
sennetti

SAGE THRASHER *Oreoscoptes montanus* | L 8½" (22 cm)

Yellow eye, white wing bars, white-cornered tail. Grayish above, boldly streaked below. **Worn** late-summer birds show much less streaking, can resemble Bendire's Thrasher. Juvenile has streaked head and back. **Song** is a long series of warbled phrases. **Calls** include a *chuck* and a high *churr*.

RANGE: Found in sagebrush plains. Casual vagrant to eastern North America.

BENDIRE'S THRASHER *Toxostoma bendirei* | L 9¾" (25 cm)

Breast mottled; bill shorter and usually less curved than in Curve-billed Thrasher; base of lower mandible pale. White tail tips are similar to *oberholseri* race of Curve-billed. Distinctive arrowhead-shaped spots on breast are not present in **worn** summer plumage. **Song** is a sustained, melodic warbling, each phrase repeated one to three times.

RANGE: Uncommon; found in open farmlands, grasslands, and brushy desert. Casual to southern California coast in late summer, fall, and winter.

CURVE-BILLED THRASHER *Toxostoma curvirostre* | L 11" (28 cm)

Breast mottled; bill all-dark, longer, heavier, and usually more strongly curved than in Bendire's Thrasher. Breast spots indistinct in the westernmost race, *palmeri*. Race from extreme southeastern Arizona to south Texas, *oberholseri*, shows clearer spotting below; has pale wing bars; conspicuous white tips on tail. **Juveniles** have shorter bills. Distinctive **call**, a sharp upslurred *whit-wheet*, sometimes three-noted (*palmeri*) or even-pitched *whit-whit* (*oberholseri*). **Song** is elaborate and melodic, and includes low trills and warbles.

RANGE: Common in canyons, semiarid brushlands. Casual (*palmeri*) in southeastern California, the Great Plains, and upper Midwest.

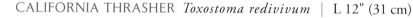

CALIFORNIA THRASHER *Toxostoma redivivum* | L 12" (31 cm)

Dark above, with pale eyebrow, dark eye, dark cheeks. Pale throat contrasts with dark breast; belly and undertail coverts tawny buff. Darker overall than Crissal Thrasher. **Calls** are a low, flat *chuck* and *chur-erp*. **Song** is loud and sustained, with mostly guttural phrases, often repeated once or twice. Imitates other species and sounds.

RANGE: Common in chaparral-covered foothills.

CRISSAL THRASHER *Toxostoma crissale* | L 11½" (29 cm)

Large and slender, with a distinctive chestnut undertail patch and a dark malar streak. **Song** is varied and musical, its cadence more leisurely than in Curve-billed Thrasher. **Calls** include a repeated *chideery* and a whistled *toit-toit-toit*.

RANGE: Very secretive, hiding in underbrush. Found mainly in dense mesquite and willows along streams and washes; sometimes on lower mountain slopes.

LE CONTE'S THRASHER *Toxostoma lecontei* | L 11" (28 cm)

Palest of the thrashers, with pale grayish brown upperparts, darker tail; tawny undertail coverts. Bill and eye are dark. **Song**, heard chiefly at dawn and dusk, is loud, melodious. **Calls** include an ascending, whistled *tweeep*.

RANGE: Prefers arid, sparsely vegetated habitats. Uncommon over most of range.

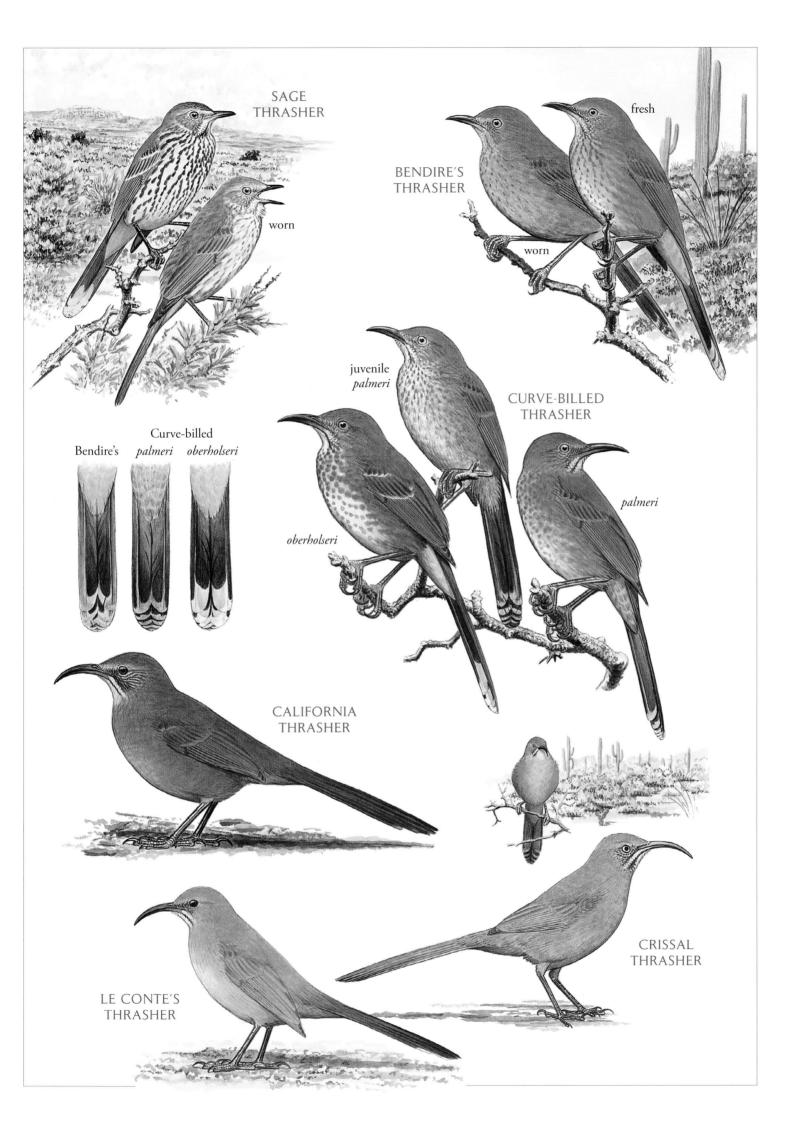

SAGE
THRASHER

worn

BENDIRE'S
THRASHER

fresh

worn

juvenile
palmeri

CURVE-BILLED
THRASHER

Bendire's

Curve-billed

palmeri   *oberholseri*

oberholseri

palmeri

CALIFORNIA
THRASHER

CRISSAL
THRASHER

LE CONTE'S
THRASHER

BULBULS (Family Pycnonotidae)

Noisy, active Old World family of the tropics and subtropics. **Species:** *118 World, 1 N.A.*

RED-WHISKERED BULBUL *Pycnonotus jocosus* | L 7" (18 cm)

Red ear spot and undertail coverts are distinctive; note conspicuous crest, long tail. **Juvenile** lacks red ear spot; undertail coverts are paler.

RANGE: Asian and African species. Escaped cage birds first noted in early 1960s in Miami, Florida; now established as a small population in suburbs and parklands south of Miami. Some also in Los Angeles area.

STARLINGS (Family Sturnidae)

Widespread Old World family. Chunky and glossy birds; most species are gregarious and bold.
Species: *115 World, 4 N.A.*

CRESTED MYNA *Acridotheres cristatellus* | L 9¾" (25 cm)

Identified by bushy crest on forehead, yellow bill and legs, white wing patch.

RANGE: Asian species, introduced in Vancouver, British Columbia, in the 1890s; population numbered in the thousands in the 1920s, but declined to fewer than 100 by early 1990s and only two by 2002; they perished in Feb. 2003. A likely factor for their extirpation was competition from European Starlings.

COMMON MYNA *Acridotheres tristis* | L 10" (25 cm)

Dark brown, with black head and white undertail coverts; yellow bill and skin around eye; white tail tip, patch at base of primaries, and wing linings distinctive in flight. Juveniles have more brownish heads. **Calls** include gurglings, whistles, and screeches.

RANGE: Southern Asian species; introduced elsewhere, including Hawaii, where it is common. Established in south Florida, where it is spreading. Found in urban areas; also open country in native range.

HILL MYNA *Gracula religiosa* | L 10½" (27 cm)

Glossy black; orange-red bill; yellow wattles and legs; white wing patch.

RANGE: Asian species, fine mimic, popular as a cage bird. A small number of escaped birds, first noted in 1960s, persists but is very local in Miami.

EUROPEAN STARLING *Sturnus vulgaris* | L 8½" (22 cm)

Adult in **breeding** plumage is iridescent black, with a yellow bill with blue base in male, pink in female. In fresh **fall** plumage, feathers are tipped with white and buff, giving a speckled appearance; bill brownish. In flight, note short, square tail, stocky body, and short, broad-based, pointed wings that appear pale gray from below. **Juvenile** is gray-brown, with brown bill. **Call** notes include squeaks, warbles, chirps, and twittering; also imitates songs of other species.

RANGE: A Eurasian species introduced in New York in 1890-91, it soon spread across the continent. Abundant, bold, aggressive, it often competes successfully with native species for nest holes. Outside nesting season, usually seen in large flocks.

juvenile

RED-WHISKERED
BULBUL

CRESTED MYNA
cristatellus

COMMON
MYNA
tristis

HILL MYNA
intermedia

Brown-headed Cowbird
(for comparison)

EUROPEAN
STARLING
vulgaris

fall

winter

breeding ♂

juvenile

ACCENTORS (Family Prunellidae)

Small Eurasian family, most species found in mountainous country. One species strays to North America. ***Species:*** *13 World, 1 N.A.*

SIBERIAN ACCENTOR *Prunella montanella* | L 5½" (14 cm)

Bright tawny buff below; back and flanks are streaked. Dark crown, with broad gray median stripe. Note buffy eyebrow that broadens behind head; dark cheek patch with buff spots below; gray patch on side of neck. **Call** is a high, thin series of *see* notes.
RANGE: Casual mainly fall and winter to Alaska and the Northwest.

WAGTAILS AND PIPITS (Family Motacillidae)

Slender-billed birds. Most species pump their tails as they walk. Wagtail flight is strongly undulating. ***Species:*** *64 World, 10 N.A.*

EASTERN YELLOW WAGTAIL *Motacilla tschutschensis* | L 6½" (17 cm)

Olive above, yellow below; tail shorter than other wagtails. In **breeding** plumage, Alaskan nesting *tschutschensis* has a speckled breast band. Asian *simillima*, recognized by some, seen on Aleutians and Pribilofs, averages greener above, yellower below. **Females** duller, **immatures** whitish below. **Call**, a loud *tsweep*, similar to Eastern Kingbird's.
RANGE: Generally common on Alaska breeding grounds. Trans-Beringian migrant; casual fall migrant on California coast.

GRAY WAGTAIL *Motacilla cinerea* | L 7¾" (20 cm)

Gray above, with greenish yellow rump, yellow below; whitish tertial edges, but lacks wing bars. **Breeding male** has black throat. **Female** and winter birds have whitish throat, paler below. Distinctly longer tail and flesh-colored legs separate Gray from Yellow Wagtail. **Call**, a metallic *chink-chink*.
RANGE: Eurasian species. Very rare spring migrant on western Aleutians; casual on Pribilofs and St. Lawrence Island; accidental south to California.

WHITE WAGTAIL *Motacilla alba* | L 7¼" (18 cm)

Breeding adult (*ocularis*) has black nape, gray back; eye line, throat, bib, and usually chin, black. In flight, shows mostly dark wings. In breeding adult male *lugens* upperparts are black, wings mostly white; chin usually white. **Breeding adult female** *lugens* similar, but duller above; **winter adults** retain distinct wing pattern. In nominate *alba* face is white in all plumages. Juveniles of all races are brownish above with two faint wing bars. **Immature** closer to adult but retains most of juvenile wing; immature *ocularis* has darker bases to median coverts than *lugens,* but separation problematic. **Calls** include a two-note *chizzik* given in flight and a whistled *chee-wee* given from perch.
RANGE: Northeast Asian *ocularis* breeds sparingly in western Alaska; *lugens* breeds coastal East Asia (south of *ocularis*), has nested and hybridized with *ocularis* in western Alaska, was formerly treated as a separate species, the Black-backed Wagtail. Nominate *alba,* breeding as close as Iceland and Greenland, is accidental on Atlantic coast.

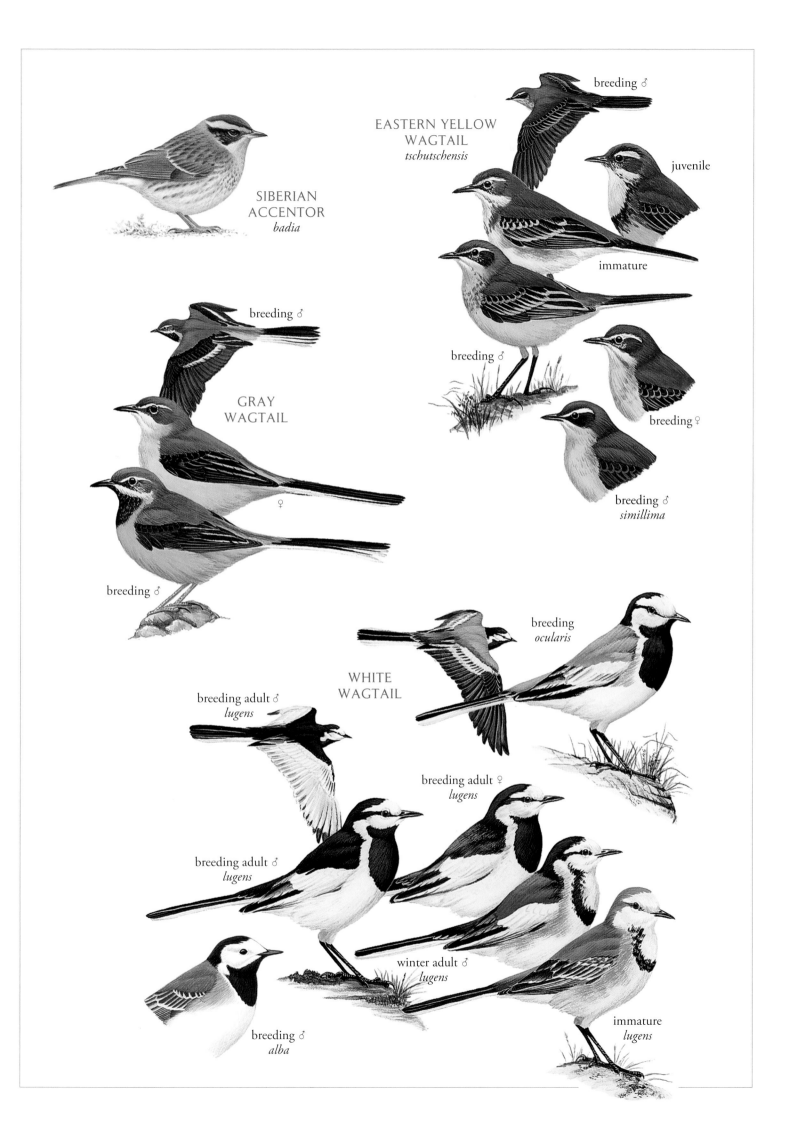

SIBERIAN
ACCENTOR
badia

EASTERN YELLOW
WAGTAIL
tschutschensis

breeding ♂

juvenile

immature

breeding ♂

breeding ♀

breeding ♂
simillima

GRAY
WAGTAIL

breeding ♂

♀

breeding ♂

WHITE
WAGTAIL

breeding
ocularis

breeding adult ♂
lugens

breeding adult ♀
lugens

breeding adult ♂
lugens

winter adult ♂
lugens

breeding ♂
alba

immature
lugens

AMERICAN PIPIT *Anthus rubescens* | L 6½" (17 cm)

Breeding birds are grayish above and faintly streaked below, except for the *alticola* race, from the Rockies and California's high mountains, which has richly colored underparts with fewer or no streaks. In **winter** American Pipit becomes browner above and more streaked below. Bill mostly dark; legs dark or tinged with pink. Tail has white outer feathers. An Asian subspecies, *japonicus*, rare in western Alaska, is more boldly streaked below, with pink legs, white wing bars; casual in fall in coastal California. **Call**, given in flight, is a sharp *pip-pit*; **song**, a rapid series of *chee* or *cheedle* notes.
RANGE: Common and widespread; nests on tundra in the far north, mountaintops farther south. Winter flocks are found in fields and on beaches.

SPRAGUE'S PIPIT *Anthus spragueii* | L 6½" (17 cm)

Dark eye prominent in pale buff face. Pale edges on rounded back feathers give a scaly look; rump is streaked. Underparts whitish, with a buffy wash and short, dark streaks on the breast. Legs pinkish. Outer tail feathers are more extensively white than in American Pipit. **Call** is a loud, squeaky *squeet,* usually given two or more times. **Song**, given continuously in high flight, is a descending series of musical *tzee* and *tzee-a* notes. Does not pump tail.
RANGE: Nests in grassy fields. Uncommon, secretive, and somewhat solitary. Very rare in fall and winter to California; accidental in eastern North America.

OLIVE-BACKED PIPIT *Anthus hodgsoni* | L 6" (15 cm)

Grayish olive back, faintly streaked. Eyebrow orange-buff in front of eye, white behind. Broken white stripe borders dark ear spot. Throat and breast rich buff, with rather large black spots on breast. Belly pure white; legs pink. **Call** is a buzzy tsee.
RANGE: Asian species. Rare migrant on western Aleutians; casual to Pribilofs and St. Lawrence Island; accidental to California and Nevada.

PECHORA PIPIT *Anthus gustavi* | L 5½" (14 cm)

Shows distinct primary projection. Resembles immature Red-throated Pipit, but compare Pechora Pipit's richly patterned back plumage, extending onto the nape; black centers with dull rufous edges contrast with white lines, or "braces," on the sides. Also a yellowish wash across the breast contrasts with the whitish belly. **Call** is a hard *pwit* or *pit*, but Pechora is often silent when flushed.
RANGE: Asian species; quite secretive. Casual in spring on the western Aleutians and St. Lawrence Island (mostly fall).

RED-THROATED PIPIT *Anthus cervinus* | L 6" (15 cm)

Note unpatterned nape. Pinkish red head and breast are distinctive in **breeding male**, less extensive in **breeding female** and fall adults. Fall **immatures** and some breeding females show no red. **Call**, given in flight, is a high, piercing *tseee*, dropping in pitch at the end. Loud, varied **song** is delivered from the ground or in song flight.
RANGE: Regular migrant on islands in Bering Sea; rare fall migrant along California coast; casual inland and in Northwest

AMERICAN
PIPIT
rubescens

winter

breeding

breeding
alticola

winter
japonicus

OLIVE-BACKED
PIPIT
yunnanensis

SPRAGUE'S
PIPIT

PECHORA
PIPIT
gustavi

RED-THROATED
PIPIT

breeding ♀

breeding ♂

immature

WAXWINGS (Family Bombycillidae)

Red, waxy tips on secondary wing feathers are often indistinct, and sometimes they are absent altogether.
All waxwings have sleek crests, silky plumage, and yellow-tipped tails. Where berries are ripening,
waxwings come to feast in amiable, noisy flocks. **Species:** *4 World, 2 N.A.*

BOHEMIAN WAXWING
Bombycilla garrulus | L 8¼" (21 cm)

Larger and grayer than Cedar Waxwing; underparts gray; undertail coverts cinnamon. White and yellow spots on wings. Individuals are sometimes seen in flocks of Cedar Waxwings. In flight, white wing patch at base of primaries is conspicuous. **Juvenile** browner above, streaked below, with pale throat. Distinctive **call**, a buzzy twittering, lower and harsher than call of Cedar Waxwing.

RANGE: Nests in open coniferous or mixed woodlands; often seen perched on top of a black spruce. Winter range varies widely and unpredictably; large flocks visit scattered locations, feeding on berries and small fruits. Irregular winter wanderer to the Northeast, usually in small numbers; annual in Maine, Maritimes, and Newfoundland.

CEDAR WAXWING
Bombycilla cedrorum | L 7¼" (18 cm)

Smaller and browner than Bohemian Waxwing; belly pale yellow; undertail coverts white. Lacks yellow spots on wings. **Juvenile** is streaked; lacks white wing patches of juvenile Bohemian. Since this species usually nests late in summer, juvenal plumage is seen well into fall. **Call** is a soft, high-pitched, trilled whistle.

RANGE: Found in open habitats where berries are available; also eats insects, flower petals, and sap. Highly gregarious in migration and winter.

SILKY-FLYCATCHERS (Family Ptilogonatidae)

This New World tropical family of slender, crested birds is closely related to the waxwings. The family's common
name describes their soft, sleek plumage and agility in catching insects on the wing. **Species:** *4 World, 2 N.A.*

PHAINOPEPLA
Phainopepla nitens | L 7¾" (20 cm)

Male is shiny black; white wing patch conspicuous in flight. In both sexes, note distinct crest, long tail, red eyes. Juvenile resembles **adult female**; but has browner eyes; both have gray wing patches. Young males acquire patchy black in fall. Distinctive **call** note is a low-pitched, whistled, querulous *wurp?* **Song** is a brief warble, seldom heard. Flight is fluttery but direct, and often very high.

RANGE: Phainopeplas nest in early spring in mesquite brushlands, feeding chiefly on insects and mistletoe berries. In late spring they move into cooler, wetter habitat and raise a second brood. Rare post-breeding wanderers north and east of mapped range. Casual to southern British Columbia and the Great Plains; accidental in eastern North America.

BOHEMIAN
WAXWING
pallidiceps

juvenile

CEDAR
WAXWING

juvenile

PHAINOPEPLA

♂

♀

♂

CHESTNUT-SIDED WARBLER
Dendroica pensylvanica | L 5" (13 cm)

Breeding male has yellow crown, black eye line, black whisker stripe; extensive chestnut on sides; **female** has greenish crown, less chestnut. Fall adults and **immatures** are lime green above, with white eye ring, whitish underparts, yellowish wing bars. **Song** is a whistled *please please pleased to meetcha*; **call**, a *chip* note like Yellow Warbler. Chestnut-sided often cocks its tail.

RANGE: Fairly common in second-growth deciduous woodlands. Rare migrant in West; casual in winter.

CAPE MAY WARBLER
Dendroica tigrina | L 5" (13 cm)

Most plumages have yellow on face, the color usually extending to sides of neck. Note also short tail; yellow or greenish rump; thin bill, slightly downcurved. **Breeding male**'s chestnut ear patch and striped underparts distinctive; wing patch white. **Female** drabber, grayer, with two narrow white wing bars. **Immature male**'s ear patch is less distinct. **Immature female** can be extremely drab, with gray face and only a tinge of yellow below and on rump; always has greenish edges on flight feathers. One **song** is a high, thin *seet seet seet seet*; **call**, a very high, thin *sip*.

RANGE: Cape May Warbler shows aggressive behavior. Breeds in black spruce forests, where often uncommon except during spruce budworm outbreaks. Rare west to Texas in migration; very rare to casual throughout the West. Winters chiefly in the West Indies; a few birds winter in southernmost Florida.

MAGNOLIA WARBLER
Dendroica magnolia | L 5" (13 cm)

Male is blackish above, with white eyebrow, white wing patch, yellow rump; broad white tail patches. Underparts yellow, streaked on breast and sides; undertail coverts white; undertail white except for black band at tip. Female has two wing bars; some **first-spring females** have dull white eye ring; often confused with rare Kirtland's Warbler (page 386). **Fall adults** and **immatures** are drabber, with grayish olive upperparts; white eye ring; faint gray band across breast. Compare immature Prairie Warbler (page 386). Magnolia Warbler does not bob tail. **Song** is a short, whistled *weety-weety-weeteo*.

RANGE: Fairly common to common in moist coniferous forests. Casual in winter in southern Florida. Rare throughout the West in migration.

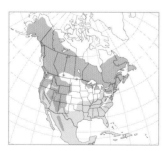

YELLOW-RUMPED WARBLER
Dendroica coronata | L 5½" (14 cm)

Yellow rump, yellow patch on side, yellow crown patch, white tail patches. In northern and eastern birds, "**Myrtle Warbler**," note white eyebrow, white throat and sides of neck, contrasting cheek patch. Western birds, "**Audubon's Warbler**," have yellow throat, except for a few immature females. Some males in the mountains of the Southwest show more black. All **females** and fall males are duller than **breeding males** but show same basic pattern. **Song**, a slow warble, usually rising or falling at the end in "Audubon's," a musical trill in one song of "Myrtle." **Call** note of "Myrtle" is lower, flatter.

RANGE: Abundant in coniferous or mixed woodlands. "Myrtle" is fairly common in winter on West Coast; "Audubon's" is casual in the East.

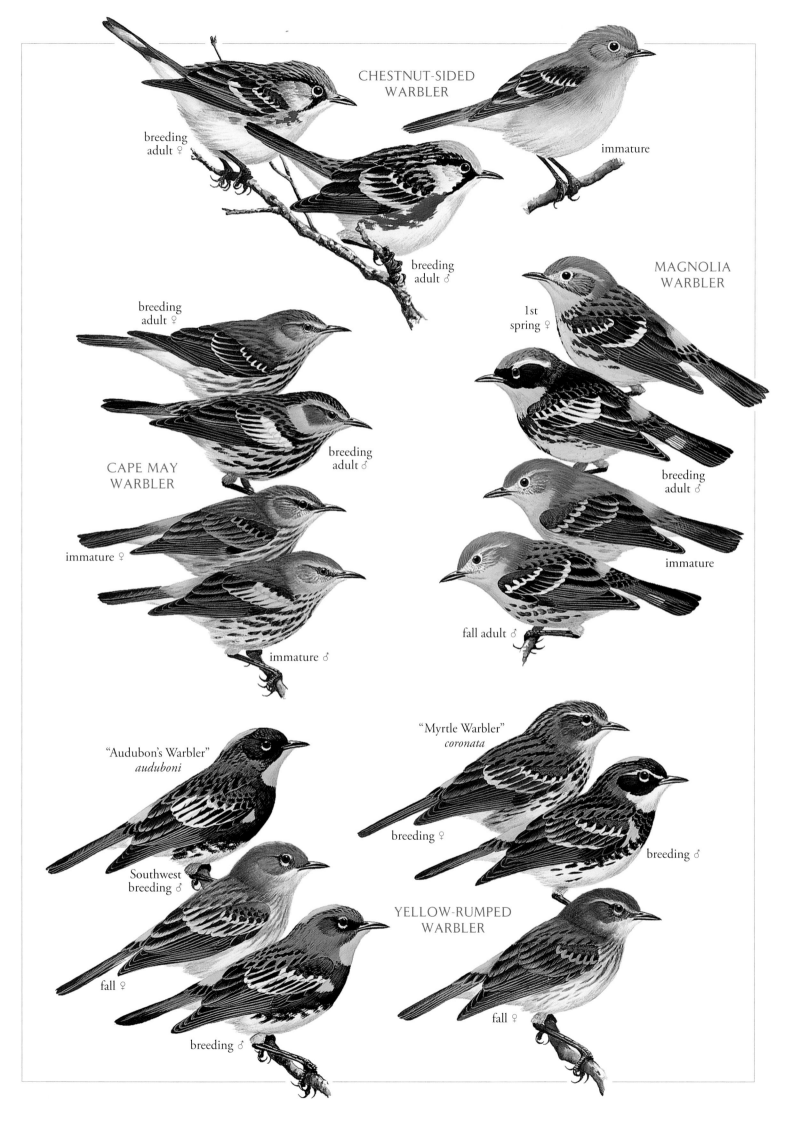

CHESTNUT-SIDED
WARBLER

breeding
adult ♀

breeding
adult ♂

immature

MAGNOLIA
WARBLER

1st
spring ♀

breeding
adult ♂

breeding
adult ♀

breeding
adult ♂

CAPE MAY
WARBLER

immature ♀

immature ♂

immature

fall adult ♂

"Myrtle Warbler"
coronata

"Audubon's Warbler"
auduboni

breeding ♀

breeding ♂

Southwest
breeding ♂

YELLOW-RUMPED
WARBLER

fall ♀

breeding ♂

fall ♀

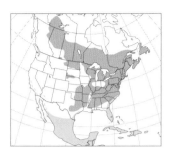

BLACK-AND-WHITE WARBLER
Mniotilta varia | L 5¼" (13 cm)

The only warbler that regularly creeps along branches and up and down tree trunks like a nuthatch. Boldly striped on head, most of body, and undertail coverts. **Male**'s throat and cheeks are black in breeding plumage; in winter, chin is white. **Female** and **immatures** have pale cheeks; female diffusely streaked on buffy flanks; buffy wash particularly bright on immatures. **Song** is a long series of high, thin *wee-see* notes. **Calls** include a sharp *chip* and high *seep-seep*.
RANGE: Common in mixed woodlands.

BLACK-THROATED BLUE WARBLER
Dendroica caerulescens | L 5¼" (13 cm)

Male's black throat, cheeks, and sides separate blue upperparts, white underparts. Bold white patch at base of primaries. Appalachian males south of Susquehanna drainage average darker above; back largely black in the case of *cairnsi* in southern Appalachians. **Female**'s pale eyebrow is distinct on dark face; upperparts brownish olive; underparts buffy; wing patch smaller, occasionally absent on immature females. Typical **song** is a slow series of four or five wheezy notes, the last note higher: *zwee zwee zwee zweeee* or a slower *zur zurr zreee*. **Call** is a single sharp *dit*, like the call of a Dark-eyed Junco.
RANGE: Inhabits deciduous forests; usually seen in lower or mid-level branches. Rare fall vagrant in the West. A few birds winter in south Florida; most migrate to the West Indies.

CERULEAN WARBLER
Dendroica cerulea | L 4¾" (12 cm)

Small, with short tail and two wide white wing bars. **Adult male** is bluish above with dark streaks; white below, with dark breast band and dark blue-gray streaking on sides. **Female** has greenish mantle, blue-green or bluish crown; pale eyebrow broadens behind the eye; breast and throat are pale yellowish. Immature male is like female, but shows some bluish and dark streaks above. **Song** is a short, fast, accelerating series of buzzy notes on one pitch, ending with a long, single buzz note.
RANGE: Sharply declining in the heart of its range. Found in tall trees in swamps, bottomlands, mixed woodlands near water. Fall migration begins from the second week of July. Range is expanding in Northeast.

BLACKBURNIAN WARBLER
Dendroica fusca | L 5" (13 cm)

Fiery orange throat, broad white wing patch, triangular ear patch, conspicuous in **adult male**. **Female** and immature male have paler throat, **immature female** paler still; note also the two white wing bars, streaked back, and bold yellow or buffy eyebrow, broader behind the eye, that curls around onto side of neck. Orange or yellow forehead stripe and white in outer tail feathers are distinct in all males, less so in females. One **song**, a short series of high notes followed by a squeaky, ascending trill, ends on a very high note.
RANGE: Fairly common in coniferous or mixed forests of northern breeding range; pine-oak woodlands in the Appalachians. Generally stays in the upper branches. Vagrants are seen rarely in coastal California in fall migration; casual elsewhere west of dashed line on map in spring and fall.

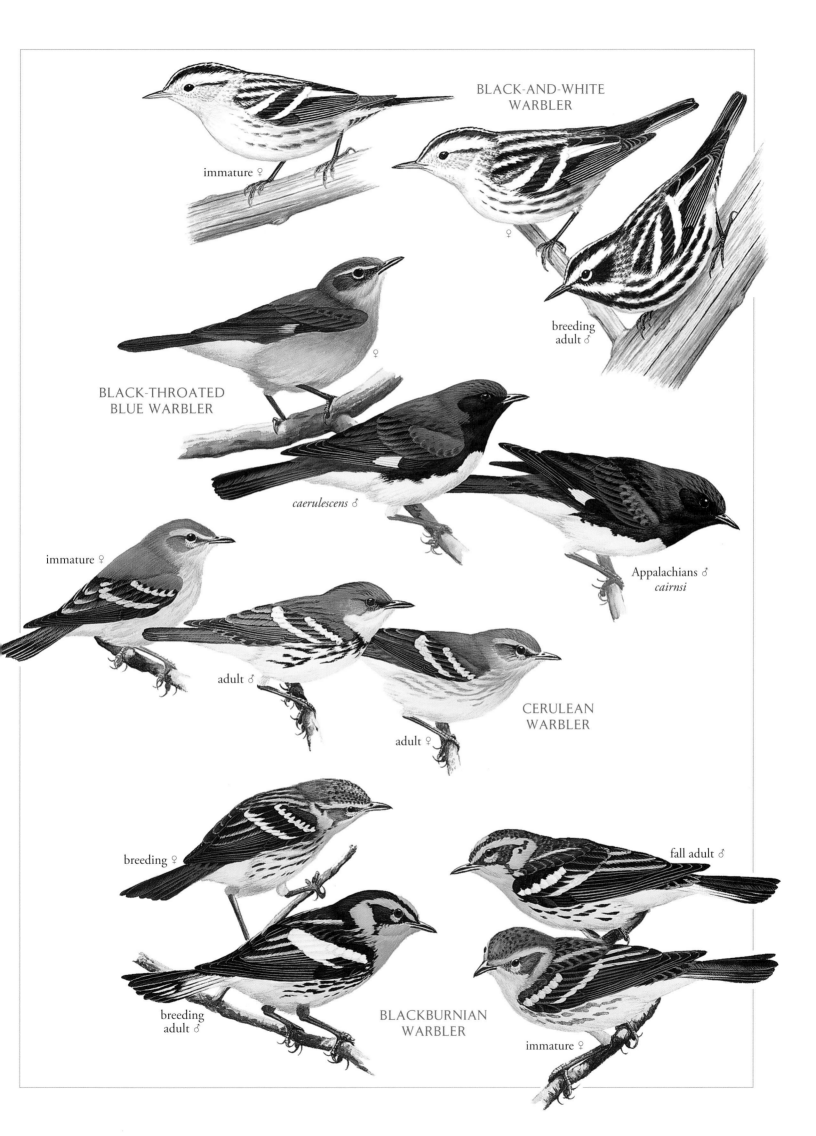

BLACK-AND-WHITE
WARBLER

immature ♀

♀

breeding
adult ♂

BLACK-THROATED
BLUE WARBLER

♀

caerulescens ♂

Appalachians ♂
cairnsi

immature ♀

adult ♂

adult ♀

CERULEAN
WARBLER

breeding ♀

fall adult ♂

breeding
adult ♂

BLACKBURNIAN
WARBLER

immature ♀

BLACK-THROATED GRAY WARBLER
Dendroica nigrescens | L 5" (13 cm)

Adult plumage is basically the same year-round: black-and-white head; gray back streaked with black; white underparts, sides streaked with black; small yellow spot between eye and bill. Lacks central crown stripe of the Black-and-white Warbler (preceding page); undertail coverts are white. Immature male resembles adult male; immature female is brownish gray above, throat white. Varied **songs** include a buzzy *weezy weezy weezy weezy-weet*.
RANGE: Inhabits woodlands, brushlands, chaparral. Rare in winter in lower Rio Grande Valley, Texas; casual otherwise in eastern North America. Very rare during migration and in winter along the Gulf Coast.

TOWNSEND'S WARBLER
Dendroica townsendi | L 5" (13 cm)

Dark crown, dark ear patch bordered in yellow. Olive above, streaked with black; yellow breast, white belly, yellowish black-streaked sides. **Adult male**'s throat and upper breast are black; **female** and immature male have streaked lower throat. **Immature female** is duller, lacks streaking on back; streaking on underparts is diffuse. Variable **song**, a series of hoarse *zee* notes. Frequently hybridizes with Hermit Warbler; **hybrids** usually have yellowish, streaked underparts of Townsend's, yellow head of Hermit.
RANGE: Found in coniferous forests. Casual in the East.

HERMIT WARBLER
Dendroica occidentalis | L 5½" (14 cm)

Yellow head, with dark markings extending from nape onto crown. **Male** has black chin and throat; in **female** and **immatures**, chin is yellowish, throat shows less or no dark color. Immature female is more olive above. **Song** is a high *seezle seezle seezle seezle zeet-zeet*.
RANGE: Fairly common in mountain forests; nests in tall conifers. During migration, also seen in lowlands. Casual in the East.

BLACK-THROATED GREEN WARBLER
Dendroica virens | L 5" (13 cm)

Bright olive green upperparts; yellow face with greenish ear patch. Underparts are white, tinged with yellow on sides of vent and often on breast. **Male** has black throat and upper breast and black-streaked sides. **Female** and **immatures** show much less black below; immature female generally has dark streaking only on sides. One **song** is a hoarse *zeee zeee zee-zo-zee*; the other, often written as *trees, trees, whispering trees*.
RANGE: Fairly common in coniferous or mixed forests in summer.

GOLDEN-CHEEKED WARBLER
Dendroica chrysoparia | **E** | L 5½" (14 cm)

Dark eye line, unmarked yellow ear patches, and lack of any yellow on underparts distinguish this species from similar Black-throated Green Warbler. **Male** black above, with black crown, black bib, black-streaked sides. **Female** and immature male duller, upperparts olive with dark streaks; chin yellowish or white; sides of throat streaked. **Immature female** shows less black on underparts. **Song**, *bzzzz layzee dayzee*, ends on a high note.
RANGE: Endangered; local in mature juniper-oak woodlands of the Edwards Plateau in central Texas. Winters from southern Mexico to Nicaragua.

BLACK-THROATED
GRAY WARBLER

♀

adult ♂

adult ♀

TOWNSEND'S
WARBLER

adult ♂

TOWNSEND'S x
HERMIT HYBRID

adult ♂

adult ♀

HERMIT
WARBLER

adult ♂

adult ♀

BLACK-THROATED
GREEN WARBLER

adult ♂

GOLDEN-CHEEKED
WARBLER

adult ♀

adult ♂

immature females

Golden-cheeked   Black-throated Green   Townsend's   Hermit

GRACE'S WARBLER
Dendroica graciae | L 5" (13 cm)

Black-streaked gray back; throat and upper breast bright yellow; rest of underparts white, with black streaks on sides; short bill; yellow eyebrow becomes white behind eye. **Female** slightly duller and browner above. Grace's Warbler's **song** is a rapid, accelerating trill.
RANGE: Inhabits coniferous or mixed forests of southwestern mountains, especially yellow pines. Usually forages high in the trees. Very rare to southern California.

YELLOW-THROATED WARBLER
Dendroica dominica | L 5½" (14 cm)

Plain gray back; large, white patch on each side of head. **Male** has black crown and face; in female, black is less extensive. Throat and upper breast bright yellow; rest of underparts white, with black streaks on sides; bold white eyebrow sometimes tinged with yellow. Eastern races *dominica* and *stoddardi* (of eastern Gulf Coast, not shown) have yellow supraloral area, unlike more westerly *albilora*; *stoddardi* and birds from Delmarva Peninsula have very long bills. **Song** is a series of clear, downslurred whistles ending with a rising note.
RANGE: Fairly common in live oak and pine woodlands, cypress, and sycamores. Usually forages high in the trees, creeping methodically along the branches. Very rare northward to southern Ontario in spring and Newfoundland in fall. Casual in West during migration.

KIRTLAND'S WARBLER
Dendroica kirtlandii | E | L 5¾" (15 cm)

Blue-gray above, strongly black-streaked on back; yellow below, streaked on sides; white eye ring, broken at front and rear; two whitish wing bars, thin and indistinct. Often confused with first-spring female Magnolia Warbler (page 380). **Adult female** is slightly duller; **immature female** brownish above. Kirtland's Warbler constantly wags its tail. **Song** is loud and lively, a variable series of low, sharp notes followed by slurred whistles.
RANGE: An endangered species: The annual breeding census counted 1,420 singing males in 2005, up from the historic low of 167 singing males in 1987. Nests in northern Michigan, where controlled plantings and fires help produce the required habitat: thickets of young jack pines. Very rare in summer outside Michigan, with records from southern Ontario and especially Wisconsin. Very rarely seen in migration. Winters in the Bahamas.

PRAIRIE WARBLER
Dendroica discolor | L 4¾" (12 cm)

Olive above, with faint chestnut streaks on back; bright yellow eyebrow, yellow patch below eye; bright yellow below, streaked with black on sides of neck and body. Two indistinct wing bars. **Female** and immature male are slightly duller. **Immature female** is duller still, grayish olive above; lack of complete eye ring or gray breast band distinguish her from fall Magnolia Warbler (page 380). Usually forages in lower branches and brush, twitching its tail. Distinctive **song**, a rising series of buzzy *zee* notes.
RANGE: Generally common in open woodlands, scrublands, overgrown fields, and mangrove swamps. Casual in the West, except in coastal California, where it is rare in fall. Also rare in fall to Maritimes and Newfoundland. Declining in upper Midwest.

YELLOW-THROATED
WARBLER

yellow-lored
dominica ♂

white-lored
albilora ♂

GRACE'S
WARBLER

♀

♂

KIRTLAND'S
WARBLER

immature ♀

adult ♀

adult ♂

adult ♀

PRAIRIE
WARBLER

adult ♂

immature ♀

BAY-BREASTED WARBLER
Dendroica castanea | L 5½" (14 cm)

Breeding male has chestnut crown, throat, and sides; black face; creamy patch at each side of neck; two white wing bars. **Female** is duller. **Fall adults** and **immatures** resemble Blackpoll Warbler and Pine Warbler. Bay-breasted is brighter green above, wing bars are thicker; underparts show little or no streaking and little yellow; flanks usually show some buff or bay color; legs usually entirely dark; undertail coverts are buffy or whitish. Short tail projection past undertail coverts for both Bay-breasted and Blackpoll, unlike Pine Warbler. **Song** consists of high-pitched double notes.
RANGE: Fairly common; nests in coniferous forests. Migrates earlier in fall than Blackpoll. Very rare in the West.

BLACKPOLL WARBLER
Dendroica striata | L 5½" (14 cm)

Solid black cap, white cheeks, and white underparts identify **breeding male**; back and sides boldly streaked with black. Compare with Black-and-white Warbler (page 382). **Female** is duller overall, variably greenish above and pale yellow below; some are gray; note streaking. **Fall adults** and immatures resemble Bay-breasted and Pine Warblers. Blackpoll is mostly pale greenish yellow below, with dusky streaking on sides; legs pale on front and back, dark on sides; undertail coverts long and usually white. **Song** is a series of high *tseet* notes.
RANGE: Common; nests in coniferous forests. Migrates later in fall than Bay-breasted Warbler. Rare migrant over much of West. Very rare in fall in most of South because much migration is off East Coast.

PINE WARBLER
Dendroica pinus | L 5½" (14 cm)

Relatively large bill; long tail projection past undertail coverts; throat color extends onto sides of neck, setting off dark cheek patch. **Male** is greenish olive above, without streaking; throat and breast yellow, with dark streaks on sides of breast; belly and undertail coverts white. **Female** is duller. **Immatures** are brownish or brownish olive above, with whitish wing bars and brownish tertial edges; male is dull yellow below, female largely white; both have brown wash on flanks. **Song** is a twittering musical trill, varying in speed.
RANGE: Common in pines in summer; also in mixed woodlands in winter. Casual to very rare in the West; rare to Newfoundland in fall.

PALM WARBLER
Dendroica palmarum | L 5½" (14 cm)

Upperparts olive. **Breeding adult** of eastern race, *hypochrysea*, has chestnut cap, yellow eyebrow, and entirely yellow underparts, with chestnut streaking on sides of breast. Fall adults and immatures lack chestnut cap and streaking; yellow is duller. Western nominate race, *palmarum*, has whitish belly and darker streaks on sides of breast; less chestnut. **Fall adults** and immatures are drab. **Song** is a rapid, buzzy trill.
RANGE: Fairly common; nests in brush at edge of spruce bogs. During migration and winter, found in woodland borders, open brushy areas, and marshes. Habitually wags its tail as it forages. Regular on West Coast in fall and winter (*palmarum*); *hypochrysea* casual to California.

fall ♂

BAY-BREASTED
WARBLER

breeding ♀

immature ♀

breeding
adult ♂

breeding ♀

breeding ♂

BLACKPOLL
WARBLER

fall

PINE WARBLER
pinus

adult ♀

immature ♀

adult ♂

immature ♂

eastern breeding
hypochrysea

PALM
WARBLER

western fall
palmarum

western breeding
palmarum

YELLOW WARBLER

Dendroica petechia | L 5" (13 cm)

Plump, yellow overall, with short tail; dark eye prominent in uniformly yellow face; reddish streaks below are distinct in **male**, faint or absent in **female**; **immatures** are duller. Much geographic variation: Northern races are greener above; southwest *sonorana* is pale with faint red streaks below; resident *gundlachi* of southernmost Florida is of the "Golden" group of the West Indian races. Note green crown and short primary projection. Resident subspecies in mangroves from Mexico south are known as "**Mangrove Warbler**"; adult males of most subspecies have chestnut heads; immatures of this and "**Golden**" are dull. **Song**, rapid, variable, is sometimes written *sweet sweet sweet I'm so sweet.*

RANGE: Favors wet habitats, especially willows and alders; open woodlands, orchards. A few "Mangroves" resident in mangroves in south coastal Texas.

MOURNING WARBLER

Oporornis philadelphia | L 5¼" (13 cm)

Lack of bold white eye ring distinguishes **adult male** from Connecticut Warbler. **Adult female** and especially **immatures** may show a thin, nearly complete eye ring, but compare with Connecticut. Immatures generally have more yellow on throat than MacGillivray's; compare also with female Common Yellowthroat (page 396). Immature males often show a little black on breast. Mourning Warblers hop rather than walk. **Call** is a flat, hollow *chip.* **Song** is a series of slurred two-note phrases followed by two or more lower phrases.

RANGE: Fairly common in dense undergrowth, thickets, moist woods; nests on the ground. Most spring migration is west of the Appalachians. Very rare in the West.

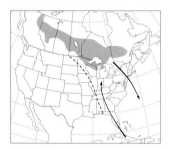

MACGILLIVRAY'S WARBLER

Oporornis tolmiei | L 5¼" (13 cm)

Bold white crescents above and below eye distinguish all plumages from male Mourning and all Connecticut Warblers. Crescents may be very hard to distinguish from the thin, nearly complete eye ring found on female and immature Mourning Warblers. Immature MacGillivray's Warblers generally have grayer throat than immature Mournings and a fairly distinct breast band above yellow belly. Field identification is often difficult. MacGillivray's hops rather than walks. **Call** is a sharp, harsh *tsik.* **Song** has two parts: a buzzy trill ending in a downslur.

RANGE: Fairly common; found in dense undergrowth; nests on the ground. Casual to East.

CONNECTICUT WARBLER

Oporornis agilis | L 5¾" (15 cm)

Large eye with bold white eye ring conspicuous on **male**'s gray hood and **female**'s brown or gray-brown hood. Eye ring is sometimes slightly broken on one side only. **Immature** has a brownish hood and brownish breast band. A large, stocky warbler, noticeably larger than Mourning and MacGillivray's Warblers. Like Mourning, long undertail coverts give Connecticut a short-tailed, plump appearance. Walks rather than hops. Loud, accelerating **song** repeats a brief series of explosive *beech-er* or *whip-ity* notes.

RANGE: Uncommon; found in spruce bogs, moist woodlands; nests on the ground; generally feeds on the ground or on low limbs. Spring migration is almost entirely west of the Appalachians. Fall migrants uncommon in the East; very rare in the West.

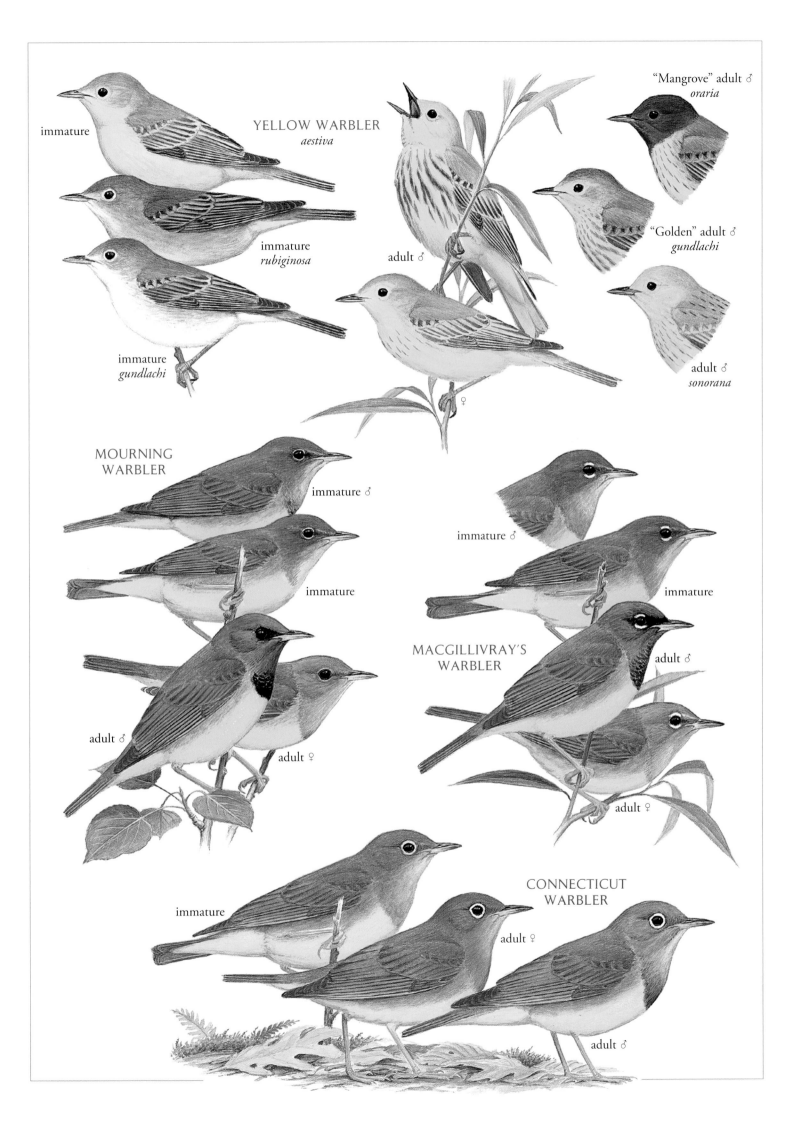

immature

YELLOW WARBLER
aestiva

immature
rubiginosa

adult ♂

immature
gundlachi

♀

"Mangrove" adult ♂
oraria

"Golden" adult ♂
gundlachi

adult ♂
sonorana

MOURNING
WARBLER

immature ♂

immature

adult ♂

adult ♀

immature ♂

immature

MACGILLIVRAY'S
WARBLER

adult ♂

adult ♀

immature

CONNECTICUT
WARBLER

adult ♀

adult ♂

KENTUCKY WARBLER
Oporornis formosus | L 5¼" (13 cm)

A short-tailed, long-legged warbler. Bold yellow spectacles separate black crown from black on face and sides of neck; underparts are entirely yellow, upperparts bright olive. Black areas are duller on **female**, olive on immature female. **Song** is a series of rolling musical notes, *churry churry churry*, much like the song of the Carolina Wren. **Call** is a low, sharp *chuck*. RANGE: Common in rich, moist woodlands; nests and feeds on the ground in dense undergrowth. Rare to Maritimes; casual to Newfoundland. Very rare vagrant to California and the Southwest.

CANADA WARBLER
Wilsonia canadensis | L 5¼" (13 cm)

Black necklace on bright yellow breast identifies **male**; note also bold yellow spectacles. In **females**, necklace is dusky and indistinct. Male is blue-gray above, females duller. All birds have white undertail coverts. **Song** begins with one or more short, sharp *chip* notes and continues as a rich and highly variable warble.
RANGE: Common in dense woodlands and brush. Usually forages in undergrowth or low branches, but also seen fly-catching. Winters in South America. Casual vagrant in the West; very rare in fall along California coast and in Newfoundland.

WILSON'S WARBLER
Wilsonia pusilla | L 4¾" (12 cm)

Olive above; yellow below, with yellow lores. Long tail is all-dark above and below, and often cocked. **Male** has black cap; in **females**, cap is blackish or absent, forehead yellowish. Yellow lores and lack of white in tail help distinguish female from female Hooded Warbler. Coloration varies geographically from bright *chryseola* of Pacific states to duller and olive-faced nominate *pusilla* of East; *pileolata* (shown) from Alaska and Rockies intermediate. Females average more black in western races. **Song** is a rapid, variable series of *chee* notes; common **call**, a sharp *chimp*.
RANGE: Fairly common, much more numerous in the West than in the East; nests in dense, moist woodlands, bogs, willow thickets, and streamside tangles.

HOODED WARBLER
Wilsonia citrina | L 5¼" (13 cm)

All ages have dark lores, unlike Wilson's Warbler; also bigger bill and larger eye. Extensive black hood identifies **male**. **Adult female** shows blackish or olive crown and sides of neck; sometimes has black throat or black spots on breast; **immature female** lacks black. Note that in both sexes tail is white below; seen from above, white outer tail feathers are conspicuous as the bird flicks its tail open. **Song** is loud, musical, whistled variations of *ta-wit ta-wit ta-wit tee-yo*. **Call** is a flat, metallic *chink*.
RANGE: Fairly common in swamps, moist woodlands; generally stays hidden in dense undergrowth and low branches. Rare migrant in the Southwest and California, where it has nested; also has nested in Colorado; casual in other western states. Rare in fall to the Maritimes; casual to Newfoundland.

KENTUCKY
WARBLER

♂

♀

immature ♀

CANADA
WARBLER

adult ♂

HOODED
WARBLER

WILSON'S
WARBLER
pileolata

♀

♂

adult ♀

adult ♂

immature ♀

WORM-EATING WARBLER

Helmitheros vermivorum | L 5¼" (13 cm)

Bold, dark stripes on buffy head; upperparts brownish olive; underparts buffy; long, spike-like bill. **Song** is a series of sharp, dry *chip* notes, like Chipping Sparrow's song but faster. Common **call**, *zeep-zeep*.

RANGE: Found chiefly in dense undergrowth on wooded slopes. Often feeds in clusters of dead leaves. Casual vagrant to California, the Southwest, and Atlantic Canada.

SWAINSON'S WARBLER

Limnothlypis swainsonii | L 5½" (14 cm)

Pale eyebrow, conspicuous between brown crown and dark eye line. Brown-olive above, grayish below. Bill very long and spiky. Uncommon and secretive. **Song** is a series of thin, slurred whistles like beginning of song of Louisiana Waterthrush; often ends with a rising *tee-oh*. **Calls** include a loud, dry *chip*.

RANGE: Found in undergrowth in swamps and canebrakes; rare and local in mountain laurel and rhododendron. Walks or shuffles on the ground and shivers while picking up dead leaves. Casual north to Ontario and Nova Scotia and west to Colorado, New Mexico, and Arizona.

OVENBIRD

Seiurus aurocapilla | L 6" (15 cm)

Russet crown bordered by dark stripes; bold white eye ring. Olive above; white below, with bold streaks of dark spots; pinkish legs. Generally seen on the ground; walks, with tail cocked, rather than hops. Typical **song** is a loud *teacher teacher teacher,* rising in volume.

RANGE: A plump warbler, common in mature forests. Rare in the West. Rare in winter along Gulf and Atlantic coasts to North Carolina.

LOUISIANA WATERTHRUSH

Seiurus motacilla | L 6" (15 cm)

Distinguished from Northern Waterthrush by contrast between white underparts and salmon-buff flanks; bicolored eyebrow, pale buff in front of eye, white and much broader behind eye; larger bill; bubblegum pink legs. A ground dweller; walks, rather than hops, bobbing its tail constantly but usually slowly. **Call** note, a sharp *chink*, is slightly flatter than that of Northern Waterthrush. **Song** begins with three or four shrill, slurred notes followed by a brief, rapid jumble.

RANGE: Uncommon; found along mountain streams in dense woodlands, also near ponds and in swamps. Rare to Southwest; casual to California and Nova Scotia.

NORTHERN WATERTHRUSH

Seiurus noveboracensis | L 5¾" (15 cm)

Distinguished from Louisiana Waterthrush by lack of contrast in color between flanks and rest of underparts; buffy eyebrow, of even width throughout or slightly narrowing behind eye; smaller bill; drabber leg color. Some birds are whiter below, with whiter eyebrow. A ground dweller; walks, rather than hops, bobbing its tail constantly and usually rapidly. **Call** note, a metallic *chink*, is slightly sharper than that of Louisiana Waterthrush. **Song** begins with loud, emphatic notes and ends in lower notes, delivered more rapidly.

RANGE: Found chiefly in woodland bogs, swamps, and thickets. Rare to uncommon migrant in California and the Southwest.

WORM-EATING
WARBLER

SWAINSON'S
WARBLER

OVENBIRD

LOUISIANA
WATERTHRUSH

paler

NORTHERN
WATERTHRUSH

COMMON YELLOWTHROAT *Geothlypis trichas* | L 5" (13 cm)

Adult male's broad black mask is bordered above by gray or white, below by bright yellow throat and breast; undertail coverts yellow. **Female** lacks black mask; has whitish eye ring. Races vary geographically in color of mask border and extent of yellow below. Southwestern race, *chryseola*, is brightest below and shows the most yellow. **Immatures** are duller and browner overall. Often cocks tail. Variable **song**; one version is a loud, rolling *wichity wichity wichity wich*. **Calls** include a raspy *chuck*.
RANGE: Common; stays low in grassy fields, shrubs, and marshes.

GRAY-CROWNED YELLOWTHROAT *Geothlypis poliocephala* | L 5½" (13 cm)

Large, with a long, graduated tail; thick, bicolored bill with curved culmen; split white eye ring; lores blackish in **males**, slaty gray in **females**. **Song**, a rich, varied warble; **call**, a rising *chee dee*. Favors grassland with scattered bushes.
RANGE: Tropical species; former resident of the Brownsville area in southern Texas; population eliminated in early 20th century. Recently, several certain records in the lower Rio Grande Valley; other reports uncertain.

FAN-TAILED WARBLER *Euthlypis lachrymosa* | L 5¾" (15 cm)

Large, with long, graduated, white-tipped tail held partly open and pumped sideways or up and down. Head pattern distinct with broken white eye ring; white lore spot; yellow crown patch. Note tawny wash on breast; long pink legs. **Song** of rich, loud slurred notes; **call**, a penetrating *schree*.
RANGE: Tropical species; casual, mainly late spring, to southeast Arizona. Found low in canyons or ravines; often walks or shuffles on ground; secretive.

GOLDEN-CROWNED WARBLER *Basileuterus culicivorus* | L 5" (13 cm)

Resembles Orange-crowned Warbler (page 376) but crown shows a distinct yellow or buffy orange central stripe, bordered in black; note also yellowish green eyebrow. **Call**, a rapidly repeated *tuck*.
RANGE: Tropical species, casual in southern Texas, chiefly in winter; accidental in eastern New Mexico (spring).

RUFOUS-CAPPED WARBLER *Basileuterus rufifrons* | L 5¼" (13 cm)

Rufous crown, bold white eyebrow, throat extensively bright yellow. Long tail, often cocked. **Song** begins with musical *chip* notes and accelerates into a series of dry, whistled warbles; **call**, a *tik*, often doubled or in a rapid series. Inhabits brush and woodlands of foothills or low mountains, generally staying low in the undergrowth.
RANGE: Mexican and Central American species, casual to Edward's Plateau and Big Bend, Texas; also southeastern Arizona.

YELLOW-BREASTED CHAT *Icteria virens* | L 7½" (19 cm)

Our largest warbler, with long tail, thick bill, and white spectacles. Lores black in **males**, gray in **females**. Unmusical **song**, a jumble of harsh, chattering clucks, rattles, clear whistles, and squawks, sometimes given in hovering display flight
RANGE: Inhabits dense thickets and brush. Rather shy. Regular straggler in fall to Maritimes and Newfoundland; rare in winter on the East Coast.

immature ♂

adult ♀

immature ♀

adult ♂

**COMMON
YELLOWTHROAT**
trichas

southwestern
adult ♂
chryseola

♀

♂

**GRAY-CROWNED
YELLOWTHROAT**
ralphi

**FAN-TAILED
WARBLER**
tephra

**GOLDEN-
CROWNED
WARBLER**
brasherii

**RUFOUS-CAPPED
WARBLER**

eastern ♂
virens

western ♂
auricollis

eastern ♀
virens

**YELLOW-BREASTED
CHAT**

AMERICAN REDSTART *Setophaga ruticilla* | L 5¼" (13 cm)

Male glossy black, with bright orange patches on sides, wings, and tail; belly and undertail coverts white. **Female** is gray-olive above, white below with yellow patches. Immature male resembles female; by **first spring**, lores are usually black, breast has some black spotting; adult male plumage is acquired by second fall. Like redstarts of the genus *Myioborus*, often fans its tail and spreads its wings when perched. Variable **song**, a series of high, thin notes usually followed by a wheezy, downslurred note.
RANGE: Common in second-growth woodlands. Rare to uncommon migrant in California and the Southwest.

PAINTED REDSTART *Myioborus pictus* | L 5¾" (15 cm)

Bright red lower breast and belly; black head and upperparts; bold white wing patch. White outer tail feathers conspicuous as the bird fans its tail. **Juvenile** acquires full adult plumage by end of summer. **Song** is a series of rich, liquid warbles; **call**, a clear, whistled *chee*.
RANGE: Found in pine-oak canyons. Very rare visitor to southern California; a scattering of records elsewhere.

SLATE-THROATED REDSTART *Myioborus miniatus* | L 6" (15 cm)

Head, throat, and back are slate black, breast dark red. Chestnut crown patch visible only at close range. Lacks white wing patch of similar Painted Redstart; white on outer tail feathers less extensive; tail strongly graduated. **Call**, a *chip* note, very different from Painted Redstart.
RANGE: Middle and South American species, casual in southeast Arizona, southeast New Mexico, and west and south Texas. Found in pine-oak canyons, forests.

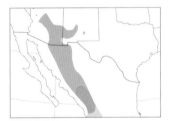

RED-FACED WARBLER *Cardellina rubrifrons* | L 5½" (14 cm)

Adult's red-and-black face pattern distinctive; back and tail gray, rump and underparts white. Immature is duller, face pinkish. **Song** is a series of varied, ringing *zweet* notes.
RANGE: A warbler of high mountains, generally found above 6,000 feet. Fairly common, especially in fir and spruce mixed with oaks. Nests on the ground. Rare to west Texas; casual to southern California.

OLIVE WARBLER (FAMILY PEUCEDRAMIDAE)

Recently placed in its own family because relationships are uncertain.

OLIVE WARBLER *Peucedramus taeniatus* | L 5¼" (13 cm)

Dark face patch broadens behind eye. Long, thin bill; two broad white wing bars; outer tail feathers extensively white. **Adult male**'s head, throat, and nape tawny brown. **Female** has olive crown, yellow face; pale yellow throat and breast. Juveniles and **first-fall** birds resemble female but are paler or whitish below; crown is gray. Young male acquires adult plumage by second fall. Typical **song** is a loud *peeta peeta peeta*, similar to call of Tufted Titmouse; **call**, a soft, whistled *phew*.
RANGE: Favors open coniferous forests at elevations above 7,000 feet. Nests and forages high in trees. A few remain on breeding grounds in winter.

AMERICAN
REDSTART

1st
spring ♂

adult ♂

PAINTED
REDSTART
pictus

juvenile

SLATE-THROATED
REDSTART

RED-FACED
WARBLER

♂

adult ♂

OLIVE
WARBLER
arizonae

adult ♀

1st spring ♂

1st fall

TANAGERS (Family Thraupidae)

Brightly colored, mostly fruit-eating, tropical birds, related to warblers. **Species:** *271 World, 6 N.A.*

SUMMER TANAGER
Piranga rubra | L 7¾" (20 cm)

Adult male is rosy red year-round. **First-spring male** usually has red head. Some **females** show overall reddish wash; most have a mustard tone, lack olive of female Scarlet Tanager; bill larger. Western birds are larger and paler; females generally grayer above. **Song** is robin-like; **call**, a staccato *ki-ti-tuck*.
Range: Common in pine-oak woods in the East, cottonwood groves in the West.

HEPATIC TANAGER
Piranga flava | L 8" (20 cm)

Large grayish cheek patch and gray wash on flanks set off brighter throat, breast, and cap in both sexes; dark bill with gray base. **Adult male** plumage is acquired by second fall; dull red plumage retained year-round. Juvenile resembles yellow-and-gray **female** but is heavily streaked overall. **Song** is robinlike. **Call** is a single low *chuck*.
Range: Inhabits mixed mountain forests.

SCARLET TANAGER
Piranga olivacea | L 7" (18 cm)

Breeding male bright red and black. In late summer, becomes splotchy green-and-red as he molts to yellow-green winter plumage. **Female** has uniformly olive head, back, and rump; whitish wing linings; bill smaller than in Summer Tanager. Immature male resembles adult male, but note brownish primaries and secondaries. Some immatures show faint wing bars. Robinlike **song** (hoarser than Summer Tanager) of raspy notes, *querit queer query querit queer*, is heard in deciduous forests. **Call** is a hoarse *chip-burr*.
Range: Very rare vagrant in the West, most in late fall.

WESTERN TANAGER
Piranga ludoviciana | L 7¼" (18 cm)

Conspicuous wing bars, often paler and thinner in **female**, upper bar yellow in **male**. Male's red head becomes yellowish and finely streaked in **winter**. Female's grayish back contrasts with greenish yellow nape and rump. Some females are duller below, grayer above. **Song** is like Scarlet Tanager's; **call**, *pit-er-ick*.
Range: Breeds in coniferous forests. Rare in winter north along coastal slope to central California. Rare on Gulf Coast, casual in the East.

FLAME-COLORED TANAGER
Piranga bidentata | L 7¼" (18 cm)

Has gray bill with visible "teeth"; blackish rear border to ear patch; streaked back; white wing bars and tertial tips; whitish tail corners. Hybrids with Western Tanager are regularly noted in southeast Arizona. **Male** of nominate west Mexican race, *bidentata*, is flaming orange, eastern *sanguinolenta* male redder. **Female** and immatures are colored like female Western Tanager. **First-spring males** have brighter yellow head; some spotting. **Song** similar to Western and Scarlet Tanagers; **call** also, but huskier, low-pitched *prreck*.
Range: Resident from western Panama to northern Mexico; casual to mountains of southeast Arizona in spring and summer; casual to west and south Texas.

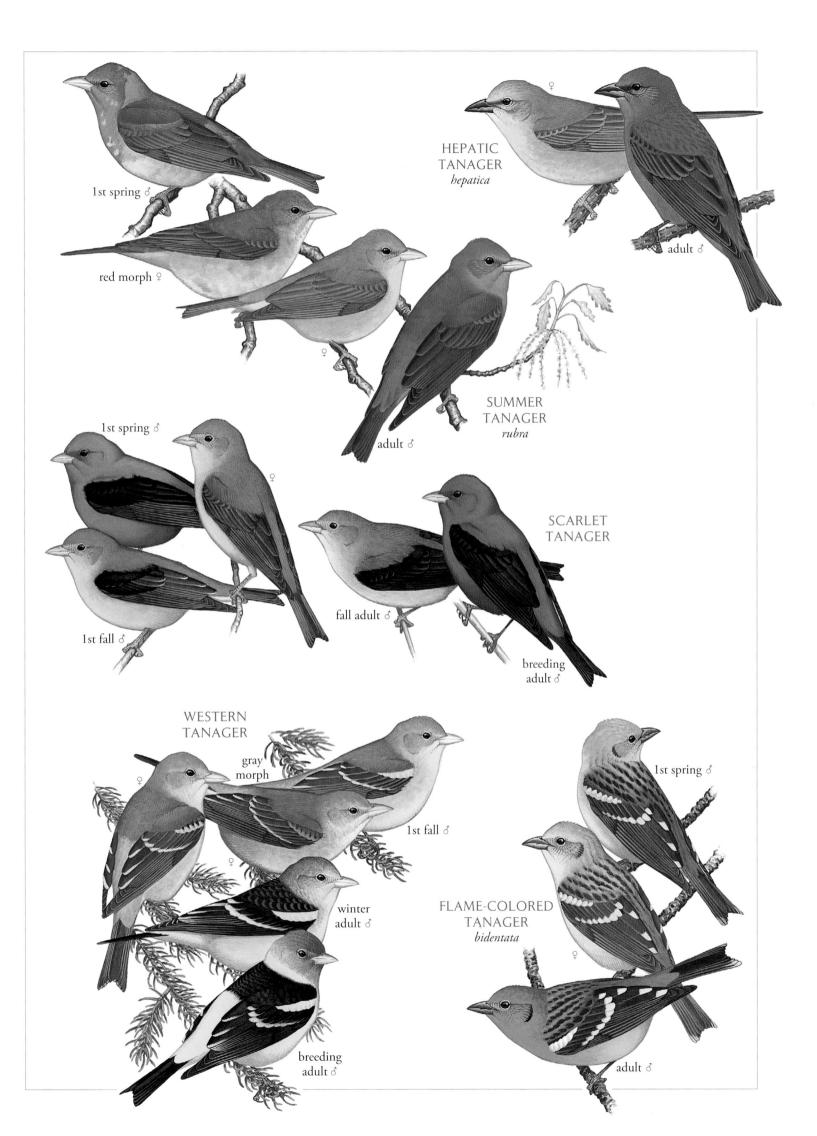

1st spring ♂

red morph ♀

HEPATIC
TANAGER
hepatica

adult ♂

♀

adult ♂

SUMMER
TANAGER
rubra

1st spring ♂

♀

1st fall ♂

fall adult ♂

breeding
adult ♂

SCARLET
TANAGER

WESTERN
TANAGER

gray
morph

♀

1st fall ♂

♀

winter
adult ♂

breeding
adult ♂

1st spring ♂

♀

FLAME-COLORED
TANAGER
bidentata

adult ♂

WESTERN SPINDALIS *Spindalis zena* | L 6¾" (17 cm)

Formerly known as Stripe-headed Tanager. **Males** are strikingly patterned: Most are black-backed (*zena*); a few unconfirmed sightings of *townsendi*, restricted to Grand Bahama and Abaco, show greenish orange back; occasionally black with dull orange edgings. Recent winter record in Key West of male *pretrei* from Cuba. **Females** of both races are grayish olive, with pale eyebrow, pale greater covert patch, and distinct white spot at base of primaries. **Call** is a thin, high *tsee*, given singly or in a series.

RANGE: West Indian species; very rare visitor, except in summer, from Bahamas to southeast Florida and Florida Keys.

BANANAQUIT (*INCERTAE SEDIS*)

Family affiliation of this species is uncertain. Prefers nectar.

BANANAQUIT *Coereba flaveola* | L 4½" (11 cm)

Note thin, downcurved bill. **Adult** has conspicuous white eyebrow, yellow rump; underparts white, with yellow breast; small white wing patch. **Juvenile** is duller.

RANGE: Tropical species; casual visitor from the Bahamas to southern Florida.

EMBERIZIDS (FAMILY EMBERIZIDAE)

All have conical bills. This large family includes the towhees, sparrows, longspurs, and Emberiza *buntings.*
Species: 308 World, 60 N.A.

WHITE-COLLARED SEEDEATER *Sporophila torqueola* | L 4½" (11 cm)

Tiny, with thick, short, strongly curved bill; rounded tail. **Adult male** has black cap; white crescent below eye; incomplete buffy collar; white wing bars; white patch at base of primaries. **Females** are paler, lack cap and collar; wing bars duller. **Song** is pitched high, then low, a variable *sweet sweet sweet sweet cheer cheer cheer*. **Calls** include a distinct, high *wink*.

RANGE: In U.S., favors canes and river bottoms.

BLACK-FACED GRASSQUIT *Tiaris bicolor* | L 4½" (11 cm)

Adult male mostly black below, dark olive above; head is black. **Female** and immatures pale gray below, gray-olive above. **Song** is a buzzing *tik-zeee*; **call**, a lisping *tst*.

RANGE: West Indian species; casual stray to south Florida.

YELLOW-FACED GRASSQUIT *Tiaris olivaceus* | L 4¼" (11 cm)

Adult male of mainland race, *pusillus*, shows extensive black on head, breast, and upper belly; golden yellow supraloral, throat, and crescent below eye; olive above. Adult **female** and immature male have traces of same head pattern; olive above. Adult male of West Indian race, *olivaceus*, shows less black. Female lacks black below. **Song** is thin, insectlike trills; **call**, a high-pitched *sik* or *tsi*.

RANGE: Tropical species. Accidental in southern Florida and southernmost Texas.

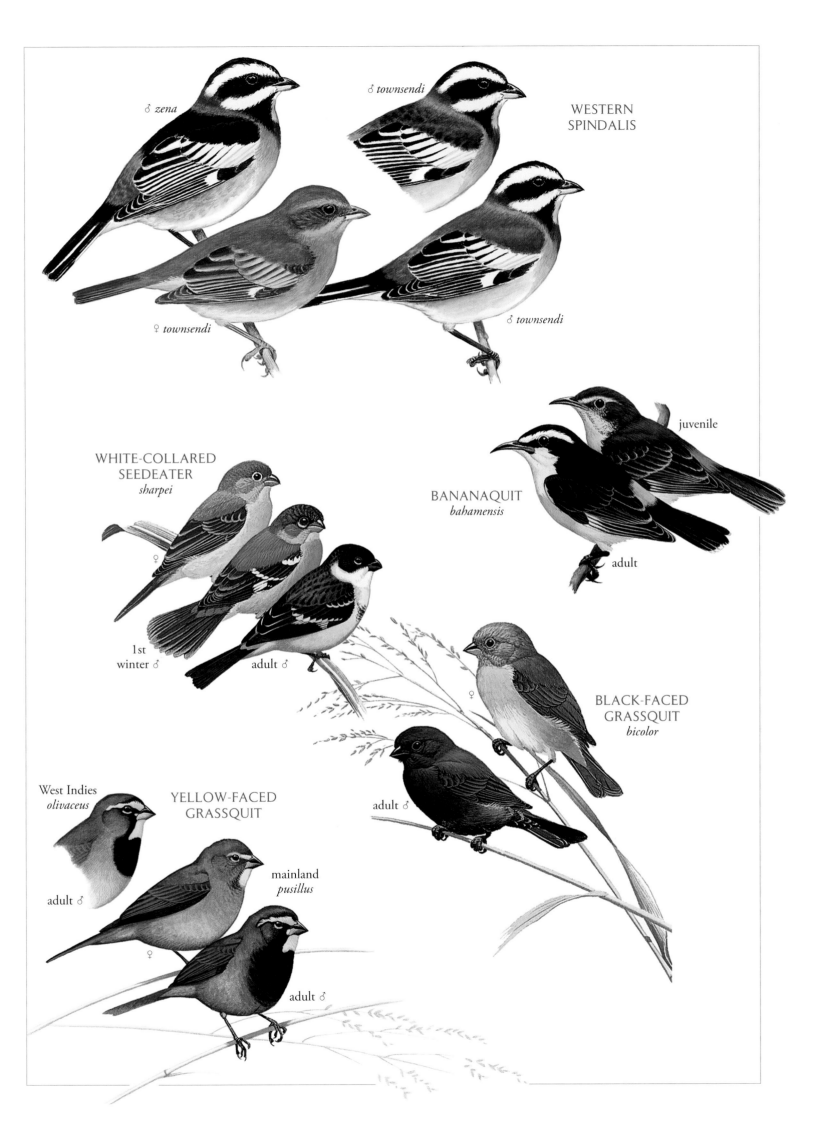

♂ *zena*

♂ *townsendi*

WESTERN
SPINDALIS

♀ *townsendi*

♂ *townsendi*

WHITE-COLLARED
SEEDEATER
sharpei

juvenile

BANANAQUIT
bahamensis

adult

♀

1st
winter ♂

adult ♂

♀

BLACK-FACED
GRASSQUIT
bicolor

adult ♂

West Indies
olivaceus

YELLOW-FACED
GRASSQUIT

adult ♂

mainland
pusillus

♀

adult ♂

OLIVE SPARROW

Arremonops rufivirgatus | L 6¼" (16 cm)

Dull olive above, with brown stripe on each side of crown. Lacks reddish cap of similar Green-tailed Towhee. **Juveniles** are buffier, with pale wing bars; faintly streaked on neck and breast. **Calls** include a dry *chip* and a buzzy *speeee*. **Song** is an accelerating series of *chip* notes.
RANGE: Tropical species, common in southernmost Texas in dense undergrowth, brushy areas, live oak.

GREEN-TAILED TOWHEE

Pipilo chlorurus | L 7¼" (18 cm)

Olive above with reddish crown, distinct white throat bordered by dark stripe and white stripe. **Juvenile** has two faint olive wing bars; plumage is streaked overall; upperparts tinged with olive; lacks reddish crown. Clear, whistled **song** begins with *weet-chur*, ends in raspy trill. **Calls** include a catlike *mew*.
RANGE: Fairly common in dense brush, chaparral, on mountainsides and high plateaus. Casual in winter throughout the East.

CALIFORNIA TOWHEE

Pipilo crissalis | L 9" (23 cm)

Brownish overall; crown slightly warmer brown than rest of upperparts. Buff throat is bordered by a distinct broken ring of dark brown spots; no dark spot on breast as in Canyon Towhee. Lores are same color as throat and contrast with cheek; undertail coverts warm cinnamon. **Juvenile** shows faint wing bars. **Call** is a sharp, metallic *chink* note; also gives some thin, lispy notes and an excited, squealing series of notes, often delivered as a duet by a pair. **Song**, accelerating *chink* notes with stutters in the middle, is heard mostly in late afternoon. With Canyon Towhee, California was formerly considered one species, Brown Towhee.
RANGE: Resident in chaparral, parks, and gardens. The race *eremophilus* (**T**) of Inyo County, California, is threatened.

CANYON TOWHEE

Pipilo fuscus | L 8" (20 cm)

Similar to California Towhee. Canyon is paler, grayish rather than brown, with shorter tail; more contrast in reddish crown gives a capped appearance; crown is sometimes raised as short crest. Larger whitish belly patch with diffuse dark spot at junction with breast; paler throat bordered by finer streaks; lores the same color as cheek; distinct buffy eye ring. Juveniles are streaked below. **Call** is a shrill *chee-yep* or *chedep*. **Song**, more musical, less metallic, than California; opens with a call note, followed by sweet slurred notes. Also gives a duet of lisping and squealing notes, like California.
RANGE: Favors arid, hilly country; desert canyons. Largely resident within range; casual southwestern Kansas (Morton County); no range overlap with California.

ABERT'S TOWHEE

Pipilo aberti | L 9½" (24 cm)

Black face; upperparts cinnamon-brown, underparts paler, with cinnamon undertail coverts. **Call** is a sharp *peek*; **song**, a series of *peek* notes.
RANGE: Common within its range, but somewhat secretive. Inhabits desert woodlands and streamside thickets, at lower altitudes than similar Canyon Towhee. Also found in suburban yards and orchards.

OLIVE
SPARROW
rufivirgatus

juvenile

spring

fall

juvenile

GREEN-TAILED
TOWHEE

juvenile

CALIFORNIA
TOWHEE

CANYON
TOWHEE

ABERT'S
TOWHEE

EASTERN TOWHEE
Pipilo erythrophthalmus | L 7½" (19 cm)

Male's black upperparts and hood contrast with rufous sides and white underparts. Distinct white patch at base of primaries and distinct white tertial edges. White in outer tail feathers is conspicuous in flight, or seen from below when bird is perched. Most have red eyes. **Females** are similarly patterned, but black areas are replaced by brown. **Juveniles** are brownish and show distinct streaks below. The nominate race is largest and shows most extensive white in tail. Wing length, and the extent of white in wings and tail, declines from the northern part of the range to the Gulf Coast, while the size of bill, legs, and feet increases. The subspecies from the Florida peninsula, *alleni*, is smaller in all measurements, paler, and duller; has less white in wings and tail; has straw-colored eyes. The *rileyi* race (not shown), from northernmost Florida to east-central North Carolina, shows intermediate characteristics, eyes being either red or straw colored; eye color particularly variable in birds from southern Georgia and coastal South Carolina. Like other towhees, Eastern scratches with its feet together. Full **song** has three parts, often rendered as *drink your tea*, or shortened to two parts: *drink tea*. Northeastern birds' **call** is a slightly upslurred *chwee*; in *alleni*, a clearer, even-pitch or upslurred *swee*. Eastern and Spotted Towhee have each been restored to full species status; formerly considered one species, Rufous-sided Towhee. The two interbreed along rivers in the Great Plains, particularly the Platte and its tributaries.
RANGE: Partial to second growth with dense shrubs and extensive leaf litter; southern races, especially *alleni*, favor coastal scrub or sand dune ridges and pinelands. Nominate race is partly migratory; casual west to Colorado and Arizona. Has declined from the northeastern part of its range by as much as 90 percent in recent decades. Other races are largely resident. Accidental in winter to Southwest.

SPOTTED TOWHEE
Pipilo maculatus | L 7½" (19 cm)

Distinguished from similar Eastern Towhee by white spotting on back and scapulars; also on tips of median and greater coverts, which forms white wing bars. In general, **females** differ less from **males** than in Eastern, with *arcticus* from Great Plains showing the greatest difference. In both sexes the amount of white spotting above and white in tail shows marked geographical variation, with *arcticus* displaying most white. Races, principally *montanus*, from the Great Basin and Rockies show less white. The Northwest coast's *oregonus* is darkest and shows least white of all the races. White increases southward to *megalonyx* of southern California and *falcinellus* (not shown) of the Central Valley region. **Song** and calls also show great geographical variation. Interior races give introductory notes, then a trill. Pacific coast birds sing a simple trill of variable speed. **Call** of *montanus* is a descending and raspy mewing. Great Plains *arcticus* and all coastal races give an upslurred, questioning *queee*.
RANGE: Some populations are largely resident while others are migratory; *arcticus* is the most migratory and is casual in eastern North America.

juvenile

Florida ♂
alleni

EASTERN
TOWHEE
erythrophthalmus

♀

♂

♀ *arcticus*

SPOTTED
TOWHEE

♂ *montanus*

♂ *arcticus*

♀ *oregonus*

♀ *montanus*

♂ *oregonus*

♀ *megalonyx*

juvenile
oregonus

♂ *megalonyx*

Eastern

Spotted

erythrophthalmus

alleni

arcticus

montanus

megalonyx

oregonus

BACHMAN'S SPARROW
Aimophila aestivalis | L 6" (15 cm)

A large sparrow with large bill, fairly flat forehead, and long, rounded, dark tail. **Adults** gray above, heavily streaked with chestnut or dark brown; sides of head buffy gray; a thin dark line extends back from eye. Breast and sides buff or gray; belly whitish. Subspecies range in overall brightness from the reddish *illinoensis* of the western part of range to the grayer and darker *aestivalis* of Florida. Birds from northeastern part of range are intermediate. **Juvenile** has a distinct eye ring; throat, breast, and sides are streaked. First-winter plumage usually retains some streaking. Quite secretive outside breeding season; best located and identified by **song**: one clear, whistled introductory note, followed by a variable trill or warble on a different pitch. Male sings from open perch; often heard in late summer.
RANGE: Inhabits dry, open woods, especially pines; scrub palmetto. Northern range has markedly declined over last several decades.

BOTTERI'S SPARROW
Aimophila botterii | L 6" (15 cm)

A large, plain sparrow with large bill, fairly flat forehead; tail long, rounded, dusky brown, lacking white tips and central barring of the very similar Cassin's Sparrow. Upperparts streaked with dull black, rust or brown, and gray; underparts unstreaked; throat and belly whitish, breast and sides grayish buff. Subspecies *arizonae* of southeastern Arizona is redder above; *texana* of extreme southern Texas is slightly grayer. **Juvenile**'s belly is buffy; breast broadly streaked, sides narrowly streaked. Best located and identified by **song**: several high sharp *tsip* or *che-lik* notes, often followed by a short, accelerating, rattly trill.
RANGE: Generally secretive; inhabits grasslands dotted with mesquite, cactus, and brush. The *texana* race is declining because of habitat loss; now uncommon and local.

CASSIN'S SPARROW
Aimophila cassinii | L 6" (15 cm)

A large, drab sparrow, with large bill, fairly flat forehead. Long, rounded tail is dark gray-brown; distinctive white tips on outer feathers are most conspicuous in flight. Gray upperparts are streaked with dull black, brown, and variable amount of rust; blackish marks form anchor marks; underparts are grayish white, usually with a few short streaks on the flanks. **Juvenile** is streaked below; paler overall than juvenile Botteri's. In fresh fall plumage, shows bolder white wing bars than similar Botteri's Sparrow, and black-centered, white-fringed tertials. Best located and identified by **song**, often given in brief, fluttery song flight: typically a soft double whistle, a loud, sweet trill, a low whistle, and a final, slightly higher note; or a series of *chip* notes ending in a trill or warbles. Also gives a trill of *pit* notes.
RANGE: Secretive; inhabits arid grasslands with scattered shrubs, cactus, and mesquite. Casual vagrant to the East and Far West.

illinoensis

BACHMAN'S
SPARROW
aestivalis

juvenile

arizonae

BOTTERI'S
SPARROW
texana

juvenile

CASSIN'S
SPARROW

juvenile

RUFOUS-WINGED SPARROW
Aimophila carpalis | L 5¾" (15 cm)

Pale gray head marked with reddish eye line and black moustachial and malar stripe on side of face; two-toned bill, with pale lower mandible; sides of crown streaked with reddish brown. Back is gray-brown, streaked with black; two whitish wing bars. Reddish lesser wing coverts distinctive but difficult to see. Underparts grayish white, without streaking. Tail long, rounded. **Juvenile**'s facial stripes are less distinct; wing bars buffier; bill dark; breast and sides lightly streaked; plumage can be seen as late as Nov. Distinctive **call** note, a sharp, high *seep*. Variable **song**, several *chip* notes followed by an accelerating trill of *chip* or *sweet* notes. RANGE: Fairly common but local; found in flat areas of tall desert grass mixed with brush and cactus. No migration or vagrancy known.

RUFOUS-CROWNED SPARROW
Aimophila ruficeps | L 6" (15 cm)

Gray head with dark reddish crown, distinct whitish eye ring, rufous line extending back from eye, single black malar stripe on each side of face. Gray-brown above, with reddish streaks; gray below; tail long, rounded. Subspecies range in overall color from paler, grayer *eremoeca*, found over most of eastern interior range, to widespread southwestern race, *scottii*, which is paler and reddish. Pacific coastal races are slightly smaller and darker and show variable amounts of reddish above. **Juvenile** is buffier above; breast and crown streaked; may show two pale wing bars. Distinctive **call**, a sharp *dear*, usually given in a series; **song**, a rapid, bubbling series of *chip* notes.
RANGE: Locally common on rocky hillsides and steep brushy or grassy slopes.

AMERICAN TREE SPARROW
Spizella arborea | L 6¼" (16 cm)

Gray head and nape crowned with rufous; rufous stripe behind eye. Gray throat and breast, with dark central spot, rufous patches at sides of breast. Back and scapulars streaked with black and rufous. Tail notched; outer feathers thinly edged in white on outer webs. Underparts grayish white with buffy sides. **Winter** birds are buffier; rufous color on crown sometimes forms a central stripe. **Juvenile** is streaked on head and underparts. Western *ochracea* is paler overall. American Tree Sparrow gives a musical *teedle-eet* **call** and also gives a thin *seet*. **Song** usually begins with several clear notes followed by a variable, rapid warble.
RANGE: Fairly common. Uncommon to rare west of Rockies. Breeds along edge of tundra, in open areas with scattered trees and brush. Winters in weedy fields, marshes, and groves of small trees.

FIELD SPARROW
Spizella pusilla | L 5¾" (15 cm)

Gray face with reddish crown, distinct whitish eye ring, bright pink bill. Back is streaked except on gray-brown rump. Breast and sides are buffy red; belly grayish white; legs pink. **Juvenile** streaked below; wing bars buffy. Birds in westernmost part of range, *arenacea*, are paler and grayer; extremes are shown here. **Song** is a series of clear, plaintive whistles accelerating into a trill; *chip* note similar to Orange-crowned Warbler.
RANGE: Fairly common in open, brushy woodlands, fields. Uncommon to rare in Maritimes; casual west of mapped range.

RUFOUS-WINGED
SPARROW
carpalis

juvenile

RUFOUS-CROWNED
SPARROW

coastal

interior
eremoeca

coastal
juvenile

AMERICAN TREE
SPARROW

juvenile
ochracea

breeding
arborea

winter
ochracea

western
arenacea

FIELD
SPARROW

eastern juvenile
pusilla

eastern
pusilla

CHIPPING SPARROW
Spizella passerina | L 5½" (14 cm)

Breeding adult identified by bright chestnut crown, distinct white eyebrow, and black line extending from bill through eye to ear; note also the gray nape and cheek with no dark moustachial stripe; gray unstreaked rump; and two white wing bars. As in all sparrows of the genus *Spizella*, the tail is fairly long and notched. **Winter adult** has browner cheek, dark lores, and streaked crown showing some rufous color. **First-winter** bird is similar but averages less rufous on the crown; breast and sides are tinged with buff. In juvenal plumage, often held into Oct, especially in the West, underparts are prominently streaked; crown usually lacks rufous; rump may show slight streaking. **Song**, rapid trill of dry *chip* notes, all on one pitch. Flight **call**, also given perched, is a high, hard *seep* or *tsik*.
RANGE: Widespread and common, Chipping Sparrows are found on lawns and in fields, woodland edges, and pine-oak forests.

CLAY-COLORED SPARROW
Spizella pallida | L 5½" (14 cm)

Brown crown with black streaks and a distinct buffy white or whitish median crown stripe. Broad, whitish eyebrow; pale lores; brown cheek outlined by dark postocular and moustachial stripes; conspicuous pale submoustachial stripe. Nape gray; back and scapulars are buffy brown, with dark streaks; rump is not streaked but color does not contrast with back as in Chipping Sparrow. Adult in fall and winter is buffier overall. **Juvenile** and **immature** birds are much buffier; gray nape and pale stripe on sides of throat stand out more; in juvenile, breast and sides are streaked. **Song** is a series of three or four insectlike buzzes. Flight **call** is a thin *sip*.
RANGE: Fairly common in brushy fields, groves, and streamside thickets. Winters primarily from Mexico south, uncommonly in southern Texas. Rare in fall, casual in winter and spring on both coasts and in Arizona. Rare in winter in south Florida.

BREWER'S SPARROW
Spizella breweri | L 5½" (14 cm)

Brown crown with fine black streaks, without clearly defined, pale median crown stripe of Clay-colored; head pattern lacks Clay-colored's strong contrast. Distinct whitish eye ring; grayish white eyebrow; ear patch pale brown with darker borders; pale lores; dark malar stripe. Upperparts buffy brown and streaked; rump buffy brown, may be lightly streaked. **Juvenile** is buffier overall, lightly streaked on breast and sides. Immature, and fall and winter adult, are somewhat buffy below. A separate subspecies, *taverneri* (not shown) of the alpine zone of the Canadian Rockies to east-central Alaska, has a slightly different song and larger bill. Juveniles are heavily streaked with blackish below. Some believe that this "Timberline Sparrow" may be a separate species. **Song** is a series of varied bubbling notes and buzzy trills at different pitches. **Call** is a thin *sip*, like call of Clay-colored Sparrow.
RANGE: Common; breeds in mountain meadows and sagebrush flats.

breeding

CHIPPING
SPARROW
passerina

winter
adult

juvenile

1st winter

immature

juvenile

breeding

CLAY-COLORED
SPARROW

BREWER'S
SPARROW
breweri

juvenile

LARK SPARROW
Chondestes grammacus | L 6½" (17 cm)

Head pattern distinctive in adults; note dark central breast spot. **Juvenile**'s colors are duller; breast, sides, and crown streaked. In all ages, white-cornered tail is conspicuous in flight. **Song** begins with two loud, clear notes, followed by a series of rich, melodious notes and trills and unmusical buzzes. **Call** is a sharp *tsip*, often a rapid series.
Range: Gregarious, found in various types of open country, often along roads. Formerly bred as far east as New York and Maryland; now rare in the East.

BLACK-CHINNED SPARROW
Spizella atrogularis | L 5¾" (15 cm)

Medium gray overall; back and scapulars rusty, with black streaks; bill bright pink; long tail. **Male** has black lores and chin; lower belly is whitish gray; long tail is all-dark. **Female** has less or no black. **Juvenile** and winter birds lack any black on face. Juveniles resemble female but show light streaks below. Plaintive **song** begins with slow *sweet sweet sweet* and continues in a rapid trill. **Call** is a high, thin *seep*.
Range: Inhabits brushy arid slopes in foothills and mountains. Rarely seen in migration.

BLACK-THROATED SPARROW
Amphispiza bilineata | L 5½" (14 cm)

Black lores and triangular black patch on throat and breast contrast with white eyebrow, white submoustachial stripe, white underparts. Upperparts plain brownish gray. **Juvenal** plumage, often held well into fall, lacks black on throat, but note bold white eyebrow; breast and back finely streaked. In all ages, extent of white on tail is greater than in Sage Sparrow. **Song** is rapid, high-pitched: two clear notes followed by a trill; **calls** are faint, tinkling notes.
Range: Fairly common in a variety of desert habitats, especially on rocky slopes; casual to eastern U.S. in fall and winter.

FIVE-STRIPED SPARROW
Aimophila quinquestriata | L 6" (15 cm)

Dark brown above; breast and sides gray; white throat bordered by black and white stripes. Dark central spot at base of breast. Juvenile lacks the streaks found on juveniles of other sparrows.
Range: West Mexican species; range barely reaches southeastern Arizona. Highly specialized habitat: tall, dense shrubs on rocky, steep hillsides, and canyon slopes. Very local; most often seen in breeding season; few winter records, but likely overlooked.

SAGE SPARROW
Amphispiza belli | L 6¼" (16 cm)

White eye ring, white supraloral, broad white submoustachial stripe bordered by dark malar stripe. Back buffy brown with dusky streaks on interior (largest and palest) race *nevadensis*. Dark central breast spot, dusky streaking on sides. **Juvenile** is duller overall but more streaked. Smaller California coastal race, *belli*, is much darker, lacks streaks on back, has stronger malar stripe. From a low perch, male **sings** a geographically variable jumbled series of rising and falling phrases. Twittering **call** consists of thin, juncolike notes. Runs on ground with tail cocked.
Range: Interior *nevadensis* and slightly darker *canescens* favor alkaline flats in sagebrush, saltbush. Coastal *belli* found in mountain chaparral. Ongoing studies may reveal that this complex represents two or three species.

LARK SPARROW
grammacus

juvenile

breeding ♀

breeding ♂

juvenile

BLACK-CHINNED
SPARROW

BLACK-THROATED
SPARROW
deserticola

juvenile

FIVE-STRIPED
SPARROW
septentrionalis

SAGE
SPARROW

coastal
belli

interior
nevadensis

interior juvenile
nevadensis

canescens

GRASSHOPPER SPARROW
Ammodramus savannarum | L 5" (13 cm)

Buffy breast and sides, usually without obvious streaking. Small and chunky, with short tail and flat head. Dark crown has a pale central stripe; note also white eye ring and, on most birds, a yellow-orange spot in front of eye. Lacks broad buffy orange eyebrow and pale blue-gray ear patch of Le Conte's Sparrow (next page). Compare also with female Orange Bishop (page 464) and Savannah Sparrow (page 420). **Juvenile**'s breast and sides are streaked with brown. **Fall** birds are buffier below but never as bright as Le Conte's. Subspecies vary in overall color from dark Florida race, *floridanus* (**E**), to reddish *ammolegus* of southeastern Arizona. Eastern *pratensis* is slightly more richly colored than western *perpallidus*, which spreads east through the Great Plains. Typical **song** is one or two high *chip* notes followed by a brief, grasshopperlike *buzz*; also sings a series of varied squeaky and buzzy notes.
RANGE: Found in pastures, grasslands, palmetto scrub, and old fields. Somewhat secretive; feeds and nests on the ground. Declining in East.

BAIRD'S SPARROW
Ammodramus bairdii | L 5½" (14 cm)

Orange tinge to head (duller on worn summer birds), usually with less distinct median crown stripe than Savannah Sparrow (page 420); note especially the two isolated dark spots behind ear patch and lack of postocular line. Widely spaced, short dark streaks on breast form a distinct necklace; also shows chestnut on scapulars. **Juvenile**'s head is paler and creamier; central crown stripe is finely streaked; white fringes give a scaly appearance to upperparts; underparts are more extensively streaked. Very secretive, especially away from breeding grounds. **Song** consists of two or three high, thin notes, followed by a single warbled note and a low trill.
RANGE: Uncommon, local, and declining. Found in grasslands and weedy fields. Casual east to Wisconsin and from coastal California. Accidental from East Coast and coastal Louisiana.

HENSLOW'S SPARROW
Ammodramus henslowii | L 5" (13 cm)

Large flat head; large gray bill. Resembles Baird's Sparrow but head, nape, and most of central crown stripe are greenish; wings extensively dark chestnut. **Juvenile** is paler, yellower, with less streaking below; compare with adult Grasshopper Sparrow. Secretive, but after being flushed several times may perch in the open for a few minutes before dropping back into cover. Distinctive **song**, a short *se-lick*, accented on second syllable.
RANGE: Uncommon, local, and declining; now occurs only casually in the Northeast; accidental to New Mexico (fall). Found in wet shrubby fields, weedy meadows, and reclaimed strip mines. In winter, found also in the understory of pine woods.

summer
perpallidus

floridanus

GRASSHOPPER
SPARROW

juvenile
pratensis

fall *pratensis*

fall *ammolegus*

Orange Bishop ♀
(for comparison)

HENSLOW'S
SPARROW

BAIRD'S
SPARROW

juvenile

juvenile

SALTMARSH SPARROW

Ammodramus caudacutus | L 5" (13 cm)

Similar to Nelson's, but bill longer and head flatter; orange-buff face triangle contrasts strongly with paler, crisply streaked underparts. Also dark markings around eye and head are more sharply defined; eyebrow streaked with black behind eye. **Juvenile**'s crown is blacker than juvenile Nelson's; cheek darker; streaks below more widespread and distinct. **Song** softer, more complex than Nelson's, with which it hybridizes in Maine. With Nelson's Sparrow, formerly treated as one species, Sharp-tailed Sparrow.

RANGE: Coastal salt marshes; winters south of Florida.

LE CONTE'S SPARROW

Ammodramus leconteii | L 5" (13 cm)

White central crown stripe, becoming orange on forehead, chestnut streaks on nape, and straw-colored back streaks distinguish Le Conte's from Saltmarsh and Nelson's Sparrows. Bright, broad, buffy orange eyebrow, grayish ear patch, thinner bill, and orange-buff breast and sides separate it from Grasshopper Sparrow (preceding page). Sides of breast and flanks have dark streaks. **Juvenal** plumage, seen on breeding grounds and in fall migration, is buffy; crown stripe tawny; breast heavily streaked. **Song** is a short, high, insectlike buzz.

RANGE: A bird of wet grassy fields, marsh edges. Fairly common but secretive; scurries through matted grasses like a mouse. Casual migrant in the Northeast and in the West.

NELSON'S SPARROW

Ammodramus nelsoni | L 4¾" (12 cm)

Distinguished from Le Conte's by gray median crown stripe; whitish or gray streaks on scapulars; gray, streakless nape. **Juvenile** has fainter median crown stripe; duller nape; variably thicker eye line; less contrast above; lacks streaking across breast. Plumage variable: *nelsoni*, of interior, has orange-buff triangle on face; streaked buffy breast contrasts with white belly; back strongly marked with black and white stripes; *subvirgatus*, of the Maritimes and coastal Maine, is duller overall; has diffuse streaking below; grayer upperparts. In *alterus* (not shown) from James and Hudson Bays, brightness is intermediate, streaks blurred. **Song**, a wheezy *p-tssssshh-uk*, ends on a lower note.

RANGE: All subspecies winter in coastal marshes from mid-Atlantic southward.

SEASIDE SPARROW

Ammodramus maritimus | L 6" (15 cm)

Long, spikelike bill with thick base, thin tip. Tail is short, pointed. Yellow supraloral patch. Dark malar stripe separates whitish throat and broad, pale stripe below cheek. Breast is white or buffy, with at least some streaking. **Juveniles** are duller, browner, than **adults**. Seaside Sparrows vary widely in overall color. Most races, like the widespread *maritimus*, are grayish olive above. The greener *mirabilis* (**E**), formerly called "Cape Sable Sparrow," inhabits a small area in southwestern Florida. Gulf Coast races such as *fisheri* have buffier breasts. The darkest race, *nigrescens*, formerly called "Dusky Seaside Sparrow," was found only near Titusville, Florida, and became extinct in June 1987. This subspecies was blackish above, heavily streaked below. Seaside Sparrow's **song** resembles that of Red-winged Blackbird but softer and buzzier.

RANGE: Fairly common in grassy tidal marshes; accidental inland. Rare in Maine; very rare to the Maritimes.

SALTMARSH
SPARROW

juvenile

LE CONTE'S
SPARROW

juvenile

subvirgatus

NELSON'S
SPARROW
nelsoni

juvenile

maritimus

fisheri

maritimus
juvenile

nigrescens

SEASIDE
SPARROW

mirabilis

FOX SPARROW

Passerella iliaca | L 7" (18 cm)

Highly variable. Most subspecies have reddish rump and tail; reddish in wings; underparts heavily marked with triangular spots merging into a larger spot on central breast. The many named subspecies are divided into four subspecies groups; may represent distinct species. The brightest, *iliaca*, and slightly duller *zaboria* ("**Red**" group) breed in the far north, from Seward Peninsula, Alaska, to Newfoundland; winter in southeastern U.S. Western mountain races have gray head and back, grayish olive base to bill; range from small-billed Rockies *schistacea* ("**Slate-colored**" group) to large-billed California *stephensi* ("**Thick-billed**" group). Dark coastal races ("**Sooty**" group), with browner rumps and tails, vary from sooty *fuliginosa* of the Pacific Northwest to paler *unalaschcensis* of southwest Alaska. **Songs** are sweet, melodic in northern, "Red" group; include harsher trills in other races. Large-billed Pacific races give a sharp *chink* call, like California Towhee; others give a *tschup* note, like Lincoln's Sparrow but louder.

RANGE: Uncommon to common; found in undergrowth in coniferous or deciduous woodlands and chaparral. Within "Sooty" group, paler northernmost races migrate the farthest south. The "Slate-colored" group migrates primarily southwest to California.

LARK BUNTING

Calamospiza melanocorys | L 7" (18 cm)

Stocky, with short tail and whitish wing patches; bill bluish gray. **Breeding male** is mostly black. **Female** is streaked below, with buffy sides and brown primaries. **Winter male** is similar, but has black primaries; immature male darker, has some black around bill and chin. Distinctive **call** is a soft *hoo-ee*. **Song** is a varied series of rich whistles and trills.

RANGE: Common; nests in dry plains and prairies, especially in sagebrush. Gregarious in migration and winter. In flight, looks short and round winged, with shallow wingbeats. Rare in fall and winter to West Coast; casual to Pacific Northwest and in the East.

SAVANNAH SPARROW

Passerculus sandwichensis | L 5½" (14 cm)

Highly variable. Eyebrow yellow or whitish; pale median crown stripe; strong postocular stripe. **Song** begins with two or three *chip* notes, followed by two buzzy trills. Distinctive flight **call**, a thin *seep*. The numerous subspecies vary geographically; extremes are shown here. West Coast races show increasingly darker color from north to south, with Alaskan and interior races paler, widespread *nevadensis* the palest; *beldingi* of southern California coastal marshes is darkest. The *rostratus* subspecies, "**Large-billed Sparrow**," which winters on the edge of Salton Sea, rarely in coastal California, is very dull with a large bill. Its **song**, markedly different from other Savannah Sparrows, has short, high introductory notes, followed by about three rich, buzzy *dzeeee* notes; **call**, a soft, metallic *zink*. In the East, the degree of darkness is reversed: Arctic races are darker than more southerly Canadian and U.S. races. Large, pale *princeps*, "**Ipswich Sparrow**," breeds on Sable Island, Nova Scotia; winters on East Coast beaches.

RANGE: Common in a variety of open habitats, marshes, grasslands.

"Red"
iliaca

"Slate-colored"
schistacea

FOX
SPARROW

"Thick-billed"
stephensi

"Sooty"
unalaschcensis

"Sooty"
fuliginosa

breeding ♂

LARK
BUNTING

early spring ♂

winter ♂

♀

nevadensis

SAVANNAH
SPARROW

"Ipswich
Sparrow"
princeps

beldingi

"Large-billed
Sparrow"
rostratus

LINCOLN'S SPARROW
Melospiza lincolnii | L 5¾" (15 cm)

Buffy wash and fine streaks on breast and sides, contrasting with whitish, unstreaked belly. Note broad gray eyebrow, whitish chin and eye ring. Briefly held **juvenal** plumage is paler overall than juvenile Swamp Sparrow. Distinguished from juvenile Song Sparrow by shorter tail, slimmer bill, and thinner malar stripe, often broken. Often raises slight crest when disturbed. Two **call** notes: a flat *tschup*, repeated in a series as an alarm call; and a sharp, buzzy *zeee*. Rich, loud **song**, a rapid, bubbling trill.
RANGE: Found in brushy bogs and mountain meadows; in winter prefers thickets.

SONG SPARROW
Melospiza melodia | L 4¾–6¾" (13–17 cm)

All subspecies have long, rounded tail, pumped in flight. All show broad grayish eyebrow and broad, dark malar stripe bordering whitish throat. Highly variable. Upperparts are usually streaked. Underparts whitish, with streaking on sides and breast that often converges in a central spot. Legs and feet are pinkish. **Juvenile** is buffier overall, with finer streaking. The numerous subspecies vary geographically in size, bill shape, overall coloration, and streaking. Eastern races typified by *melodia;* large Alaskan races, the largest resident on the Aleutians, reach an extreme in the gray-brown *maxima;* paler races such as *saltonis* inhabit southwestern deserts; *morphna* represents the darker, redder races of the Pacific Northwest; *heermanni* is one of the blackish-streaked California races. Typical **song** has three or four short clear notes followed by a buzzy *tow-wee*, then a trill. Distinctive **call** note is a nasal, hollow *chimp*; flight note is a clear, rising *seet*.
RANGE: Generally common, Song Sparrows are found in brushy areas, especially dense streamside thickets.

VESPER SPARROW
Pooecetes gramineus | L 6¼" (16 cm)

White eye ring; dark ear patch bordered in white along lower and rear edges; white outer tail feathers. Lacks bold eyebrow of Savannah Sparrow (preceding page). Distinctive chestnut lesser coverts not easily seen. Eastern nominate race is slightly darker overall than the widespread subspecies, *confinis*. **Song** is rich and melodious: two long, slurred notes followed by two higher notes, then a series of short, descending trills.
RANGE: Uncommon to fairly common in dry grasslands, farmlands, forest clearings, and sagebrush; declining in the East.

SWAMP SPARROW
Melospiza georgiana | L 5¾" (15 cm)

Gray face; rich rufous upperparts and wings; variable black streaks on back; white throat. **Breeding adult** has reddish crown, gray breast, and whitish belly. **Winter adult** is buffier overall; crown is streaked, shows gray central stripe; sides are rich buff. Briefly held **juvenal** plumage is usually even buffier; darker overall than juvenile Lincoln's or Song Sparrow; wings and tail redder. **Immature** resembles winter adult. Typical **song** is a slow, musical trill, all on one pitch. Two **call** notes: a prolonged *zeee*, softer than Lincoln's Sparrow, and an Eastern Phoebe-like *chip*.
RANGE: Nests in dense, tall vegetation in marshes and bogs. Winters in marshes and brushy fields. Generally rare in the West.

juvenile

LINCOLN'S
SPARROW

juvenile
melodia

melodia

SONG
SPARROW

morphna

heermanni

maxima

saltonis

western
confinus

eastern
gramineus

VESPER
SPARROW

breeding

SWAMP
SPARROW

winter
adult

juvenile

immature

HARRIS'S SPARROW
Zonotrichia querula | L 7½" (19 cm)

A large sparrow with black crown, face, and bib; pink bill. **Winter adult**'s crown is blackish; cheeks buffy; throat may be all-black or show white flecks or partial white band. Immature resembles winter adult but shows less black; white throat is bordered by dark malar stripe. **Song** is a series of long, clear, quavering whistles, often beginning with two notes on one pitch followed by two notes on another pitch. **Calls** include a loud *wink* and a drawn-out *tseep*. RANGE: Fairly common; nests in stunted boreal forest; winters in open woodlands and brushlands. Rare to casual in winter in rest of North America outside mapped range.

WHITE-THROATED SPARROW
Zonotrichia albicollis | L 6¾" (17 cm)

Conspicuous and strongly outlined white throat; mostly dark bill; dark crown stripes and eye line. Broad eyebrow is yellow in front of eye; remainder is either white or tan. Upperparts rusty brown; underparts grayish, sometimes with diffuse streaking. **Juvenile**'s eyebrow and throat are grayish, breast and sides heavily streaked. **Song** is a thin whistle, generally two single notes followed by three triple notes: *pure sweet Canada Canada Canada*, often heard in winter. **Calls** include a sharp *pink* and a drawn-out, lisping *tseep*. RANGE: Common in woodland undergrowth, brush, and gardens. Generally rare in the West.

WHITE-CROWNED SPARROW
Zonotrichia leucophrys | L 7" (18 cm)

Black-and-white striped crown; pink, orange, or yellowish bill; whitish throat; underparts mostly gray. **Juvenile**'s head is brown and buff, underparts streaked. **Immature** has tan and brownish head stripes; compare with immature Golden-crowned Sparrow. Nominate race *leucophrys*, mainly found in the east Canadian tundra, and *oriantha* (not shown), of the High Sierra, southern Cascades, and Rockies, have a black supraloral area and large, dark pink bill; *gambelii*, found from Alaska to Hudson Bay, has whitish supraloral and a smaller, orange-yellow bill; in coastal *nuttalli* and *pugetensis* (not shown), breast and back are browner, bill dull yellow, supraloral pale. **Song**, often heard in winter, is usually one or more thin, whistled notes followed by a twittering trill; *leucophrys* and *gambelii* give a more mournful song with no trill at the end. **Calls** include a loud *pink* and sharp *tseep*. RANGE: Generally common in woodlands, grasslands, and roadside hedges.

GOLDEN-CROWNED SPARROW
Zonotrichia atricapilla | L 7" (18 cm)

Yellow patch tops black crown; back brownish, streaked with dark brown; breast, sides and flanks grayish brown. Bill is dusky above, pale below. Yellow is less distinct on **immature**'s brown crown. Briefly held **juvenal** plumage has dark streaks on breast and sides. **Winter adults** are duller overall; amount of black on crown varies. **Song** is a series of three or more plaintive, whistled notes: *oh dear me*. **Calls** include a soft *tseep* and a flat *tsick*. RANGE: Fairly common in stunted boreal bogs and in open areas near tree line, especially in willows. Winters in dense woodlands, tangles, and brush; casual in East.

HARRIS'S
SPARROW

winter
adults

breeding

immature

WHITE-
THROATED
SPARROW

tan-striped
morph

juvenile

white-striped
morph

adult
gambelii

WHITE-CROWNED
SPARROW

juvenile
gambelii

adult
nuttalli

adult
leucophrys

immature
leucophrys

juvenile

GOLDEN-CROWNED
SPARROW

winter
adult

immature

breeding

DARK-EYED JUNCO

Junco hyemalis | L 6¼" (16 cm)

Variable; most races have a gray or brown head and breast sharply set off from white belly. White outer tail feathers are conspicuous in flight. **Male** of the widespread "**Slate-colored Junco**" group of subspecies has a dark gray hood; upperparts are entirely gray or have varying amount of brown at center of back. **Female** is brownish gray overall. **Juveniles** of all races are streaked. "Slate-colored" winters mostly in eastern North America; uncommon in the West. Male "**Oregon Junco**" of the West has slaty to blackish hood, rufous-brown to buffy brown back and sides; females have duller hood color. Of the eight subspecies in the "Oregon" group (two are resident in northern Baja California), the more southerly races are paler. "Oregon" types winter mainly in the West; very rare during winter in the East. "**Pink-sided Junco**," *mearnsi* (considered within "Oregon" group)—breeding in the central Rockies and wintering from western Great Plains to the foothills of the Southwest and northern Mexico, rarely to southern California—has very broad, bright pinkish cinnamon sides that sometimes meet across the breast, blue-gray hood, and blackish lores. The "**White-winged Junco**" race, *aikeni*—breeding in the Black Hills area and wintering largely in the Front Range south to north-central New Mexico, rarely on Great Plains and casually to Southwest and California—is mostly pale gray above, usually with two thin, white wing bars; also larger, with a bigger bill and more white on tail. In the "**Gray-headed Junco**" of the southern Rockies, the pale gray hood is barely darker than the underparts; back is rufous. It winters on the western Great Plains and in the foothills of the Southwest and northern Mexico, rarely to California; accidental in Midwest. In much of Arizona and New Mexico, largely resident "Gray-headed," *dorsalis*, has an even paler throat and a large, bicolored bill, black above and bluish below. Intergrades between some races are frequent. Dark-eyed Junco's **song** is a musical trill on one pitch, often heard in winter. Varied **calls** include a sharp *dit* and, in flight, a rapid twittering. Some songs and calls of "Gray-headed" *dorsalis* are more suggestive of Yellow-eyed Junco.

RANGE: Breeds in coniferous or mixed woodlands. In migration and winter, found in a wide variety of habitats.

YELLOW-EYED JUNCO

Junco phaeonotus | L 6¼" (16 cm)

Bright yellow eyes, set off by black lores. Pale gray above, with a bright rufous back and rufous-edged greater wing coverts and tertials; underparts paler gray. **Juveniles** are similar to juveniles of gray-headed races of Dark-eyed Junco; eye is brown, becoming pale before changing to yellow of adult; look for rufous on wings. **Song** is a variable series of clear, thin whistles and trills. **Calls** include a high, thin *seep*, similar to call of Chipping Sparrow.

RANGE: Yellow-eyed Junco is found on coniferous and pine-oak slopes, generally above 6,000 feet. Resident within its range. Some move to lower altitude in winter.

"Oregon"
thurberi

♀

♂

"Slate-colored"
hyemalis

♀

♂

juvenile

"White-winged"
aikeni

♂

DARK-EYED
JUNCO

"Pink-sided"
mearnsi

♂

"Slate-colored" ♂

"Gray-headed"
races

dorsalis

caniceps

juvenile

YELLOW-EYED
JUNCO
palliatus

CHESTNUT-COLLARED LONGSPUR

Calcarius ornatus | L 6" (15 cm)

White tail marked with blackish triangle. Very short primary projection; primary tips barely extend to base of tail. **Breeding adult male**'s black-and-white head, buffy face, and black underparts are distinctive; a few have chestnut on underparts. Lower belly and undertail coverts whitish. Upperparts black, buff, and brown, with chestnut collar, whitish wing bars. **Winter males** are paler; feathers edged in buff and brown, obscuring black underparts. Male has small white patch on shoulder, often hidden; compare with Smith's Longspur (next page). Breeding adult female resembles **winter female** but is darker, usually shows some chestnut on nape. Juvenile's pale feather fringes give upperparts a scaled look; tail pattern and bill shape distinguish juvenile from juvenile McCown's Longspur. Fall and winter birds have grayer bills than McCown's. **Song**, heard only on breeding grounds, is a pleasant rapid warble, given in song flight or from a low perch. Distinctive **call**, a two-syllable *kittle*, repeated one or more times. Also gives a soft, high-pitched rattle and a short *buzz* call.
RANGE: Fairly common; nests in moist upland prairies. Somewhat shy; generally found in dense grass; gregarious in fall and winter. Casual during migration to eastern North America and Pacific Northwest; more regularly to California.

MCCOWN'S LONGSPUR

Calcarius mccownii | L 6" (15 cm)

White tail marked by dark inverted-T shape. Note also stouter, thicker-based bill than bills of other longspurs. Primary projection slightly longer than in Chestnut-collared Longspur; in perched bird, wings extend almost to tip of short tail. **Breeding adult male** has black crown, black malar stripe, black crescent on breast; gray sides. Upperparts streaked with buff and brown, with gray nape, gray rump; chestnut median coverts form contrasting crescent. **Breeding adult female** has streaked crown; may lack black on breast and show less chestnut on wing. In **winter adults**, bill is pinkish with dark tip; feathers are edged with buff and brown. Winter adult female is paler than female Chestnut-collared, with fewer streaks on underparts and a broader buffy eyebrow. Some winter males have gray on rump; variable blackish on breast; retain chestnut median coverts. When with Horned Larks, look for McCown's chunkier, shorter-tailed shape, slightly darker plumage, mostly white tail, thicker bill, and undulating flight. **Juvenile** is streaked below; pale fringes on feathers give upperparts a scaled look; paler overall than juvenile Chestnut-collared. **Song**, heard only on breeding grounds, is a series of exuberant warbles and twitters, generally given in song flight. **Calls** include a dry rattle, a little softer and more abrupt than Lapland Longspur; also gives single finchlike notes.
RANGE: Locally fairly common but range has shrunk significantly since the 19th century. Nests in dry shortgrass plains; in winter, also found in plowed fields and dry lake beds, often amid large flocks of Horned Larks. Very rare visitor to interior California, Nevada. Casual in coastal California and southern Oregon; accidental to the East Coast.

winter ♂

winter ♀

breeding
males

CHESTNUT-COLLARED
LONGSPUR

tail from above

breeding ♂

breeding ♀

MCCOWN'S
LONGSPUR

tail from above

winter ♀

winter ♂

juvenile

SMITH'S LONGSPUR
Calcarius pictus | L 6¼" (16 cm)

Outer two feathers on each side of tail are almost entirely white. Bill is thinner than in other longspurs. Note long primary projection, a bit shorter than Lapland, but much longer than Chesnut-collared or McCown's (preceding page); shows rusty edges to greater coverts and tertials. **Breeding adult male** has black-and-white head, rich buff nape and underparts; white patch on shoulder, often obscured. **Breeding adult female** and all **winter** plumages are duller, crown streaked, chin paler. Dusky ear patch bordered by pale buff eyebrow; pale area on side of neck often breaks through dark rear edge of ear patch. Underparts are pale buff with thin reddish brown streaks on breast and sides. Females have much less white on lesser coverts than males. Typical **call** is a dry, ticking rattle, harder and sharper than call of Lapland and McCown's Longspurs. **Song**, heard in spring migration and on the breeding grounds, is delivered only from the ground or a perch. It consists of rapid, melodious warbles, ending with a vigorous *wee-chew*.
RANGE: Generally uncommon and secretive, especially in migration and winter. Nests on open tundra and damp, tussocky meadows. Winters in open, grassy areas; sometimes seen with Lapland Longspurs. Regular spring migrant in the Midwest, east to western Indiana. Casual vagrant to East Coast from Massachusetts to South Carolina, and to California.

LAPLAND LONGSPUR
Calcarius lapponicus | L 6¼" (16 cm)

Outer two feathers on each side of tail are partly white, partly dark. Note also, especially in winter plumages, the reddish edges on the greater coverts and on the tertials. The reddish edges of the tertials form an indented, or notched, shape. **Breeding adult male**'s head and breast are black and well outlined: a broad white or buffy stripe extends back from eye and down to sides of breast; nape is reddish brown. **Breeding adult female** and all **winter** plumages are duller; note bold dark triangle outlining plain buffy ear patch; dark streaks (female) or patch (male) on upper breast; dark streaks on side. On all winter birds, note broad buffy eyebrow and buffier underparts; belly and undertail are white, unlike Smith's Longspur; also compare head and wing patterns. **Juvenile** is yellowish and heavily streaked above and on breast and sides. **Song**, heard only on the breeding grounds, is a rapid warbling, frequently given in short flights. **Calls** include a musical *tee-lee-oo or tee-dle* and, in flight, a dry rattle distinctively mixed with whistled *tew* notes.
RANGE: Fairly common, Lapland Longspurs breed on Arctic tundra, winter in grassy fields, grain stubble, and on shores. Often found amid flocks of Horned Larks and Snow Buntings; look for Lapland's darker overall coloring and smaller size.

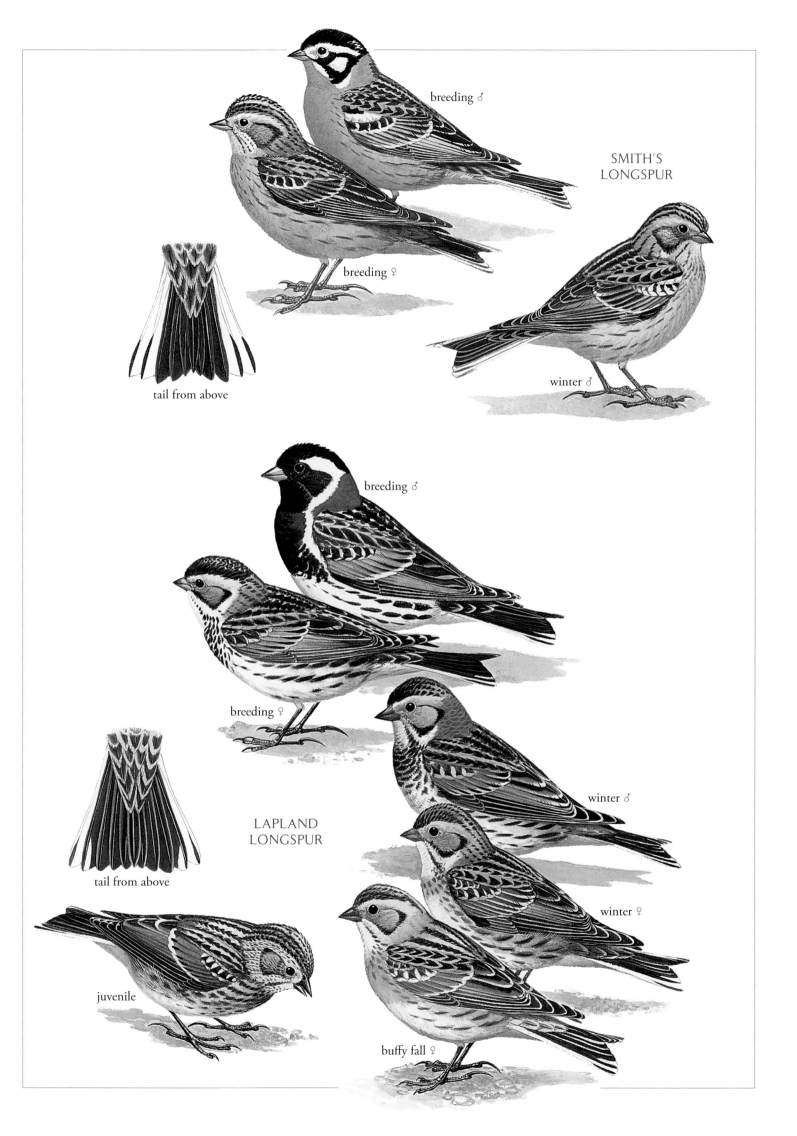

breeding ♂

SMITH'S
LONGSPUR

breeding ♀

winter ♂

tail from above

breeding ♂

breeding ♀

winter ♂

LAPLAND
LONGSPUR

winter ♀

tail from above

juvenile

buffy fall ♀

SNOW BUNTING
Plectrophenax nivalis | L 6¾" (17 cm)

Black-and-white breeding plumage acquired by end of spring by wear. Bill is black in summer, orange-yellow in winter. In all seasons, note long black-and-white wings. **Males** usually show more white overall than **females**, especially in the wings. **Juvenile** is grayish and streaked, with buffy eye ring; very similar to juvenile McKay's Bunting. First-winter plumage, acquired before migration, is darker overall than adult. **Calls** include a sharp, whistled *tew*; a short buzz; and a musical rattle or twitter. **Song**, heard only on the breeding grounds, is a loud, high-pitched musical warbling.
RANGE: Fairly common; breeds on tundra, rocky shores, and talus slopes. During migration and winter, found on shores, especially sand dunes and beaches, in weedy fields and grain stubble, and along roadsides, often in large flocks that may include Lapland Longspurs and Horned Larks (page 328).

MCKAY'S BUNTING
Plectrophenax hyperboreus | L 6¾" (17 cm)

Adult breeding plumage mostly white, with less black on wings and tail than Snow Bunting; **female** shows a white panel on greater coverts. **Winter** plumage is edged with rust or tawny brown, but male is whiter overall than Snow Bunting; female very similar to male Snow Bunting. Juvenile is buffy gray and streaked, with gray head, prominent buffy eye ring; very similar to juvenile Snow Bunting. **Calls** and **song** similar to Snow Bunting.
RANGE: McKay's is known to breed only on Hall and St. Matthew Islands in the Bering Sea. A few sometimes present in late spring on St. Lawrence Island; summer rarely on Pribilofs. Rare to uncommon in winter along west coast of Alaska; casual in winter south on coast to Oregon, in interior of Alaska and on Aleutians. Some authorities think McKay's may be a subspecies of Snow Bunting.

YELLOW-BREASTED BUNTING
Emberiza aureola | L 6" (15 cm)

Has the white outer tail feathers characteristic of most *Emberiza* buntings. **Breeding male** has rufous-brown upperparts, bright yellow underparts, white patch on lesser and median wing coverts. East Asian race, *ornata*, which has reached Alaska, has black on forehead and base of breast band. Winter adult male usually shows features of breeding plumage. **Female** and immatures have striking head pattern with median crown stripe; ear patches with dark border and pale spot in rear; yellowish underparts with sparse streaking; unmarked belly. **Call** is a *tzip*, similar to Little Bunting.
RANGE: A declining, mostly Asian species; casual to Alaska, mostly on western Aleutians.

GRAY BUNTING
Emberiza variabilis | L 6¾" (17 cm)

A large, heavy-billed bunting; shows no white in tail. **Breeding male** is gray overall, prominently streaked with blackish above. Winter males are a bit browner above, paler below. **Adult female** is brown; chestnut rump is conspicuous in flight. **Immature male** resembles adult female above but is mostly gray below with some gray on the head; immature plumage is largely held through first spring. **Call** is a sharp *zhii*.
RANGE: Asian species, casual spring vagrant on western Aleutians.

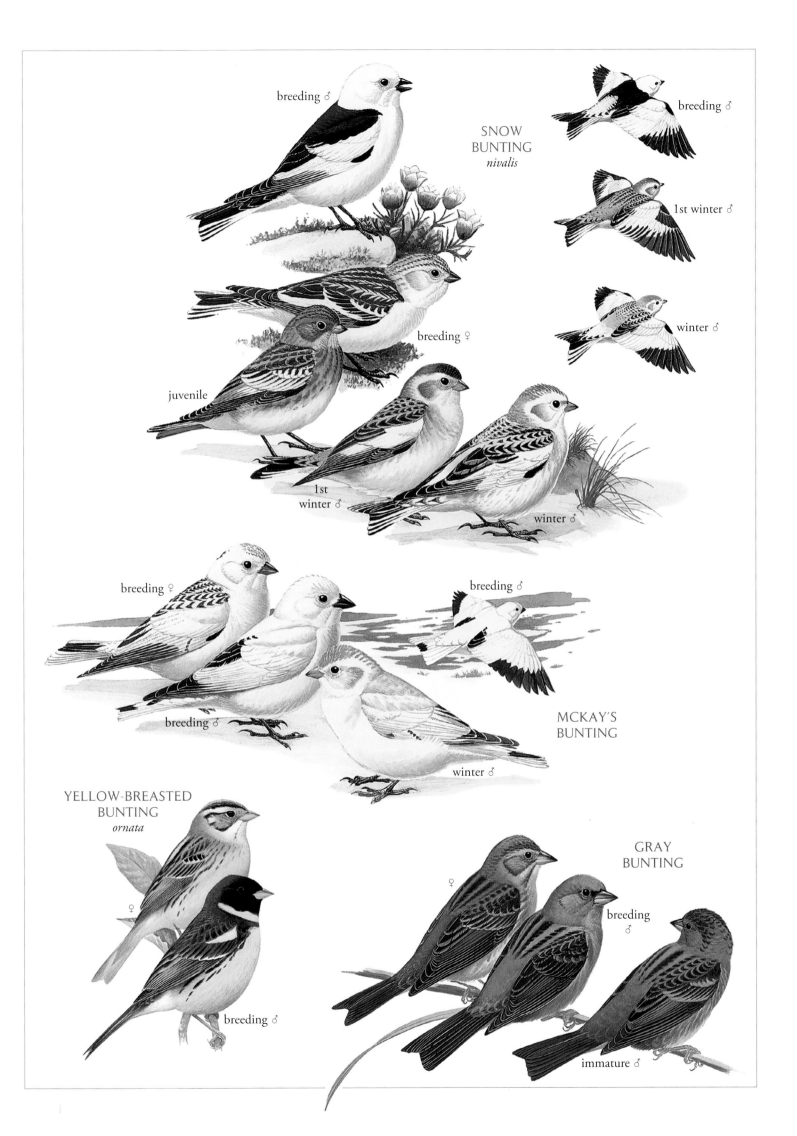

breeding ♂

SNOW
BUNTING
nivalis

breeding ♂

1st winter ♂

winter ♂

juvenile

breeding ♀

1st
winter ♂

winter ♂

breeding ♀

breeding ♂

MCKAY'S
BUNTING

breeding ♂

winter ♂

YELLOW-BREASTED
BUNTING
ornata

♀

breeding ♂

GRAY
BUNTING

♀

breeding
♂

immature ♂

REED BUNTING
Emberiza schoeniclus | L 6" (15 cm)

All records are of pale East Asian *pyrrhulina*, which resembles Pallas's Bunting in female and winter plumages. Reed Bunting has solid chestnut lesser wing coverts. Note also heavy, gray bill with curved culmen; cinnamon wing bars; dark lateral crown stripes, paler median crown stripe. **Breeding male** has black head and throat, a broad white submoustachial stripe, and white nape; upperparts streaked black and rust; rump gray. Underparts white with thin reddish streaks along sides and flanks. **Female** has pale brownish rump, broad buffy white eyebrow; compare with female Pallas's. **Fall male** resembles female Reed, but shows black on throat and a more distinct collar. Active and conspicuous, Reed Bunting often flicks its tail, showing white outer tail feathers. **Calls**, a *seeoo*, falling in pitch; flight note, a hoarse *brzee*. RANGE: Eurasian species, casual vagrant on westernmost Aleutians in late spring; one fall record on St. Lawrence Island.

PALLAS'S BUNTING
Emberiza pallasi | L 5" (13 cm)

Compare with Reed Bunting. Note Pallas's smaller size, smaller two-toned bill with straighter culmen (except **breeding male**, which has black bill); grayish lesser wing coverts; less rufous wing bars; shorter tail. **Female** has more indistinct eyebrow and lateral crown stripes than female Reed; lacks median crown stripe. **Call**, a *cheeep*, recalling Eurasian Tree Sparrow, very unlike Reed's call.
RANGE: Asian species, accidental spring vagrant at Point Barrow, Alaska; St. Lawrence Island; western Aleutians.

LITTLE BUNTING
Emberiza pusilla | L 5" (13 cm)

Small, with short legs, short tail, a small triangular bill, bold creamy white eye ring, chestnut ear patch, and two thin, pale wing bars. Underparts whitish and heavily streaked; outer tail feathers white. In **breeding** plumage, shows chestnut crown stripe bordered by black stripes. Many **males** have chestnut on chin. **Immatures** and winter adults have chestnut crown, tipped and streaked with buff and black; compare especially with female Rustic and Reed Buntings. **Call** note is a sharp *tsick*.
RANGE: Eurasian species, very rare in fall to St. Lawrence Island. Casual on western Aleutians and in California.

RUSTIC BUNTING
Emberiza rustica | L 5¾" (15 cm)

Has a slight crest, whitish nape spot, and prominent pale line extending back from eye. **Male** has black head; upperparts bright chestnut with buff and blackish streaks on back; outer tail feathers white. Underparts are white, with chestnut breast band and streaks on sides. **Female** and fall and winter males have brownish head pattern, with pale spot at rear of ear patch. Female may be confused with Little Bunting; note Rustic's larger size, heavier bill with pink lower mandible, diffuse rusty streaking below, and lack of eye ring. **Call** note is a hard, sharp *jit* or *tsip*. **Song**, a soft, bubbling warble.
RANGE: Eurasian species; uncommon spring migrant on western and central Aleutians, rare in fall; very rare on other islands in Bering Sea; casual elsewhere from West Coast.

fall adult ♂

REED
BUNTING
pyrrhulina

♀

breeding ♂

PALLAS'S
BUNTING
polaris

♀

breeding ♂

immature

LITTLE
BUNTING

breeding ♂

RUSTIC
BUNTING

♀

breeding ♂

CARDINALS, SALTATORS, AND ALLIES (Family Cardinaliade)

In North America, these seedeaters include Northern Cardinal, certain grosbeaks,
*Passerina and other buntings, and Dickcissel. **Species: 42 World, 13 N.A.***

ROSE-BREASTED GROSBEAK
Pheucticus ludovicianus | L 8" (20 cm)

Large size; very large, triangular bill; upper mandible paler than Black-headed Grosbeak. **Breeding male** has rose red breast, white underparts, white wing bars, white rump. Rose red wing linings show in flight. Brown-tipped **winter** plumage is acquired before migration. **Female**'s streaked plumage and yellow wing linings resemble female Black-headed, but underparts are more heavily and extensively streaked. Similar **first-fall male** is buffier above, with buffy wash across breast; often has a few red feathers on breast; red wing linings distinctive. Rich, warbled **songs** of both species are nearly identical. Rose-breasted's **call**, a sharp *eek*, is squeakier than Black-headed's.
RANGE: Common in wooded habitat along watercourses. Rare throughout West in migration.

BLACK-HEADED GROSBEAK
Pheucticus melanocephalus | L 8¼" (21 cm)

Large, with a very large, triangular bill, upper mandible darker than in Rose-breasted Grosbeak. **Male** has cinnamon underparts, all-black head. In flight, both sexes show yellow wing linings. **Female** plumage is generally buffier above and below than female Rose-breasted, with less streaking below. First-fall male Black-headed is rich buff below, with little or no streaking. **Songs** and **calls** of the two species are nearly identical, but Black-headed's call is lower-pitched. Black-headed hybridizes occasionally with Rose-breasted in range of overlap on the Great Plains.
RANGE: Common in open woodlands and forest edges. Casual during migration and winter to the Midwest and East.

CRIMSON-COLLARED GROSBEAK
Rhodothraupis celaeno | L 8½" (22 cm)

Stubby, mostly black bill; long tail; black on head variable. **Adult male** collar and much of underparts an intense shade of red; upperparts darker. **Adult female** is olive above; has thin yellowish wing bars; yellow-green rear collar; yellowish olive underparts. Immatures show less black than female; male shows some red and black patches by first spring. Often skulks on or near ground; often raises rear crown feathers. **Song**, a variable warble; **call**, a penetrating, rising and falling *seeiyu*.
RANGE: Endemic to northeast Mexico; casual to south Texas, mainly in winter.

YELLOW GROSBEAK
Pheucticus chrysopeplus | L 9¼" (24 cm)

Male distinguished by large size, massive bill, yellow plumage. **Females** similar to male but duller; crown streaked; immature male has yellower head than female; like adult male by second fall. Compare with Evening Grosbeak (page 462). **Call** and **song** are like Black-headed's.
RANGE: Mexican species, casual vagrant to southeastern Arizona in late spring to early summer, chiefly in open woodlands and river courses of low mountains.

ROSE-BREASTED
GROSBEAK

breeding
adult ♂

winter
adult ♂

breeding
adult ♂

1st fall ♂

1st spring ♂

♀

BLACK-HEADED
GROSBEAK

♀

breeding adult ♂

1st fall ♂

adult ♀

CRIMSON-
COLLARED
GROSBEAK

adult ♂

adult ♂

adult ♀

YELLOW
GROSBEAK

NORTHERN CARDINAL

Cardinalis cardinalis | L 8¾" (22 cm)

Conspicuous crest; cone-shaped reddish bill. **Male** is red overall, with black face. **Female** is buffy brown or buffy olive, tinged with red on wings, crest, and tail. **Juvenile** browner overall, dusky bill; juvenile female lacks red tones. Bill shape helps distinguish female and juveniles from similar Pyrrhuloxia. **Song** is a loud, liquid whistling with many variations, including *cue cue cue* and *cheer cheer cheer* and *purty purty purty*. Both sexes sing almost year-round. Common **call** is a sharp *chip*.

RANGE: Abundant throughout the East, inhabits woodland edges, swamps, streamside thickets, and suburban gardens. Nonmigratory, but this species has expanded its range northward during the 20th century.

PYRRHULOXIA

Cardinalis sinuatus | L 8¾" (22 cm)

Thick, strongly curved, pale bill helps distinguish this species from female and juvenile Northern Cardinal. **Male** is gray overall, with red on face, crest, wings, tail, and underparts. **Female** shows little or no red. **Song** is a liquid whistle, thinner and shorter than Northern Cardinal's; **call** is a sharper *chink*.

RANGE: Fairly common in thorny brush, mesquite thickets, desert, woodland edges, and ranchlands. Casual to southern California.

DICKCISSEL

Spiza americana | L 6¼" (16 cm)

Yellowish eyebrow, thick bill, and chestnut wing coverts are distinctive. **Breeding male** has black bib under white chin, bright yellow breast. **Female** lacks black bib, but has some yellow on breast; chestnut wing patch muted. **Winter adult male**'s bib is less distinct. **Immatures** are duller overall than adults, breast and flanks lightly streaked; female may show almost no yellow or chestnut. Dickcissel's common **call**, often given in flight, is a distinctive electric-buzzer *bzrrrt*. **Song**, a variable *dick dick dickcissel*.

RANGE: Breeds in open weedy meadows, grainfields, and prairies. Abundant and gregarious, especially in migration when pccurs in large flocks, but numbers and distribution vary locally from year to year outside core breeding range. Irregular east of the Appalachians; occasional breeding is reported outside mapped range. Rare migrant to both coasts; more common in the East, where a few winter.

BLUE GROSBEAK

Passerina caerulea | L 6¾" (17 cm)

Wide chestnut wing bars, large heavy bill, and larger overall size distinguish **male** from male Indigo Bunting (next page). **Females** of these two species also similar; compare bill shape, wing bars, and overall size. Juvenile resembles female; in first fall, some **immatures** are richer brown than female. **First-spring male** shows some blue above and below; resembles adult male by second winter. In poor light, Blue Grosbeak resembles Brown-headed Cowbird (page 448); note Blue Grosbeak's bill shape and wing bars; also the habit of twitching and spreading its tail. Listen for distinctive **call**, a loud, explosive *chink*. **Song** is a series of rich, rising and falling warbles.

RANGE: Fairly common; found in low, overgrown fields, streamsides, woodland edges, and brushy roadsides. Uncommon to rare in fall north to New England and Maritime Provinces.

NORTHERN
CARDINAL
cardinalis

♂

♀

juvenile ♂

PYRRHULOXIA
fulvescens

♀

♂

DICKCISSEL

breeding ♂

breeding ♀

winter adult ♂

BLUE
GROSBEAK
caerulea

breeding
adult ♂

immature

♀

immature ♀

immature ♂

1st spring ♂

INDIGO BUNTING
Passerina cyanea | L 5½" (14 cm)

Breeding male deep blue. Smaller than Blue Grosbeak (preceding page); bill much smaller; lacks wing bars. In **winter** plumage, blue is obscured by brown and buff edges. **Female** is brownish, with diffuse streaking on breast and flanks. Young birds resemble female. **Song**, a series of varied phrases, usually paired.

RANGE: Common in woodland clearings and borders. Uncommon to rare to Maritimes and Newfoundland; rare to Pacific states.

LAZULI BUNTING
Passerina amoena | L 5½" (14 cm)

Adult male bright turquoise above and on throat; cinnamon across breast; thick white upper wing bars. **Female** is grayish brown above, rump grayish blue; whitish underparts with buffy wash across breast. Juveniles resemble female but have distinct fine streaks across breast; immature male is mostly blue by **first spring**. Winter adult male's blue color is obscured by brown and buff edges. **Song** is a series of varied phrases, sometimes paired; faster and less strident than Indigo Bunting's song.

RANGE: Found in open deciduous or mixed woodlands and chaparral, especially in brushy areas near water. Occasionally hybridizes with Indigo.

PAINTED BUNTING
Passerina ciris | L 5½" (14 cm)

Adult male's gaudy colors are retained year-round. **Female** is bright green above, paler yellow-green below. **Juvenile** is much drabber; look for telltale hints of green above, yellow below. Fall molt in eastern nominate race takes place on breeding grounds; western *pallidior* molts on winter grounds. First-winter male resembles adult female; by spring, may show tinge of blue on head, red on breast. **Song** is a rapid series of varied phrases, thinner and sweeter than song of Indigo Bunting. **Call** is a loud, rich *chip*.

RANGE: Locally common in low thickets, streamside brush, and woodland borders. Casual vagrant north on Atlantic coast to New York; west to Midwest, California. Declining in Southeast.

VARIED BUNTING
Passerina versicolor | L 5½" (14 cm)

Breeding male's plumage is colorful in good light; otherwise appears black. In **winter**, colors are edged with brown. **Female** is plain gray-brown or buffy brown above, slightly paler below; resembles female Indigo Bunting but lacks streaks and all wing markings; note also that Varied Bunting's culmen is slightly more curved. First-spring male resembles female. **Song** is similar to song of Painted Bunting.

RANGE: Locally common in thorny thickets in washes and canyons, often near water.

BLUE BUNTING
Cyanocompsa parellina | L 5½" (14 cm)

Smaller than Blue Grosbeak; lacks wing bars. Found in brushy fields and woodland edges. **Adult male** is blackish blue overall, paler blue on crown, cheeks, shoulder, and rump. Immature male similar, but with brownish cast to wings. Contrasting colors and thick, strongly curved bill distinguish male from male Indigo Bunting. Separated by female Blue Bunting's richer, uniform color, and lack of streaking below.

RANGE: Tropical species, very rare and irregular winter visitor to southern Texas.

INDIGO
BUNTING

winter
adult ♂

fall

♀

1st spring ♂

breeding
adult ♂

breeding
adult ♂

LAZULI
BUNTING

1st spring ♂

♀

PAINTED
BUNTING

juvenile

adult ♂

♀

VARIED
BUNTING

breeding
adult ♂

winter
adult ♂

♀

BLUE
BUNTING

♀

adult ♂

BLACKBIRDS (Family Icteridae)

Strong, direct flight and pointed bills mark this diverse group. **Species:** *98 World, 25 N.A.*

BOBOLINK
Dolichonyx oryzivorus | L 7" (18 cm)

Breeding male entirely black below; hindneck is buff, fading to whitish by midsummer; scapulars and rump white. Male in **spring** migration shows pale edgings. **Breeding female** is buffy overall, with dark streaks on back, rump, and sides; head is striped with dark brown. Juvenile resembles female, but lacks streaking below; has indistinct spotting on throat and upper breast. All **fall** birds resemble female, but are rich yellow-buff below—especially, on average, the immatures. In all plumages, note sharply pointed tail feathers. Bobolinks nest primarily in hayfields, weedy meadows, where male's loud, bubbling *bob-o-link* **song**, often given during display flight, is heard in spring and summer. Flight **call** heard year-round is a repeated, whistled *ink*.
RANGE: Most birds migrate east of the Great Plains. Rare in fall on West Coast.

EASTERN MEADOWLARK
Sturnella magna | L 9½" (24 cm)

Black V-shaped breast band on yellow underparts is characteristic of both meadowlark species after post-juvenal molt. In fresh **fall** plumage, birds are more richly colored overall, with partly veiled breast band and rich buffy flanks. On Eastern females, yellow does not reach submoustachial area, and barely does so on males. In widespread northern nominate race, dark centers are visible on central tail feathers, uppertail coverts, secondary coverts, and tertials. Southeastern *argutula* is smaller and darker, especially those from Florida. Southwestern *lilianae* is pale, like Western Meadowlark, but note more extensively white tail. South Texas birds, *hoopesi*, are intermediate in color. **Song** is a clear, whistled *see-you see-yeeer*; distinctive **call** a high, buzzy *drzzt*, given in a rapid series in flight.
RANGE: Generally common in fields and meadows; has declined in the East in recent decades.

WESTERN MEADOWLARK
Sturnella neglecta | L 9½" (24 cm)

Plumages similar to those of Eastern Meadowlark, but in **spring** and summer yellow extends well into the submoustachial area, especially in males; yellow often veiled in **fall**. Lack of dark centers to feathers of upperparts helps to separate from the more easterly races of Eastern, in areas where ranges overlap. Also, in fresh fall and winter plumage, upperparts, flanks, and undertail region are much paler. Distinguished from pale Eastern *lilianae* by mottled cheeks, more mottled postocular and lateral crown stripes, and less white in tail. Northwestern *S. n. confluenta* is darker above and can show dark feather centers like Eastern. **Song** is a series of bubbling, flutelike notes of variable length, usually accelerating toward the end. Sharp *chuck* note; rattled flight **call** similar to Eastern, but lower-pitched; also gives a whistled *wheet*. Eastern and Western Meadowlarks hybridize in Midwest.
RANGE: Westerns are gregarious in winter; large flocks often gather along roadsides, while Easterns usually prefer taller cover.

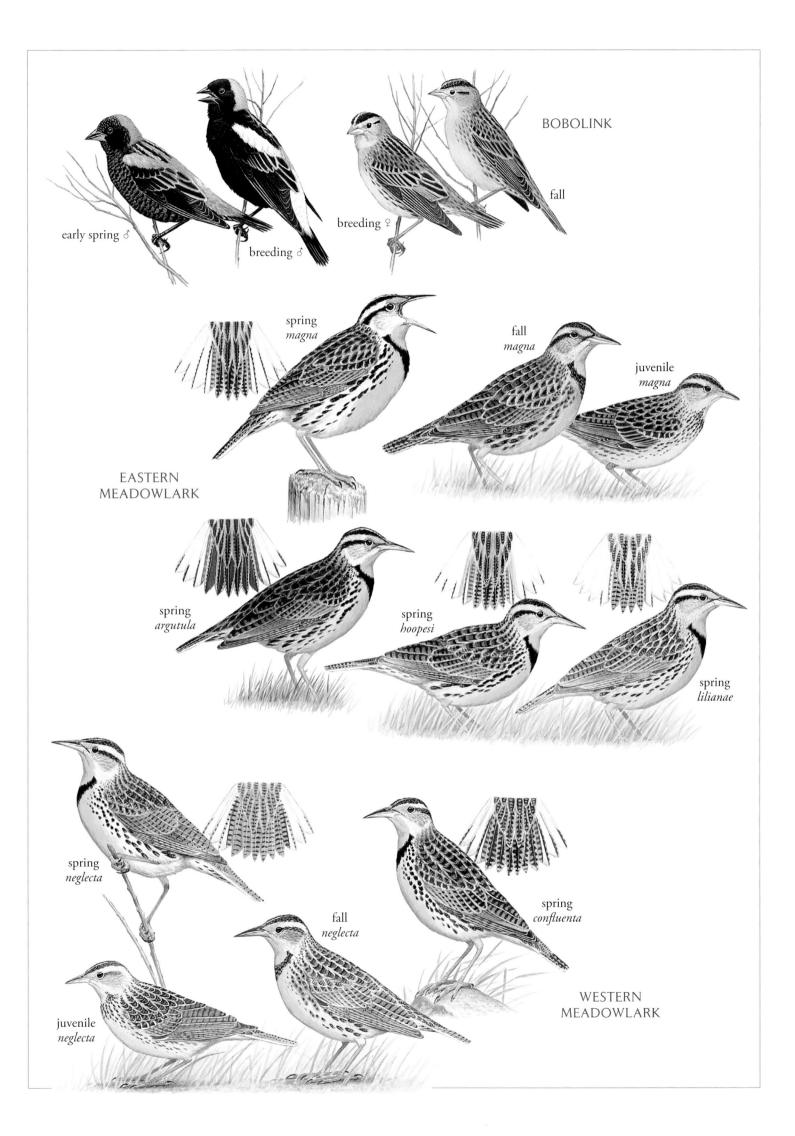

BOBOLINK

early spring ♂

breeding ♂

breeding ♀

fall

spring
magna

fall
magna

juvenile
magna

EASTERN
MEADOWLARK

spring
argutula

spring
hoopesi

spring
lilianae

spring
neglecta

fall
neglecta

spring
confluenta

juvenile
neglecta

WESTERN
MEADOWLARK

YELLOW-HEADED BLACKBIRD
Xanthocephalus xanthocephalus | L 9½" (24 cm)

Adult male's yellow head and breast and white wing patch contrast sharply with black body. **Adult female** is dusky brown, lacks wing patch; eyebrow, lower cheek, and throat are yellow or buffy yellow; belly streaked with white. **Juvenile** is dark brown with buffy edgings on back and wing; head mostly tawny. **Immature male** resembles female but darker; wing coverts tipped with white; acquires adult plumage by following fall. **Song** begins with a harsh, rasping note, ends with a long, descending buzz. **Call** note is a rich *croak*.
RANGE: Locally common throughout most of range; uncommon and very local in the Midwest. Prefers freshwater marshes or reedy lakes; often seen foraging in nearby grasslands or farmlands. Rare fall and winter visitor to the East Coast. Casual in spring and fall as far north as southern Alaska.

RED-WINGED BLACKBIRD
Agelaius phoeniceus | L 8¾" (22 cm)

More rounded wings and usually stouter bill than Tricolored Blackbird. Glossy black male has red shoulder patches broadly tipped with buffy yellow. In perched birds, red patch may not be visible; only the buffy or whitish border shows. **Females** are dark brown above, heavily streaked below; sometimes show a red tinge on wing coverts or pinkish wash on chin and throat. **First-year male** plumage is distinguished from female Tricolored by reddish shoulder patch. Males in races of California's Central Valley and central coast region nearly or totally lack the buffy band behind red shoulder patch; known collectively as "**Bicolored Blackbird**." Females have darker bellies, more like female Tricolored; but note chestnut-buff edging on feathers of upperparts, except when worn away; more rounded wings; stouter bill, except for *aciculatus* of Kern Basin in south-central California, which has a bill like Tricolored. Red-winged Blackbird's **song** is a liquid, gurgling *konk-la-reee*, ending in a trill. Most common **call** is a *chack* note.
RANGE: This abundant, aggressive species is often found in immense flocks in winter. Generally nests in thick vegetation of freshwater marshes, sloughs, and dry fields; forages in surrounding fields, orchards, and woodlands.

TRICOLORED BLACKBIRD
Agelaius tricolor | L 8¾" (22 cm)

More pointed wings and bill than Red-winged Blackbird. Glossy (slightly grayish sheen) black **male** has dark red shoulder patches, often hidden, broadly tipped with white; tips are buffy white in fresh fall plumage. **Females** usually lack any red on shoulder and never show pinkish on throat; plumage is sooty brown and streaked overall; darker than female Red-winged Blackbird, particularly on belly; note more pointed wings and bill. In fresh fall plumage, all Tricolored Blackbirds have grayish buff edging on feathers of upperparts, unlike chestnut-buff of Red-winged. Distinction between females is more difficult when feathers are worn. Tricolor's variety of **calls** are much like Red-winged's, but its harsh, braying *on-ke-kaaangh* **song** lacks Red-winged's liquid tones.
RANGE: Gregarious; found year-round in large flocks in open country and dairy farms; nests in large colonies in marshes.

spring adult ♂

♀

juvenile

immature ♂

YELLOW-HEADED
BLACKBIRD

spring
adult ♂

1st year ♂

adult ♀

RED-WINGED
BLACKBIRD

immature ♀

♀

"Bicolored
Blackbird"

adult ♂

adult ♂

TRICOLORED
BLACKBIRD

breeding ♂

♀

COMMON GRACKLE
Quiscalus quiscula | L 12½" (32 cm)

Long, keel-shaped tail; pale yellow eyes. Plumage appears all-black at a distance. In good light, **males** show glossy purplish blue head, neck, and breast. Widespread race *versicolor*, called "Bronzed Grackle," occurs in most of New England and west of the Appalachians; it has a bronze back, blue head, and purple tail. Smaller "Purple Grackle," *quiscula* of the Southeast, has a narrow bill, purple head, bottle green back, and blue tail. An intergrade population from the mid-Atlantic (*stonei*) shows variable head color and purplish back with iridescent bands of variable color. Females are smaller and duller than males. **Juveniles** are sooty brown, with brown eyes. Common Grackle's **song** is a short, creaky *koguba-leek*; **call** note, a loud *chuck*.

RANGE: Abundant and gregarious, roaming in mixed flocks in open fields, marshes, parks, and suburban areas. Casual in Pacific states and Alaska, primarily in late spring.

BOAT-TAILED GRACKLE
Quiscalus major | ♂ L 16½" (42 cm) ♀ L 14½" (37 cm)

Large grackle with a very long, keel-shaped tail; smaller overall size, duller eye color, and more rounded crown than Great-tailed Grackle. **Adult male** is iridescent blue-black. **Adult female** is tawny brown with darker wings and tail. Male and female eye color is mostly brown in nominate race of coastal Texas and Louisiana and *westoni* of Florida; *alabamensis* of coastal Mississippi to northwest Florida and the largest race, *torreyi*, on the Atlantic coast, have a yellow iris. **First-fall male** is black but lacks iridescence; **juvenile** shows a hint of spotting or streaking on breast. Immatures resemble respective adults by mid-fall. **Calls** include a quiet *chuck* and a variety of rough squeaks, rattles, and other chatter. Most common **song** is a series of harsh *jeeb* notes.

RANGE: This common, noisy grackle seldom strays beyond coastal saltwater marshes except in Florida, where it also inhabits inland lakes and streams. Nests in small colonies. Range is slowly expanding northward on the Atlantic coast.

GREAT-TAILED GRACKLE
Quiscalus mexicanus | ♂ L 18" (46 cm) ♀ L 15" (38 cm)

A large grackle with very long, keel-shaped tail and golden yellow eyes. **Adult male** is iridescent black with purple sheen on head, back, and underparts. **Adult female**'s upperparts are brown; underparts cinnamon-buff on breast to grayish brown on belly; shows less iridescence than male. **Juveniles** resemble adult female but are even less glossy and show some streaking on underparts. Immature males are duller, with shorter tails and darker eyes than adults by mid-fall. Females west of central Arizona are smaller and paler below than races to the east. In narrow zone of range overlap, Great-tailed Grackle is distinguished from Boat-tailed by bright yellow eyes, larger size, and flatter crown. Varied **calls** include clear whistles and loud *clack* notes.

RANGE: Common, especially in open flatlands with scattered groves of trees and in marshes and wetlands. Casual far north of breeding range; rapidly expanding north and west in western U.S.

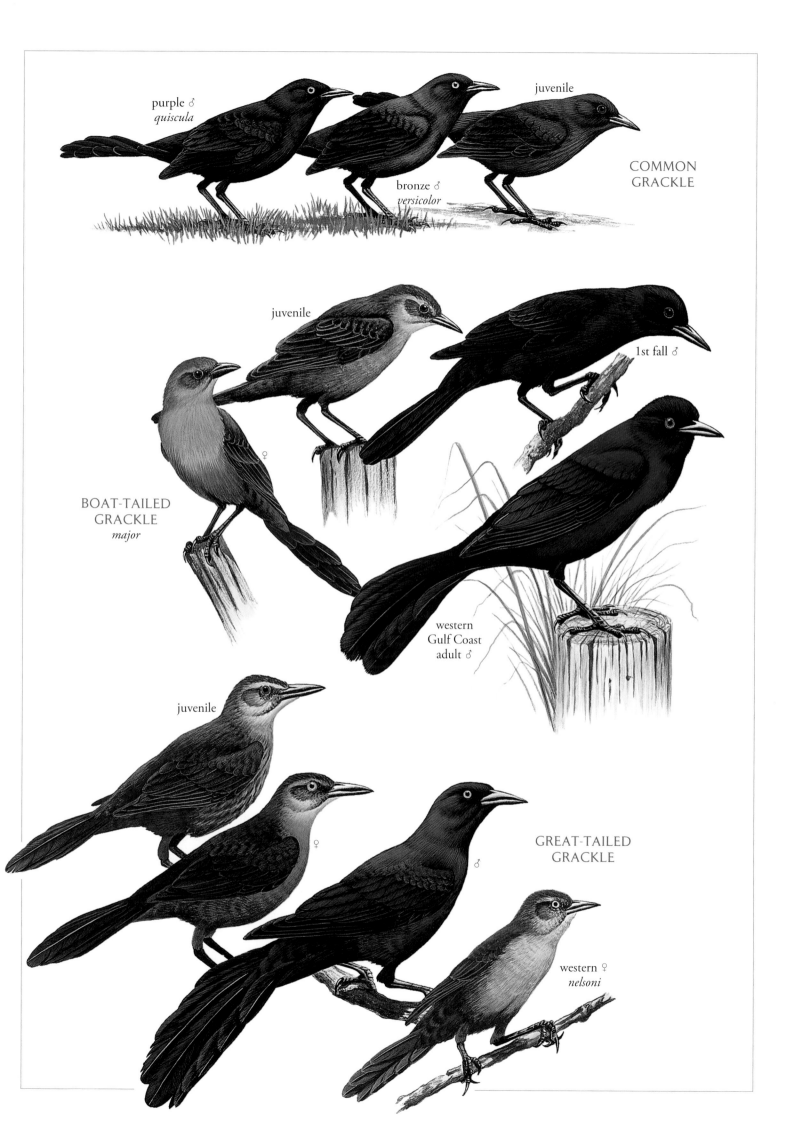

purple ♂
quiscula

bronze ♂
versicolor

juvenile

COMMON
GRACKLE

juvenile

1st fall ♂

BOAT-TAILED
GRACKLE
major

western
Gulf Coast
adult ♂

juvenile

♀

♂

GREAT-TAILED
GRACKLE

western ♀
nelsoni

RUSTY BLACKBIRD
Euphagus carolinus | L 9" (23 cm)

Adults and fall immatures have yellow eyes. Fall adults and immatures are broadly tipped with rust; tertials and wing coverts edged with rust. **Fall female** has broad, buffy eyebrow, buffy underparts, gray rump. **Fall male** is darker; eyebrow fainter. The rusty feather tips wear off by spring, producing the dark **breeding** plumage. Juveniles have dark eyes. **Call** is a harsh *tschak*; **song**, a high, squeaky *koo-a-lee*.

RANGE: Overall uncommon and declining in wet woodlands and swamps; nests in shrubs or conifers near water. Gregarious in fall and winter. Very rare in West.

BREWER'S BLACKBIRD
Euphagus cyanocephalus | L 9" (23 cm)

Male has yellow eyes; **female**'s are usually brown. Male is black year-round, with purplish gloss on head and neck, greenish gloss on body and wings. **Immature males** show variable buffy feather edgings, but never on tertials or wing coverts, as in Rusty Blackbird; note also the shorter, thicker bill. Female and juveniles are gray-brown. Typical **call** is a harsh *check*; **song**, a wheezy *que-ee or k-seee*.

RANGE: Common in open habitats; gregarious. Casual in winter to East Coast.

SHINY COWBIRD
Molothrus bonariensis | 7½" (19 cm)

Sleeker, with longer tail, flatter head, and longer, more pointed bill than Brown-headed. **Male** blackish with purple gloss on head, breast, and back. **Female** and juveniles resemble female Brown-headed except for shape, darker color, and more prominent eyebrow. **Song**, whistled notes followed by trills.

RANGE: Mainly South American species, which spread through the West Indies, arriving in south Florida in 1985. Uncommon in coastal south Florida; rare to eastern Gulf Coast; casual to Texas and North Carolina; accidental to Oklahoma, Maine, and Maritimes.

BROWN-HEADED COWBIRD
Molothrus ater | L 7½" (19 cm)

Male's brown head contrasts with metallic green-black body. **Female** is gray-brown above, paler below. **Juvenile** is paler above, more heavily streaked below; pale edgings give its back a scaled look. Young males molting to adult plumage in late summer are a patchwork of buff, brown, and black. Southwestern *obscurus* is distinctly smaller than eastern nominate *ater*; Rockies and Great Basin *artemisiae* is largest. Feeds with tail cocked up. All cowbirds lay their eggs in nests of other species. Male's **song** is a squeaky gurgling. **Calls** include a harsh rattle and squeaky whistles.

RANGE: Common; found in woodlands, farmlands, and suburbs.

BRONZED COWBIRD
Molothrus aeneus | L 8¾" (22 cm)

Red eyes distinctive at close range. Bill larger than Brown-headed Cowbird. **Adult male** is black with bronze gloss; wings and tail blue-black; thick ruff on nape and back gives a hunchbacked look. **Adult female** of the Texas race, *aeneus*, is duller than the male; **juveniles** are dark brown. In southwestern *loyei*, females and juveniles are gray. Call is a harsh, guttural *chuck*. **Song** is wheezy and buzzy.

RANGE: Locally common in open country, brushy areas, and wooded mountain canyons; forages in flocks.

RUSTY
BLACKBIRD

fall ♀

fall ♂

breeding ♀

breeding ♂

immature ♂

♂

♀

BREWER'S
BLACKBIRD

♀

SHINY
COWBIRD

♂

juvenile

♀

♂

BROWN-HEADED
COWBIRD

♂

southwestern
♀ *loyei*

BRONZED
COWBIRD
aeneus

juvenile

♂

♀

ORCHARD ORIOLE
Icterus spurius | L 7¼" (18 cm)

Adult male is chestnut overall, with black hood. During winter, **female** is olive above, yellowish below. Immature male resembles female; acquires black bib and, sometimes, traces of chestnut by first spring. Smaller size, lack of orange tones or whitish belly, and thinner, more curved bill distinguish female and immature male from Baltimore and Bullock's Orioles. Compare especially with *nelsoni* race of Hooded Oriole. **Calls** include a sharp *chuck*. **Song** is a loud, rapid burst of whistled notes, downslurred at the end.
RANGE: Locally common in suburban shade trees and orchards. Rare vagrant to Arizona, California, the Maritimes.

HOODED ORIOLE
Icterus cucullatus | L 8" (20 cm)

Bill long and slightly curved. **Breeding male** is orange or orange-yellow; note black patch on throat. Western birds (*nelsoni*) are yellower; the two Texas races—*sennetti* (shown) and *cucullatus*—are orange. All **winter adult males** have buffy brown tips on back, forming a barred pattern; compare with Streak-backed Oriole (next page). Hooded **female** and immature male lack pale belly of Bullock's Oriole; bill is more curved. Compare *nelsoni* also with female and immature male Orchard Orioles, which are smaller and purer lemon yellow below, with smaller bill. Immature male acquires black patch on throat by winter. **Calls** include a whistled, rising *wheet*; **song** is a series of whistles, trills, and rattles.
RANGE: Common in varied habitats, especially near palms. Breeding has expanded northward on West Coast.

BALTIMORE ORIOLE
Icterus galbula | L 8¼" (21 cm)

Adult male has black hood and back, bright orange rump and underparts; large orange patches on tail. **Adult females** are brownish olive above and orange below, with varying amounts of black on head and throat; those with maximum black (shown) resemble first-spring males. Extent and intensity of color on underparts of **fall immatures** is highly variable; has distinctly contrasting wing bars and palish lores; no eye line or yellowish eyebrow. Common **call** is a rich *hew-li*; also gives a series of rattles. **Song** is a musical, irregular sequence of *hew-li* and other notes.
RANGE: Common breeder in deciduous woodland over much of the East. Some winter in the South. Rare vagrant to West.

BULLOCK'S ORIOLE
Icterus bullockii | L 8¼" (21 cm)

Formerly considered same species as Baltimore Oriole; some interbreeding on Great Plains. **Adult male** has less black on head: crown, eye line, throat patch; note bold white patch on wing, entirely orange outer tail feathers. **Females** and **immatures** have yellow throat and breast, unlike Baltimore's extensive orange; note Bullock's dark eye line, less contrasting wing bars. Most birds show dark "teeth" intruding into white of median covert bar. By **first spring**, males have black lores, throat. **Song**, mix of whistles and harsher notes; **call**, a harsh *cheh* or series of same.
RANGE: Breeds where shade trees grow. Small numbers winter in coastal California. Casual vagrant to the East; most reports are of dull, immature Baltimores.

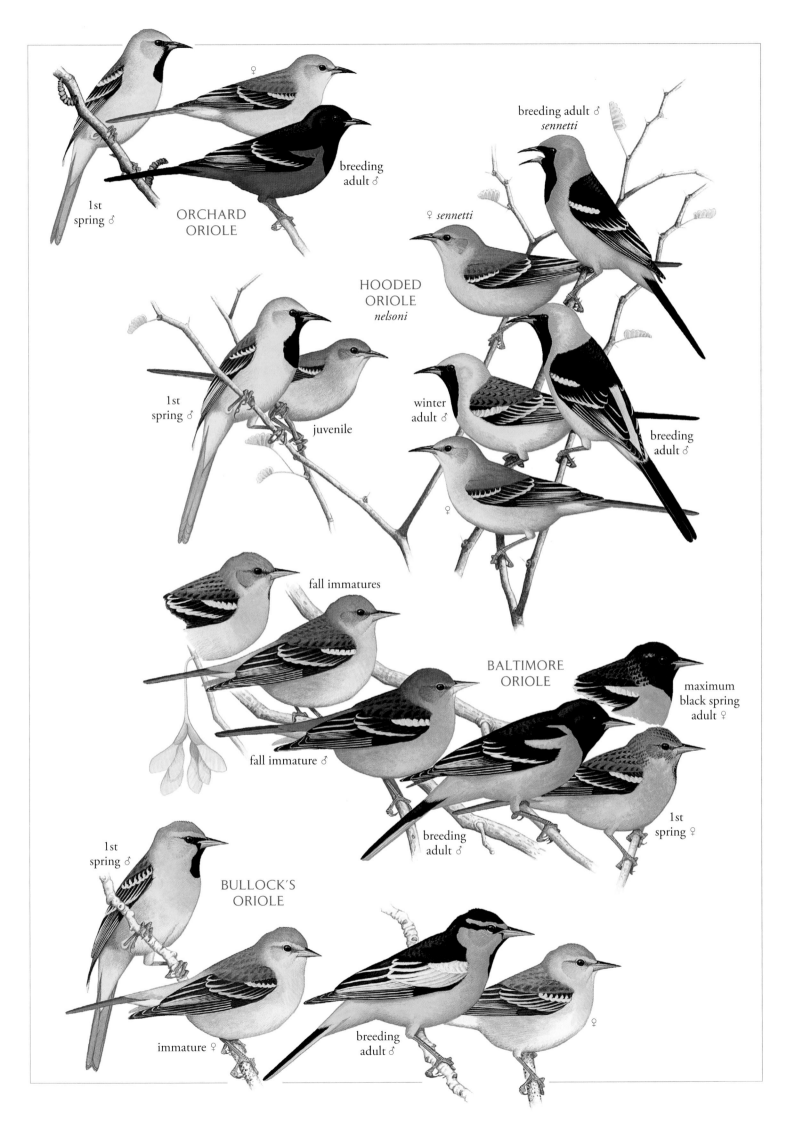

1st
spring ♂

**ORCHARD
ORIOLE**

♀

breeding
adult ♂

breeding adult ♂
sennetti

♀ *sennetti*

**HOODED
ORIOLE**
nelsoni

1st
spring ♂

juvenile

winter
adult ♂

breeding
adult ♂

♀

fall immatures

**BALTIMORE
ORIOLE**

maximum
black spring
adult ♀

fall immature ♂

breeding
adult ♂

1st
spring ♀

1st
spring ♂

**BULLOCK'S
ORIOLE**

immature ♀

breeding
adult ♂

♀

BLACK-VENTED ORIOLE *Icterus wagleri* | L 8½" (22 cm)

Long, narrow bill and long, graduated tail. **Adult** has solid black head, back, undertail coverts, tail, and wings, except for yellow shoulders; the border between breast and belly is chestnut. **First-spring** has black lores and chin; streaked back. Juvenile lacks black bib. Black-vented Oriole's **call** is a nasal *nyeh*, often repeated.
RANGE: Resident from central Nicaragua to northern Mexico; accidental to south and west Texas, southeast Arizona.

STREAK-BACKED ORIOLE *Icterus pustulatus* | L 8¼" (21 cm)

Distinguished from winter Hooded Oriole (preceding page) by broken streaks on upper back; deeper orange head; and much thicker-based, straighter bill. **Female** is duller than **male**. Immatures resemble adult female. *Wheet* **call** is softer than Hooded Oriole's call and does not rise in pitch. Chatter calls resemble those of Baltimore Oriole.
RANGE: Mexican species, casual in fall and winter in southeastern Arizona, southern California; accidental to New Mexico, east Texas, and Wisconsin.

ALTAMIRA ORIOLE *Icterus gularis* | L 10" (25 cm)

Distinguished from Hooded Oriole (preceding page) by larger size, much thicker-based bill, and, in **adult**, by orange shoulder patch. Lower wing bar whitish. **Immatures** are duller than adults, shoulder patch yellow; like adults by second fall. **Calls** include a low, raspy *ike ike ike*; **song** is a series of clear, varied whistles.
RANGE: Found in southernmost Texas in tall trees and willows.

AUDUBON'S ORIOLE *Icterus graduacauda* | L 9½" (24 cm)

Male has greenish yellow back. Female is slightly duller. Rather secretive; often seen foraging on ground. **Song** is a series of soft, tentative, three-note warbles.
RANGE: Tropical species, resident but uncommon in south Texas.

SPOT-BREASTED ORIOLE *Icterus pectoralis* | L 9½" (24 cm)

Adults have an orange or yellow-orange patch on shoulders; black lores and throat; dark spots on upper breast; extensive white on wings. **Juveniles** are yellower overall; **immatures** may lack breast spots. **Song** is a long, loud series of melodic whistles.
RANGE: Middle American species, introduced and now established in southern Florida. Prefers suburban gardens. Florida population has declined over past three decades.

SCOTT'S ORIOLE *Icterus parisorum* | L 9" (23 cm)

Adult male's black hood extends to back and breast; rump, wing patch, and underparts bright lemon yellow. Adult female is olive and streaked above, dull greenish yellow below; throat shows variable amount of black. **Immature male**'s head is mostly black by first spring. **Females** and immatures larger, grayer, more streaked above; have straighter bill than female Hooded Oriole (preceding page). Common **call** note is a harsh *shack*; **song** is a mixture of rich, whistled phrases.
RANGE: Found in arid and semiarid habitats. Casual east to Minnesota, Wisconsin, Louisiana. A few winter in southern California.

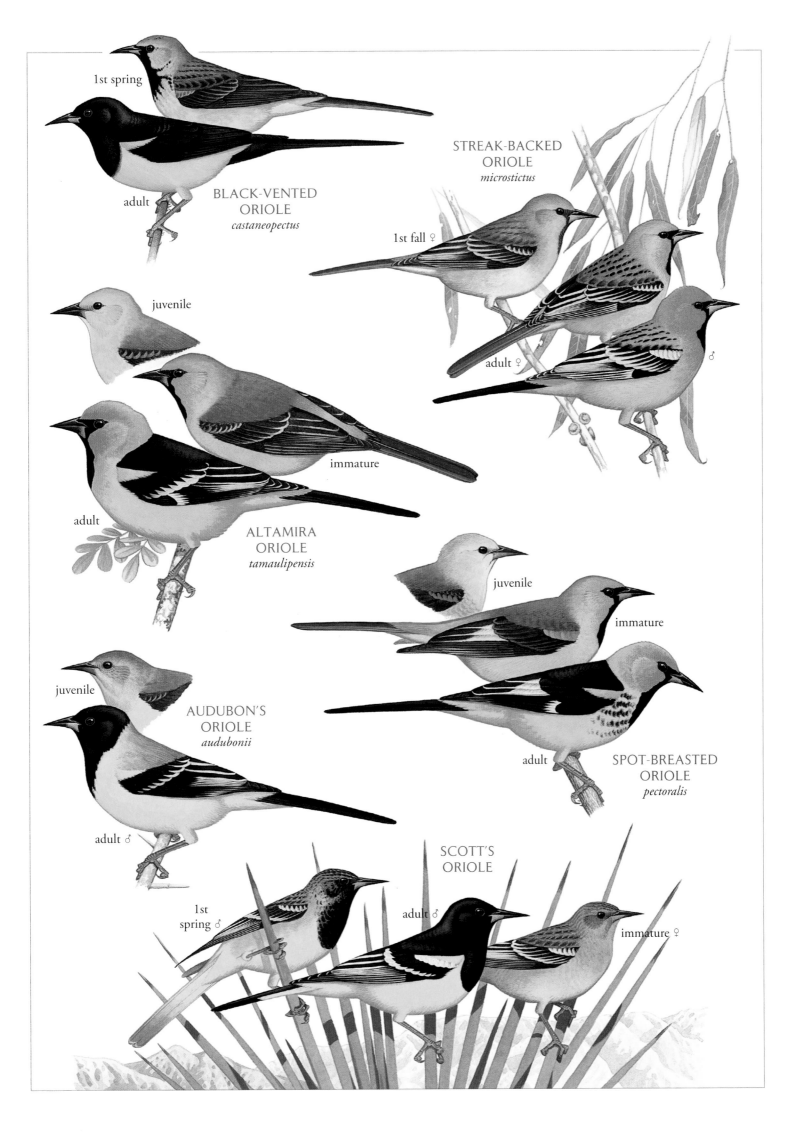

1st spring

adult

BLACK-VENTED ORIOLE
castaneopectus

STREAK-BACKED ORIOLE
microstictus

1st fall ♀

adult ♀

♂

juvenile

immature

adult

ALTAMIRA ORIOLE
tamaulipensis

juvenile

immature

adult

SPOT-BREASTED ORIOLE
pectoralis

juvenile

AUDUBON'S ORIOLE
audubonii

adult ♂

SCOTT'S ORIOLE

1st spring ♂

adult ♂

immature ♀

FRINGILLINE & CARDUELINE FINCHES AND ALLIES
(FAMILY CARDINALIADE)

Seedeaters with undulating flight. Many nest in the North; in fall, flocks of "winter finches" may roam south.
***Species:** 168 World, 23 N.A.*

ORIENTAL GREENFINCH
Chloris sinica | L 6" (15 cm)

Adult male has greenish face and rump, dark gray nape and crown, bright yellow wing patch and undertail coverts. **Adult female** is paler, with a brownish head. **Juvenile** has same yellow areas as adults but is streaked overall.
RANGE: Asian species. Casual migrant, mainly in spring, on outer Aleutians.

BRAMBLING
Fringilla montifringilla | L 6¼" (16 cm)

Adult male has tawny orange shoulders, spotted flanks; head and back fringed with buff in fresh fall plumage that wears down to black by spring. **Female** and juvenile have mottled crown, gray face, striped nape. Flight **call**, a nasal *check-check-check*; also gives a nasal *zwee*.
RANGE: Eurasian species; fairly common but irregular migrant on Aleutians; rare on Pribilofs and St. Lawrence Island; casual in fall and winter in Canada and northern U.S.

GRAY-CROWNED ROSY-FINCH
Leucosticte tephrocotis | 5½-8¼" (14-21 cm)

Dark brown, with gray on head; pink on wings and underparts; underwings silvery. Female less pink, **juveniles** grayish. All have yellow bill in **winter**, black by spring. Western *littoralis*, "**Hepburn's Rosy-Finch**," and much larger, darker Pribilofs *umbrina* and Aleutians *griseonucha* show more gray on face than widespread nominate race and closely allied *dawsoni* from Sierra Nevada and White Mountains. **Call**, a high, chirping *chew*, often given in courtship flight.
RANGE: Descends from higher elevations in winter. Some races are migratory.

COMMON CHAFFINCH
Fringilla coelebs | L 6" (15 cm)

Has white patches on lesser coverts, base of primaries, wing bar; outer tail feathers white. **Male**'s crown and nape are blue-gray; shows pinkish below, pinkish brown above. **Female**'s head is mostly gray with brown lateral stripes. **Call** a metallic *pink-pink*; also a *hweet*.
RANGE: Palearctic species; casual to northeastern North America; reports elsewhere possibly escaped cage birds.

BROWN-CAPPED ROSY-FINCH
Leucosticte australis | L 6" (15 cm)

Plumages, behavior, and voice like Gray-crowned. Lacks gray head band of other North American Rosy-Finches. **Male** rich brown; darker crown; extensive pink on underparts. **Female** much drabber; some young female Gray-crowneds can be very similar.
RANGE: Uncommon in high-elevation breeding range; lower in winter.

BLACK ROSY-FINCH
Leucosticte atrata | L 6" (15 cm)

Plumages, behavior, and voice like Gray-crowned. Darkest Rosy-Finch. **Male** is blackish; in fresh plumage, scaled with silver-gray; has gray head band; shows extensive pink. **Female** is blackish gray with little pink.
RANGE: Casual to California, and Arizona.

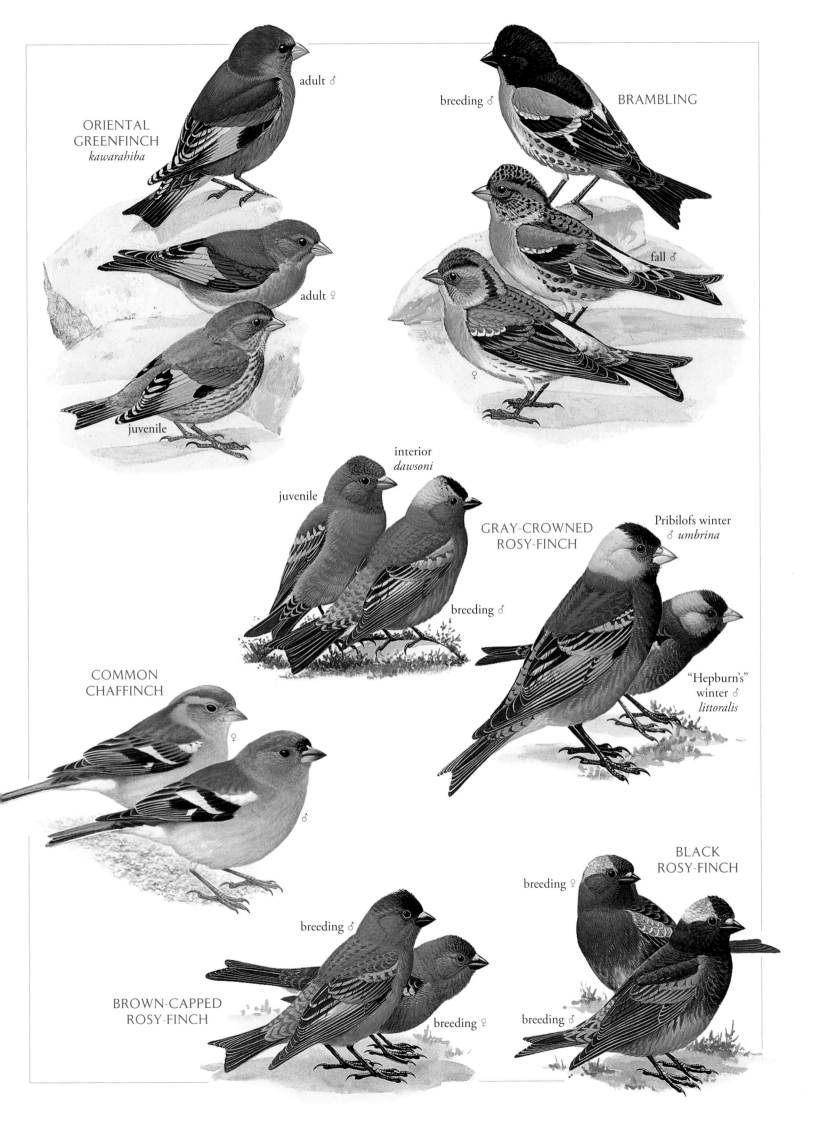

ORIENTAL
GREENFINCH
kawarahiba

adult ♂

adult ♀

juvenile

breeding ♂

BRAMBLING

fall ♂

♀

interior
dawsoni

juvenile

GRAY-CROWNED
ROSY-FINCH

Pribilofs winter
♂ *umbrina*

breeding ♂

"Hepburn's"
winter ♂
littoralis

COMMON
CHAFFINCH

♀

♂

BLACK
ROSY-FINCH

breeding ♀

breeding ♂

BROWN-CAPPED
ROSY-FINCH

breeding ♂

breeding ♀

PURPLE FINCH
Carpodacus purpureus | L 6" (15 cm)

Not purple, but rose red over most of **adult male**'s body, brightest on head and rump. Rose color is acquired in second fall. Back is streaked; tail notched. Pacific coast race, *californicus*, is buffier below and more diffusely streaked than the widespread *purpureus*, especially in females. **Adult female** and immatures are heavily streaked below; closely resemble Cassin's Finch. Ear patch and whitish eyebrow and submoustachial stripe are slightly more distinct in Purple Finch; bill slightly stubbier and more curved; undertail coverts are often not streaked. Compare also with female House Finch. **Calls** include a musical *chur-lee* and, in flight, a sharp *pit*, a bit sharper in *californicus*. **Song** is a rich warbling, longer and more variable in *purpureus*; shorter than Cassin's song, lower and less strident than House Finch song.
RANGE: Fairly common; found in coniferous or mixed woodland borders, suburbs, parks, and orchards; in the Pacific states, inhabits coniferous forests, oak canyons, and lower mountain slopes.

CASSIN'S FINCH
Carpodacus cassinii | L 6¼" (16 cm)

Crimson of **adult male**'s cap ends sharply at brown-streaked nape. Throat and breast paler than Purple Finch; streaks on sides and malar stripe more distinct. Red hues begin to appear late in second summer. Tail strongly notched. Undertail coverts always distinctly streaked, unlike many Purples. **Adult female** and immatures otherwise closely resemble Purple Finch. Cassin's facial pattern is slightly less distinct; culmen is straighter and longer; has longer primary projection. Cassin's gives a dry *kee-up* or *tee-dee-yip* **call**. Lively **song**, a variable warbling, longer and more complex than song of Purple Finch, especially *californicus*.
RANGE: Fairly common in upper mountain forests, evergreen woodlands. Casual in winter east of range and to West Coast.

HOUSE FINCH
Carpodacus mexicanus | L 6" (15 cm)

Male has brown cap; front of head, bib, and rump are typically red but can vary to orange or occasionally yellow. Bib is clearly set off from streaked underparts. Tail is squarish. **Adult female** and juveniles are streaked with brown overall; lack distinct ear patch and eyebrow of Purple and Cassin's Finches. Young males acquire adult coloring by first fall. Lively, high-pitched **song** consists chiefly of varied three-note phrases; includes strident notes, unlike Purple Finch's song; usually ends with a nasal *wheer*. **Calls** include a whistled *wheat*.
RANGE: Abundant; found in semiarid lowlands and slopes up to about 6,000 feet. Introduced in the East in the 1940s, where its range is fast expanding; range has expanded rapidly also in the West; especially numerous in towns.

COMMON ROSEFINCH
Carpodacus erythrinus | L 5¾" (15 cm)

Strongly curved culmen. Lacks distinct eyebrow. **Adult male** has red head, breast, and rump. **Female** and immatures diffusely streaked above and below except on pale throat. **Call** is a soft, nasal *djuee*.
RANGE: Eurasian species; very rare migrant, chiefly in spring, on the western Aleutians and other western Alaska islands.

PURPLE FINCH

♀ *californicus*

♀ *purpureus*

Pacific coast
adult ♂
californicus

eastern adult ♂
purpureus

CASSIN'S FINCH

adult ♀

adult ♂

HOUSE
FINCH
frontalis

typical ♂

variant ♂

♀

COMMON
ROSEFINCH
grebnitskii

♀

adult ♂

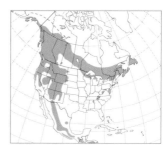

RED CROSSBILL
Loxia curvirostra | L 5½–7¾" (14–20 cm)

Bill with crossed tips identifies both crossbill species. Red Crossbill's dark brown wings lack the bold white bars of White-winged Crossbill. Plumage highly variable. Most **males** are reddish overall, brightest on crown and rump, but may be pale rose or scarlet or largely yellow; always have red or yellow on throat. Most **females** are yellowish olive; may show patches of red; throat is always gray, except in a small northern subspecies where yellow extends to center, but not sides, of throat. **Juveniles** are boldly streaked; a few juveniles and a very few adult males show white wing bars, the upper bar thinner than the lower. Immatures are like the respective adult but juvenile wing is retained. All birds except adult males have olive edges on wings. Subspecies vary widely in size, including bill size; extremes are shown here. All have their "home range." Distinct differences in vocalizations have led some authorities to believe that there may be a half dozen or more cryptic separate species in the Red Crossbill complex. All have large heads and short, notched tails. **Calls**, given chiefly in flight, vary from one subspecies to another. **Song** begins with several two-note phrases followed by a warbled trill.

RANGE: Fairly common, Red Crossbills inhabit coniferous woods. May nest at any time of year, especially in southern range. Highly irregular in their wanderings, dependent upon cone crops. Any race may turn up almost anywhere. Irruptive migrant. Has bred as far south as Georgia.

WHITE-WINGED CROSSBILL
Loxia leucoptera | L 6½" (17 cm)

All ages have black wings with white tips on the tertials; two bold, broad white wing bars. Upper wing bar is often hidden by scapulars. **Adult male** is bright pink overall, paler in winter. **Immature male** is largely yellow, with patches of red or pink. **Adult female** is mottled with yellowish olive or grayish; rump pale yellow; underparts grayish olive, with yellow wash on breast and sides. **Juvenile** is heavily streaked; wing bars thinner than in adults. White-winged Crossbill's distinctive flight **call** is a rapid series of harsh *chet* notes. Variable **song**, often delivered in display flight, combines harsh rattles and musical warbles.

RANGE: Inhabits coniferous woods. Highly irregular in its wanderings, dependent upon spruce cone crops. Irruptive migrant.

PINE GROSBEAK
Pinicola enucleator | L 9" (23 cm)

Large and plump, with long tail. Bill is dark, stubby, strongly curved. Two white wing bars, sometimes tinged with pink in adult male. **Male**'s gray plumage is tipped with red on head, back, and underparts; pinker in fresh fall plumage. **Female** and immatures are grayer overall; head, rump, and underparts variably yellow or reddish; some females and immature males are **russet**. Typical flight **call** is a whistled *pui pui pui*; alarm call, a musical *chee-vli*. Location call shows considerable geographic variation. **Song** is a rather short, musical warble.

RANGE: Fairly common; inhabits open coniferous woods. In winter, found also in deciduous woods, orchards, and suburban shade trees. Usually unwary and approachable. Irruptive winter migrant in the East.

variant ♂

juvenile

RED
CROSSBILL

northern
minor ♀

typical ♀

southwestern ♂
stricklandi

typical ♂

immature ♂

juvenile

WHITE-WINGED
CROSSBILL
leucoptera

♀

winter
adult ♂

♀

adult ♂

PINE
GROSBEAK
leucura

russet variant

PINE SISKIN
Spinus pinus | L 5" (13 cm)

Prominent streaking; yellow at base of tail and in flight feathers conspicuous in flight; bill thinner than in other finches. **Juvenile**'s overall yellow tint is lost by late summer. **Calls** include a rising *tee-ee* and, in flight, a harsh, descending *chee*. **Song** is similar to that of American Goldfinch but much huskier.

RANGE: Gregarious; may flock with goldfinches in winter. Found in coniferous and mixed woods in summer; forests, shrubs, and fields in winter. Winter range is erratic. Casual to Bering Sea islands.

AMERICAN GOLDFINCH
Spinus tristis | L 5" (13 cm)

Breeding adult male is bright yellow with black cap; black wings have white bars, yellow shoulder patch; uppertail and undertail coverts white, with black-and-white tail. **Female** is duller overall, olive above; lacks black cap and yellow shoulder patch. White undertail coverts distinguish female from most Lesser Goldfinches. **Winter adults** and immatures are either brownish or grayish above; male may show some black on forehead. **Juvenal** plumage, held into Nov., has cinnamon-buff wing markings and rump. **Song** is a lively series of trills, twitters, and *swee* notes. Distinctive flight **call**, *per-chik-o-ree*.

RANGE: Common and gregarious; found in weedy fields, open second-growth woodlands, and roadsides, especially in thistles and sunflowers. Casual north to southeast and south-coastal Alaska.

LESSER GOLDFINCH
Spinus psaltria | L 4½" (11 cm)

All birds have a white wing patch at base of primaries. Entire crown black on **adult male**; back varies from black in eastern part of range to greenish in western birds. Most **adult females** are dull yellow below; except for a few extremely pale birds, they lack the white undertail coverts typical of American Goldfinch. **Immature male** lacks full black cap. Juveniles resemble adult female. **Call**, a plaintive, kittenlike *tee-yee*. **Song** is somewhat similar to that of American Goldfinch.

RANGE: Common in dry, brushy fields, woodland borders, and gardens. Range expanding northward. Casual in Great Plains; accidental in the East.

LAWRENCE'S GOLDFINCH
Spinus lawrencei | L 4¾" (12 cm)

Wings extensively yellow; upperparts grayish in breeding plumage; large yellow patch on breast. **Male** has black face and yellowish tinge on back. **Winter** birds are browner above, duller below. **Juvenile** is faintly streaked, unlike other goldfinches. **Call** is a bell-like *tink-ul*. Mixes *tink* notes into jumbled, melodious **song**. Lawrence's and Lesser Goldfinches often mimic other species' songs.

RANGE: Fairly common in spring and early summer; may sometimes flock with other goldfinches, but generally prefers drier interior foothills and mountain valleys; also western fringe of desert near watercourses. Erratic but usually uncommon at other seasons. Irregular fall movements to southwestern U.S. Casual as far east as western Texas.

PINE SISKIN
pinus

juvenile

breeding ♀

AMERICAN
GOLDFINCH
tristis

breeding ♂

winter
adult ♂

winter ♀

juvenile

LESSER
GOLDFINCH
hesperophila

♀

pale ♀

black-backed
adult ♂
psaltria

green-backed
adult ♂

LAWRENCE'S
GOLDFINCH

winter ♀

immature ♂

winter ♂

juvenile

breeding ♂

COMMON REDPOLL
Acanthis flammea | L 5¼" (13 cm)

Red or orange-red cap or "poll," black chin. Closely resembles Hoary Redpoll, but usually has distinct streaks on flanks, rump, undertail coverts; bill is slightly larger. **Male** usually has bright rosy breast and sides. Both sexes paler, buffier in winter. **Juveniles** lack red cap until late-summer molt; males acquire pinkish breast by end of second summer. Extent of interbreeding between Common and Hoary Redpolls is unknown. When perched, Common gives a *swee-ee-eet* **call**; flight call, a dry rattling. **Song** combines trills and twittering.
RANGE: Fairly common; breeds in subarctic forests and tundra scrub. Unwary. Forms large winter flocks; frequents brushy, weedy areas, also catkin-bearing trees like alder and birch.

HOARY REDPOLL
Acanthis hornemanni | L 5½" (14 cm)

Closely resembles Common Redpoll but is usually frostier and paler overall, with a slightly smaller bill. Streaking below and on rump minimal or absent. **Male**'s breast is usually paler and pinker than on Common; color does not extend to cheeks or sides. The race *hornemanni*, of Canadian Arctic islands and Greenland, is larger and paler than more widespread *exilipes*. **Calls** and **song** similar to Common.
RANGE: Fairly common; nests on or near the ground above Arctic tree line. Rare sightings, especially of *exilipes*, occur south of Canada in winter, almost always with Commons.

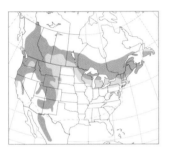

EVENING GROSBEAK
Coccothraustes vespertinus | L 8" (20 cm)

Stocky, noisy finch. Big bill pale yellow or greenish by spring, whitish by fall; prominent white patch on inner wing. Yellow forehead and eyebrow on **adult male**; dark brown and yellow body. Grayish tan **female** has thin, dark malar stripe, white-tipped tail; second wing patch, on primaries, is conspicuous in flight. **Juveniles** have brown bills; female resembles adult female; male yellower overall, wing and tail like adult male. Loud, strident **call**: *clee-ip* or *peeer*.
RANGE: Breeds in mixed woods; in the West, mainly in mountains. In winter frequents woodlots, shade trees, and feeders; numbers and range limits vary greatly.

HAWFINCH
Coccothraustes coccothraustes | L 7" (18 cm)

Stocky; yellowish brown above; pinkish brown below; has black throat and lores; shows conspicuous white band on extended wing. Big bill is blue-black in spring, yellowish in fall. Female resembles **male**, but is duller; has grayish secondaries and inner primaries. Walks with parrotlike waddle. **Call** is a loud, explosive *ptik*.
RANGE: Eurasian species. Rare spring stray on western Aleutians; casual off other islands of western Alaska.

EURASIAN BULLFINCH
Pyrrhula pyrrhula | L 6½" (17 cm)

Cheeks, breast, and belly intense reddish pink in **male**, brown in **female**. Black cap and face, gray back, prominent whitish bar on wing, distinct white rump. In profile, top of head and bill form unbroken curve. Juvenile resembles female, but with brown cap. **Call** is a soft, piping *pheew*.
RANGE: Eurasian species. Casual migrant on Aleutians and St. Lawrence Island; casual in winter on Alaskan mainland.

juvenile

breeding ♀

winter ♀

COMMON
REDPOLL
flammea

breeding ♂

winter ♂

HOARY
REDPOLL

winter ♀
exilipes

winter ♂
hornemanni

winter ♂
exilipes

EVENING
GROSBEAK
vespertinus

♂

♀

breeding ♂

juvenile
♂

breeding ♀

breeding ♂

HAWFINCH
japonicus

EURASIAN
BULLFINCH
cassinii

♂

♀

OLD WORLD SPARROWS (Family Passeriade)

Old World family. Gregarious; two species have become established in North America. **Species:** *40 World, 2 N.A.*

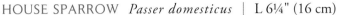

HOUSE SPARROW *Passer domesticus* | L 6¼" (16 cm)

Breeding male has gray crown, chestnut nape, black bib, black bill. Fresh **fall** plumage is edged with gray, obscuring these markings; bill becomes brownish. **Female** is best identified by the combination of streaked back, buffy eye stripe, and unstreaked breast. Juveniles resemble adult female.

RANGE: Common and aggressive, House Sparrows are omnipresent in populated areas. Gregarious in winter. Also known as English Sparrow.

EURASIAN TREE SPARROW *Passer montanus* | L 6" (15 cm)

Brown crown, black ear patch, black throat distinguish **adult**. Compare with House Sparrow's gray crown and more extensively black throat. **Juvenile** has dark mottling on crown, dark gray throat and ear patch.

RANGE: Old World species, introduced and now locally common in parks, suburbs, and farmlands within mapped range. Accidental in Indiana, Wisconsin, Manitoba, Kentucky, Ontario. Gregarious all year.

WEAVERS (Family Ploceidae)

Large, primarily African family. Breeding males are often highly colored. Known for elaborate woven nests. **Species:** *108 World, 1 N.A.*

ORANGE BISHOP *Euplectes franciscanus* | L 4" (10 cm)

Breeding male is bright orange-red with black cap, breast, and belly; long tail coverts obscure tail. **Females**, immatures, and winter birds are streaked above. Compare especially with Grasshopper Sparrow (page 416); note Orange Bishop's thicker, pinkish bill; short, blunt tail, often flicked open. Complex **song**, high and buzzy. **Calls** include a sharp *tsip* and a mechanical *tsik tsik tsk*.

RANGE: Native to sub-Saharan Africa; widely introduced. Established in Los Angeles (1980s) and Phoenix (1998) areas, where it favors weedy areas, especially river bottoms.

ESTRILDID FINCHES (Family Estrildidae)

Large, Old World family found from Africa to Australia and South Pacific islands. Most are small, with pointed tails. Related to weavers. **Species:** *108 World, 1 N.A.*

NUTMEG MANNIKIN *Lonchura punctulata* | L 4½" (11 cm)

Small, with heavy bill, pointed tail. **Adults** rich reddish brown above; scaled below; bill black. **Juveniles** are tan; bill slate gray. **Song**, *tiks* and whistles, is nearly inaudible; **call**, a loud *kibee*. Favors grassy, weedy areas.

RANGE: Widespread in Southeast Asia, found in greater Los Angeles area.

HOUSE
SPARROW
domesticus

breeding ♂

fall ♂

♀

EURASIAN TREE
SPARROW
montanus

juvenile

adult

displaying

ORANGE
BISHOP
franciscanus

breeding ♂

NUTMEG
MANNIKIN
punctulata

adult

juvenile

♀

ACCIDENTALS AND EXTINCT SPECIES

*These 71 species have been recorded for North America, but for nearly all there are
fewer than three records in the past two decades or five records in the last hundred years.
Four species that have gone extinct (EX) in the past two centuries are also included.*

adult
anser

GRAYLAG GOOSE

Anser anser | L 29–33" (74–84 cm) WS 59–66" (150–168 cm)

Palearctic species that has become widely domesticated (page 50). The nominate European
race is a common summer resident on Iceland; recently recorded from Greenland. From
24 Apr. to 2 May 2005, one landed and remained on a ship some 120 miles southeast of
St. John's, Newfoundland. Largest and bulkiest of gray geese with heavy head, neck, and
bill. Pink legs and orange (in nominate *anser*) bill. Head and neck gray, uniform with rest
of body, unlike other gray geese, which have a darker head and neck. In flight, striking pale
gray forewing and contrasting pale gray underwing coverts.

adult

LESSER WHITE-FRONTED GOOSE

Anser erythropus | L 22–26" (55-66 cm)

Palearctic species. Closely resembles Greater White-fronted Goose (page 20) and plumages
similar, but smaller and more stocky, with shorter neck and stubbier bill. The yellow orbital
ring is conspicuous. It has a darker neck than Greater White-fronted (except for the *elgasi*
and *flavirostris* subspecies of Greater). The wings extend beyond the tail when folded. Specimen record from Attu Island, Alaska, on 5 June 1994. Population is declining, especially
from the western Palearctic.

adult ♂

LABRADOR DUCK

Camptorhynchus labradorius | **EX** | L 22½" (57 cm)

Extinct. An endemic North American species. Note the unique bill shape that broadens
towards tip. Much of **adult male**'s head, neck, and chest white; remainder of body plumage blackish. Adult females, immatures, and eclipse males more grayish-brown overall, the
throat being whiter than the head. Never common, it was best known from the winter
grounds on the mid-Atlantic coast, especially the southern shore of Long Island where the
last definite record (specimen) was obtained in 1875; only 54 museum species are known
to exist. Breeding grounds unknown, perhaps Labrador, perhaps farther north. Accidental
inland in spring from Montreal, Quebec (1862).

adult

LIGHT-MANTLED ALBATROSS

Phoebetria palpebrata | L 31–35" (79–89 cm) WS 72–86" (183–218 cm)

A circumpolar species of the southern oceans that breeds on subantarctic islands. A very
graceful flyer. **Adult** has dark head with prominent white eye crescents and strikingly pale
mantle and body. The long, dark, wedge-shaped tail is distinctive. At close range, note
bluish line of skin on the lower mandible (sulcus). Juvenile is similar to adult but browner
overall, with less prominent eye crescents and gray sulcus. One individual was well documented at Cordell Bank, off northern California, on 17 July 1994. The origin of this record
has been questioned by the ABA Checklist Committee.

WANDERING ALBATROSS

Diomedea exulans | L 42–53" (107–135 cm) WS 100–138" (254–351 cm)

Circumpolar in southern oceans, breeding on subantarctic islands. A polytypic species (5 to 7 subspecies), some authorities recognize up to five species. Huge size and immense wingspan (reaches over 11 feet). Massive pinkish bill and white underwings with narrow dark trailing edge and primary tips. **Adult** has extensively white back and wings. Juvenile is dark chocolate brown with conspicuous white face. Complex plumage maturation takes up to 15 years. Becomes white first on the mantle, body, and head, eventually spreading to upperwing coverts. One onshore record at Sea Ranch, Sonoma County, California, 11 to 12 July 1967. Five European records.

adult ♀

BLACK-BELLIED STORM-PETREL

Fregetta tropica | L 8" (20 cm) WS 18" (46 cm)

A widespread southern ocean species, one was well photographed off Manteo, North Carolina, on 31 May 2004. Black-and-white coloration distinctive on all individuals, but diagnostic black line up through white belly (separating it from another southern oceans species, White-bellied Storm-Petrel, *F. grallaria*) can be hard to see. Note long legs and feet project past tail. Foraging behavior distinctive: splashes breast into water, then springs forward pushing off with long legs.

RINGED STORM-PETREL

Oceanodroma hornbyi | L 8¼–9" (21–23 cm)

South American species of the Humboldt Current from Chile to southern Ecuador; casual to Colombia. Nesting grounds unknown, but possibly in the central Andes. Recent well-documented record off San Miguel Island, California, on 2 Aug. 2005. Note large size and striking plumage pattern, including blackish cap and breast band; tail is deeply forked.

NAZCA BOOBY

Sula granti | L 32" (81 cm) WS 62" (158 cm)

Recently split from the Masked Booby, this eastern tropical Pacific endemic ranges north to Mexico. One immature landed on a ship off northern Baja California and rode to San Diego (29 May 2001). Other records of immatures off southern and central California remain problematic. Similar to Masked Booby (page 100) in all plumages; subtle shape differences include shorter and thinner bill, shorter legs, and longer wings and tail. Note **adult**'s orange-pink bill and more orange (not yellow) iris. Juvenile averages paler than Masked, and pale collar is less marked or absent.

adult

GREAT FRIGATEBIRD

*Fregata mino*r | L 37" (95 cm) WS 85" (216 cm)

Extensive breeding range in Indian and Pacific Oceans. Closely resembles Magnificent Frigatebird (page 98). Adult male distinguished from Magnificent by russet bar on upper wing coverts, pink feet and often by whitish scallops on axillars. Note **adult female**'s dark head with pale gray throat, rounder (less tapered) black belly patch, and red orbital ring. When fresh, juvenile has rusty wash to head and chest, and pink feet. Specimen from Perry, Oklahoma (3 Nov. 1975). Two photographed records of adults off California (male in Monterey Bay on 13 Oct. 1979, and female Southeast Farallon Island on 14 Mar. 1992).

adult ♀

adult ♂

adult

adult

breeding
adult

dark-morph
adult

LESSER FRIGATEBIRD

Fregata ariel | L 30" (76 cm) WS 73" (185 cm)

Widespread in southwest and central Pacific and Indian Ocean; a few colonies in south Atlantic. Our smallest frigatebird, in all plumages a white spur extends from the flanks into the axillars. Juvenile has pale, rusty head. Recorded at least once from North America, an **adult male** photographed at Deer Island, Maine, on 3 July 1960; a photographed bird from Michigan in fall of 2005 is under review.

YELLOW BITTERN

Ixobrychus sinensis | L 15" (38 cm) WS 21" (53 cm)

Widespread Asian species. One specimen record from Attu Island, Alaska (17 to 22 May 1989). In **adults** (sexes similar), head and neck are buffy, cap and tail are black, and neck is streaked. Juvenile is more streaked overall. In flight, note black primary coverts and flight feathers. An Asian congener, Schrenck's Bittern (*I. xobrychus eurhythmus*), though scarce, is highly migratory and could occur in North America. It is slightly larger than Yellow Bittern, more cinnamon dorsally, and has a slaty gray trailing edge to its wing.

GRAY HERON

Ardea cinerea | L 33–40" (84–102 cm) WS 61–69" (155–175 cm)

Widespread Old World species; one found on Newfoundland coast in Oct. 1999, subsequently died in a rehabilitation center. There is a sight record from St. Paul Island, Alaska, 1 Aug. 1999. Similar to Great Blue Heron (page 114), but smaller, with shorter legs and neck. In all plumages lacks rufous thighs of Great Blue; in flight leading edge of wing shows prominent white area, rather than rufous.

CHINESE EGRET

Egretta eulophotes | L 27" (65 cm) WS 41" (104 cm)

Threatened Asian species, breeding on islands off Korea, China, and perhaps the Russian Far East; winters in the Philippines and Borneo, some on coastal mainland of Southeast Asia. One specimen record from Agattu Island, Alaska, 16 June 1974. Note shorter legs than Little Egret (page 112). In **breeding** plumage has shaggy crest, turquoise lores, entirely orange-yellow bill, and black legs with yellow feet. Nonbreeding birds lack crest; legs and feet are yellowish green and bill mostly dark.

WESTERN REEF-HERON

Egretta gularis | L 23½" (60 cm) WS 37½" (95 cm)

Old World species; casual to the West Indies. A **dark morph** was present on Nantucket Island, Massachusetts, from 26 Apr. to 13 Sept. 1983; another dark morph summered (2005) at Stephenville Crossing, western Newfoundland. Structurally resembles closely related Little Egret (page 112), but has slightly thicker neck; thicker-based bill is longer and a little more curved. Two color morphs. Much more numerous dark morph is slaty gray overall with white chin and throat; lores and bill dusky yellow; legs black and feet yellow. White morph resembles Little Egret, but note slight structural differences; immatures often have scattered dark feathers.

CHINESE POND-HERON
Ardeola bacchus | L 18" (46 cm) WS 34" (86 cm)

Migratory East Asian species that occurs rarely, but with increasing frequency, to Japan and Korea. A breeding-plumaged adult was on St. Paul Island, Alaska, 4 to 9 Aug. 1996. Members of this genus are short and stocky and most have entirely white wings, rump, and tail (especially visible in flight); yellow legs and feet. **Breeding male** has bright chestnut head, neck, and upper breast, slaty lower breast, and blue-based bill. Breeding female lacks slaty lower breast. Immatures and nonbreeding adult much duller with streaked neck and not separable from several other congeners from south and Southeast Asia.

breeding
adult ♂

CRANE HAWK
Geranospiza caerulescens | L 18–21" (46–53 cm) WS 36–41" (91–104 cm)

Neotropical species from northeastern and northwestern Mexico to South America. One wintered at Santa Ana National Wildlife Refuge, southern Texas, from 20 Dec. 1987 to 9 Apr. 1988. Distinctive, long profile with small head, long orange legs, and long banded tail; iris reddish. Northeastern subspecies (*nigra*) is darkest. In flight, note white crescent at base of primaries. Juvenile has some whitish in face and undertail, whitish barring below, and duller soft parts.

adult

COLLARED FOREST-FALCON
Micrastur semitorquatus | L 20" (52 cm) WS 31" (79 cm)

Neotropical species found from northeastern and northwestern Mexico to South America. Recorded once in south Texas at Bentsen-Rio Grande Valley State Park, from 22 Jan. to 24 Feb. 1994 (light-morph adult). Distinctive structure: very short, rounded wings (wing tips barely reach base of tail), long graduated tail, and long legs. Three color morphs: more numerous **light morph** is black above, white below; note black crescent on white cheek and thin white bars on tail; juvenile similar but browner above, also barred and more buffy below. In buff morph, white areas replaced with buff; dark morph is rare. Usually seen within forest canopy; often located by loud calls.

light-morph
adult

RED-FOOTED FALCON
Falco vespertinus | L 11" (27 cm) WS 29" (73 cm)

A medium-sized falcon that breeds in eastern Europe and western Asia and winters in southern and southwestern Africa. Regular, especially in spring, to northwestern Europe; casual to Iceland. One **first-summer male** was present at Martha's Vineyard, Massachusetts, 8 to 24 Aug. 2004. Adult male is slaty gray overall with rufous thighs and undertail. Adult female has buffy crown and underparts, dark moustache, pale sides of neck; adults have red legs and feet. Juvenile similar but browner and more streaked, duller legs and feet. Often hovers.

1st
summer ♂

PAINT-BILLED CRAKE
Neocrex erythrops | L 7¼–8" (18–20 cm)

Found eastern Panama through South America and Galápagos Islands. Specimens from Brazos County, Texas, on 17 Feb. 1972 (*erythrops*) and near Richmond, Virginia, on 15 Dec. 1978 (*olivascens*). Olive-brown above with gray forecrown and face; throat whitish; sides, flanks, and undertail coverts barred; red legs and yellow-green bill with bright orange base.

adult

SPOTTED RAIL
Pardirallus maculatus | L 10–11¼" (25-28 cm)

Resident on Cuba and Isle of Pines, Hispaniola, and from central Mexico to South America and the Galápagos Islands. Specimens (*insolitus*) from Beaver County, Pennsylvania, on 12 Nov. 1976 and from Brown County, Texas, on 9 Aug. 1977. Rather large, blackish rail. **Adult** is spotted with white on head and upperparts; remainder of underparts banded with white; long, slender greenish yellow bill has red spot at base; iris, legs, and feet red. Juveniles are polymorphic, have duller legs and bill, and brown iris. Dark morph is plain dark brown above, sooty gray below; pale morph has grayish brown throat and breast with fine white bars on breast; barred morph has gray throat with white spots, breast and belly barred with white.

DOUBLE-STRIPED THICK-KNEE
Burhinus bistriatus | L 16½" (42 cm)

Resident from northeastern Mexico to Brazil; recent nesting record from Great Inagua, Bahamas. A specimen record on 5 Dec. 1961 from the King Ranch, Kleberg County, Texas. A bird at Yuma, Arizona, was transported. Crepuscular and terrestrial, with ploverlike gait of runs and abrupt stops. Adults with dark lateral crown stripe and bold white supercilium; dark-tipped yellow bill and yellow legs. Juvenile slightly duller. In flight, upper wing two-toned; prominent broken white bar on primaries.

winter

GREATER SAND-PLOVER
Charadrius leschenaultii | L 8½" (32 cm)

Old World species. Only North American record is one that wintered at Bolinas Lagoon, California, from 29 Jan. to 8 Apr. 2001. Overall closely resembles Lesser Sand-Plover (page 158), but bill distinctly longer; legs longer and not as dark, more greenish yellow. In flight, wing stripe broader and feet project beyond tail. In breeding plumage, colored breast band is not as dark or extensive.

adult

COLLARED PLOVER
Charadrius collaris | L 5" (14 cm)

Resident from northern Mexico to South America. Only North America record, 9 to 11 May 1992 at Uvalde, Texas. Small with disproportionately long legs, small thin black bill, pinkish legs; lacks white collar around nape. Adult with dark forecrown, often with rusty border; auriculars, nape, and sides of breast often with rusty fringes, especially in males; narrow but complete black breast band, often with distinct rusty fringes at sides, especially in male. Juvenile has incomplete breast band and pale rusty edges above.

breeding
adult

EURASIAN OYSTERCATCHER
Haematopus ostralegus | L 16½" (42 cm)

Palearctic species, breeding west to Iceland; rare migrant to Greenland (over 30 records). Two spring records from Newfoundland: Fox Island, Tors Cove, 22 to 25 May 1994; and Eastport, from 3 Apr. to 2 May 1999. Similar to American Oystercatcher (page 160), but has black back and red iris. In flight white wing bar is bolder and more extensive, and white extends up back. Nonbreeding birds have a white bar across throat.

BLACK-WINGED STILT
Himantopus himantopus | L 13" (33 cm)

Widespread in Old World. Two spring records for Alaska from western Aleutians: Nizki Island from 24 May to 3 June 1983; and Shemya Island, 1 to 9 June 2003. One specimen record from St. George Island, Alaska, 15 May 2003. Similar to Black-necked Stilt (page 160), but paler on rear neck and lacks white spot above eye. **Adults** vary from having entirely white head and neck to being darker; adult males have glossy black backs, females browner.

adult

SLENDER-BILLED CURLEW
Numenius tenuirostris | E | L 15" (39 cm)

Critically endangered, possibly extinct. Only known nests found near Tara, north of Omsk, southwestern Siberia, Russia, early in 20th century. Wintered in western Mediterranean region and coastal northwest Africa, where several recorded from one Moroccan location until 1995; last record from Northumberland, United Kingdom, 4 to 7 May 1998, ironically the United Kingdom's first record. In North America a specimen from Crescent Beach, Ontario, from "about 1925." Size of Whimbrel but patterned like Eurasian Curlew (page 173), but with slender bill and black heart-shaped spots, not chevrons, on sides and flanks.

adult

EURASIAN WOODCOCK
Scolopax rusticola | L 13" (33 cm)

Widespread Old World species. Casual to North America, where most records are old and from the Northeast; older records are from Newfoundland (1862), Quebec (twice in 1862), Pennsylvania (1886, 1890), New Jersey (1859), and Alabama (1889). The last record, and the only one accepted from the 20th century, was one at Goshen, New Jersey, 2 to 9 Jan. 1956, although one from Ohio (specimen lost) in 1935 may have been this species. All dated records fall between early Nov. and early Mar. Distinctly larger than similar American Woodcock (page 190); also duller and heavily barred below.

ORIENTAL PRATINCOLE
Glareola maldivarum | L 9" (23 cm) WS 23½–25½" (60–65 cm)

Asian species. Winters south to Australia. Recorded twice in Alaska: a specimen from Attu Island, 19 to 20 May 1985; one at Gambell, St. Lawrence Island, 5 June 1986. Short-tailed pratincole with no white trailing edge to wing and chestnut underwings. Collared Pratincole (*G. pratincola*), breeding in the western Palearctic, has occurred once on Barbados and could occur in eastern North America. Collared has longer tail than Oriental and has a white trailing edge to wing.

breeding
adult

GRAY-HOODED GULL
Chroicocephalus cirrocephalus | L 16" (41 cm) WS 43" (109 cm)

A native to Africa and South America. One adult was photographed on 26 Dec. 1998 at Apalachicola, Florida. **Breeding adult** has pale gray hood with darker border, long dark red bill, and long red legs; wing pattern distinctive. In winter loses hood and has dark ear spot and smudge around eye and dark tip to bill. First-year has similar outer wing pattern to adult, but a diagonal brown bar across the inner wing and a dark secondary bar, and a dark tail band. The pinkish yellow bill has a dark tip.

breeding
adult

WHISKERED TERN

Chlidonias hybridus | L 9½–10" (24–25 cm) WS 26½–28½" (67–72 cm)

Widespread Old World species. Two North American records, both adults. In 1993 one at Cape May, New Jersey, 12 to 15 July, later moved to Delaware shore from 19 July to 24 Aug. Another at Cape May, 8 to 12 Aug. 1998. Like congeners, secures food by picking it off surface. **Adults** have short stout dark red bill and medium length red legs; dark gray underparts set off contrasting white cheeks and undertail. In winter, head and underparts white with thin black postocular patch, blackish bill and legs; compare head pattern to "ear muff" effect of White-winged Tern (page 224).

GREAT AUK

Pinguinus impennis | **EX** | L 30" (81 cm)

Extinct. North Atlantic species known in North America from three nesting colonies on islands off Quebec and Newfoundland, the largest on Funk Island off Newfoundland. Wintered within breeding range and south to Massachusetts, casually to South Carolina. Extirpated from Funk Island about 1800; last definite record was two clubbed on Eldey Stack, Iceland, on 3 June 1844. Flightless. Resembled a large Razorbill with similarly shaped bill. Distinct, white circular patch in lores. Winter plumage imperfectly known.

SCALY-NAPED PIGEON

Patagioenas squamosa | L 14" (36 cm)

Resident throughout most of the West Indies. Two old specimen records from Key West: 24 Aug. 1898 and 6 May 1929. A large dark pigeon; in good light, head and upper breast are dark maroon; feathers on sides of neck more reddish, tipped black, forming diagonal lines giving scaled appearance; dark red bill with yellow tip; orange-red iris; orange orbital ring.

PASSENGER PIGEON

Ectopistes migratorius | **EX** | L 15¾" (40 cm)

Extinct. Believed to have been the most abundant bird species in North America—in migration, flocks said to have numbered in the millions. Formerly found in eastern North America, casually in the West. By 1870s relegated to scattered breeding locations. Last records were a specimen from Ohio in 1900 and a reliable sight record in Missouri in 1902. Last individual died in captivity in a Cincinnati zoo on 1 Sept. 1914. Destruction of old-growth deciduous forests and overhunting, especially in breeding colonies, led to its demise. Resembled a large Mourning Dove. **Adult male** bluish gray above and pinkish below; female browner above and paler below; juvenile similar to female.

CAROLINA PARAKEET

Conuropsis carolinensis | **EX** | L 13½" (34 cm)

Extinct. Only native breeding North American psittacid. Formerly resident in Southeast, especially along rivers. Last certain records were from Florida and Kansas in 1904; a reliable sight record from Missouri in 1905 and perhaps another in 1912. Last captive died on 21 Feb. 1918. **Adult** had green body with yellow patches on shoulder, thighs, and vent, a yellow head, and reddish orange face. Immature entirely green, except for orange patch on forehead.

breeding adult

adult

adult ♂

adult ♂

adult

ORIENTAL SCOPS-OWL

Otus sunia | L 7½" (19 cm) WS 21" (53 cm)

A small, nocturnal, insectivorous owl of eastern Asia. Multiple subspecies, northern ones are migratory. Two records (*japonicus*) of rufous morphs from Aleutian Islands, Alaska: a dried wing found on Buldir Island on 5 June 1977 and one found alive on Amchitka Island on 20 June 1979, subsequently died (specimen). Three color morphs: gray-brown, reddish gray, and **rufous**. Fine dark streaks on head; breast streaked vertically and horizontally with thin crossbars. Short ear tufts. Northern subspecies may be specifically distinct, calls differ.

rufous morph
japonicus

MOTTLED OWL

Ciccaba virgata | L 14" (36 cm) WS 33" (84 cm)

A medium-size and very vocal nocturnal owl found in a variety of woodland habitats from northwest and northeast Mexico to South America. A road-killed specimen was salvaged in front of Bentsen-Rio Grande Valley State Park, southern Texas, on 23 Feb. 1983. Note round head with no ear tufts and streaked underparts; brown facial disk with bold white eyebrows and whiskers is distinctive. Larger Barred Owl (page 258) has prominent barring across upper breast and paler facial disk.

STYGIAN OWL

Asio stygius | L 17" (43 cm) WS 42" (107 cm)

A medium-size, forest-dwelling nocturnal owl occurring from northern Mexico to South America; also Hispaniola and Gonâve Island, West Indies. Two winter records of birds found roosting and photographed at Bentsen-Rio Grande Valley State Park: 9 Dec. 1994 and 26 Dec. 1996. Deep chocolate brown overall with close-set ear tufts and blackish facial disk with contrasting white forehead; underparts show distinct dark streaks and crossbars. Compare to Long-eared Owl (page 256), which is browner and has a rufous facial disk.

GRAY NIGHTJAR

Caprimulgus indicus | L 11–12¾" (28-32 cm)

Asian species, formerly known as Jungle Nightjar. One desiccated specimen (*jotaka*) salvaged on Buldir Island, Alaska, on 31 May 1977. Overall color is grayish brown, patterned with black, buff, and grayish white. Note long wing-tip projection. Adult male has large white subterminal patch on inner primaries and white tips to all but central pair of tail feathers; on female, wing patch and tail tips more buffy. The two more northerly and migratory subspecies (*jotaka* and *hazarae*) have different vocalizations and are treated as a distinct species; *C. jotaka,* by some authors.

♂ *jotaka*

ANTILLEAN PALM-SWIFT

Tachornis phoenicobia | L 4¼" (11 cm)

A tiny Caribbean swift that is resident in the Greater Antilles (except Puerto Rico). Two were present and photographed at Key West, Florida, from 7 July to 13 Aug. 1972. Distinctive are dark cap, dark sides, and thin line across breast contrasting with white throat, belly, and rump; tail has shallow fork. Batlike flight with rapid wingbeats and short glides and twists; generally flies low, among trees and palms.

CINNAMON HUMMINGBIRD
Amazila rutila | L 4–4½" (10–12 cm)

Resident in lowlands from Sinaloa and Yucatán Peninsula, south to Costa Rica. Two records from Southwest: 21 to 23 July 1992 at Patagonia, Arizona; 18 to 21 Sept. 1993 at Santa Teresa, New Mexico. Adults have a cinnamon tail and underparts and a black-tipped red bill. Immature is similar but upperparts edged cinnamon when fresh, and upper mandible is mostly dark.

adult ♂

BUMBLEBEE HUMMINGBIRD
Atthis heloisa | L 2¾–3" (7–8 cm)

Endemic to montane forests of Mexico north of the Isthmus of Tehuantepec. Two specimens taken on 2 July 1896 in Ramsey Canyon, Arizona. That two would be taken on the same date from one locality with no records since seems unlikely, but the specimens are extant and the records have not yet been refuted. The question is whether the collector was actually in the present-day Ramsey Canyon or was farther south in Mexico. A tiny hummingbird with a short bill and a short, rounded or double-rounded, rufous-based tail with white tips. **Adult males** have an elongated magenta-rose gorget. Females and immatures closely resemble female-type Calliope Hummingbirds, which have darker tails that fall shorter than or equal to wing tips; on Bumblebee, tail extends beyond wing tips.

adult
saturata

EURASIAN HOOPOE
Upupa epops | L 10½" (27 cm)

Widespread Old World species. One North American record, a specimen (*saturata*) from Old Chevak, Yukon-Kuskokwim Delta, Alaska, 2 to 3 Sept. 1975. Unmistakable: pinkish brown coloration; long crest (sometimes raised); long, thin, slightly decurved bill. Striking black-and-white wing pattern in flight; wingbeats slow and floppy. The two races from equatorial Africa and farther south and from Madagascar are each treated as separate species by some authors.

adult

EURASIAN WRYNECK
Jynx torquilla | L 6" (17 cm)

Widespread in Old World. Two fall records from Alaska: a specimen (*chinensis*) from Cape Prince of Wales, 8 Sept. 1945, and one photographed at Gambell, St. Lawrence Island, 2 to 5 Sept. 2003. A dead bird found in southern Indiana was believed to have been artificially transported. Patterned in browns and grays, the Eurasian Wryneck is quite unlike a woodpecker, except for the sharply pointed bill. Note dark mask, dark vertical band on sides of back, and the long and sparsely barred tail. Often forages on ground, but also perches on branches in a somewhat horizontal posture.

GREENISH ELAENIA
Myiopagis viridicata | L 5½" (14 cm)

Resident in Mexico from southern Durango and southern Tamaulipas, south to northern Argentina. Recorded once in North America on upper Texas coast at High Island, 20 to 23 May 1984. Overall greenish above with a contrasting grayish head and dark eye stripe with distinct, pale supercilium. Primaries edged with olive; bright yellowish on secondaries; short primary projection. Grayish throat and olive breast contrast with yellow belly. Distinctive call note, a high, thin, and descending *seei-seeur*.

CARIBBEAN ELAENIA
Elaenia martinica | L 5½" (14 cm)

Resident on many Caribbean Islands, including throughout Lesser Antilles, but absent from Greater Antilles, except for Puerto Rico. Recorded once (photographs) from Santa Rosa Island, Escambia County, Florida, on 28 Apr. 1984. The AOU, while accepting the record as an elaenia, does not consider the record definitive to species. Overall plumage plain. Short crest rarely raised to show white-based crown feathers. Indistinct pale gray supercilium; dusky lores; pale yellowish wing bars and edges to secondaries; lower mandible flesh-colored at base. **Call,** a clear whistled *wheee-u.*

SOCIAL FLYCATCHER
Myiozetetes similis | L 6¾–7¼" (17-18 cm)

Common from northeastern and northwestern Mexico to northeastern Argentina. Only one fully documented record at Bentsen-Rio Grande Valley State Park, Texas, 7 to 14 Jan. 2005. A medium-sized, colorful flycatcher vaguely suggestive of the much larger Greater Kiskadee, but black bill is much smaller. **Adult** also lacks rufous in the wings and tail, which eliminates Middle American races of Great Kiskadee (page 310), although juvenile Social Flycatcher does show rufous edges. The reddish orange central crown patch is concealed by dark gray. Distinctive **call** is a loud *che cheechee cheechee cheechee*; also a harsh *cree-yooo.*

MASKED TITYRA
Tityra semifasciata | L 9" (23 cm)

Common from northwestern and northeastern Mexico to Brazil. One record from south Texas at Bentsen-Rio Grande Valley State Park, from 17 Feb. to 10 Mar. 1990. Large and chunky; **males** are pale gray above and whitish below with contrasting black on face, most of wings, and thick subterminal tail band. Bare skin on face and base of thick bill is pinkish red. Female is similar, but is darker and duller; lacks black face. Distinctive **call** is a double, nasal grunt, *zzzr-zzzrt.*

YUCATAN VIREO
Vireo magister | L 6" (15 cm)

Resident on Yucatán Peninsula and its offshore islands; also on Grand Cayman and islands off Honduras. One record from Galveston County, Texas, from 28 April to 27 May 1984. Overall brownish above with dark eye line, but no dark lateral crown stripe as in Red-eyed Vireo (page 318). Dull whitish below with grayish brown flanks, short primary projection, and very large, heavy bill. **Call** a nasal *benk*, often strung together in a series.

CUBAN MARTIN
Progne crytoleuca | L 7½" (19 cm)

Breeds Cuba and Isle of Pines; wintering grounds unknown, but presumably South America. Only acceptable record is specimen taken on 9 May 1895 at Key West, Florida. Adult male closely resembles Purple Martin (page 330), but has relatively longer and more deeply forked tail; in hand, note concealed white feathers on belly. **Female** resembles female Purple, but with unmarked white belly and undertail coverts; lacks grayish collar. Separation from Caribbean (*P. dominicensis*) and Sinaloa Martins (*P. sinaloae*), with which it sometimes is treated as conspecific, is difficult; Cuban is darker overall with dark shaft smudges on the undertail coverts and usually with some shaft streaking on the breast and sides.

adult

adult ♂

adult ♀

adult ♀

adult ♀

adult *fusca*

adult

GRAY-BREASTED MARTIN
Progne chalybea | L 6¾" (17 cm)

Breeds in Mexico from southern Sinaloa and southern Tamaulipas, south to Argentina. Withdraws from northeastern portion of range in winter. Two old specimen records for south Texas: 25 Apr. 1880 at Rio Grande City; 18 May 1889 from Hidalgo County; all other reports unsubstantiated. Similar to female Purple Martin (page 330), but smaller with a less deeply forked tail; also browner on forehead, a less well-defined collar, and paler underparts.

SOUTHERN MARTIN
Progne elegans | L 7" (18 cm)

An austral migrant from South America. One specimen record from Key West, Florida, on 14 Aug. 1890. Adult male resembles Purple Martin (page 330), but smaller with slightly longer and more forked tail, and in hand, lacks concealed white patch on sides and flanks. **Female** is darker below than female Purple Martin.

BROWN-CHESTED MARTIN
Progne tapera | L 6½" (16 cm)

South American species. Southern subspecies, *fusca,* is an austral migrant to northern South America. Two North American records: a specimen (*fusca*) from Monomoy Island, Massachusetts, 12 June 1983, and one photographed at Cape May, New Jersey, 6 to 15 Nov. 1997. Smaller than Purple Martin with brownish upper parts, white below with brown sides, and brown band across breast.

MANGROVE SWALLOW
Tachycineta albilinea | L 5¼" (13 cm)

A small swallow found from coastal slopes of central Sonora and southern Tamaulipas, Mexico, south through Panama; isolated population in Peru. An adult was at the Viera Wetlands in Brevard County, Florida, 18 to 25 Nov. 2000. **Adults** have an iridescent greenish crown and back; auriculars and lores black; narrow white line, usually meets across forehead; partial white collar; and distinct white rump. Juvenile is like adult, but brownish above. Compare also to Tree and Violet-green Swallows and Common House-Martin (page 330). Within normal range, seldom found far from water.

WILLOW WARBLER
Phylloscopus trochilus | L 4½" (11 cm)

A highly migratory Old World species. Only record for North America at Gambell, St. Lawrence Island, Alaska, 25 to 30 Aug. 2002, was thought to pertain to *yakutensis* (illustrated), but was greener above and more yellowish below. There is an additional specimen record from northeast Greenland. Like the similar Arctic Warbler (page 348) in overall coloration and long primary projection, but smaller, with smaller bill and a plainer wing. Whistled *hoo-eet* **call** note is very unlike Arctic's buzzy note. Another *Phylloscopus,* the widespread Chiffchaff (*P. collybita*), could occur in North America. It is very similar to Willow Warbler, but has a shorter primary projection and a different wing formula (primary spacing). Eastern breeding populations of Chiffchaff average grayer or browner than European populations. Typically leg color is a good indicator (pale on Willow, dark on Chiffchaff), but the Gambell bird had dark legs as it can in the easterly breeding subspecies of Willow Warbler.

adult
yakutensis

WOOD WARBLER
Phylloscopus sibilatrix | L 5" (13 cm)

Breeds in the western Palearctic; winters in tropical Africa. Two fall records for western Alaska: one collected on Shemya Island, western Aleutians, 9 Oct. 1978; one photographed on St. Paul Island, 7 Oct. 2004. A colorful *Phylloscopus* with yellow throat and upper breast contrasting sharply with white lower breast and belly; the yellow supercilium is well defined by a complete dark eye line; yellow-green edges to flight feathers and greater coverts; and dark centered tertials with sharply defined pale edges. Note very long primary projection, which contributes to short-tailed appearance.

YELLOW-BROWED WARBLER
Phylloscopus inornatus | L 4½" (11 cm)

Primarily breeds in the northeastern Palearctic region. Two photographed fall records from Gambell, St. Lawrence Island, Alaska: 23 to 24 Sept. 1999 and 30 Aug. 2002. Distinctly patterned with bold supercilium and two pale wing bars and dark band at base of secondaries; tertials with sharply defined, pale edges. Note small and mostly dark bill. **Call** a distinctive upslurred *swee-eet,* not too unlike call of male Pacific-slope Flycatcher. Most authorities have now split the less migratory and more southerly breeding subspecies (*humei* and *mandellii*) as a separate species, Hume's Leaf -Warbler (*P. humei*).

LESSER WHITETHROAT
Sylvia curruca | L 5½" (14 cm)

Old World species. Only record was one photographed at Gambell, St. Lawrence Island, Alaska, 8 to 9 Sept. 2002. Distinctive, with a rather long tail, outer rectrices tipped with white; gray crown with a dark mask, a warm brown back and wings, and whitish underparts with a tan wash on the sides and flanks. **Call** is a hard *tik,* often repeated. Multiple subspecies groups are treated by Old World authorities as up to four species.

MUGIMAKI FLYCATCHER
Ficedula mugimaki | L 5¼" (13 cm)

Highly migratory species of eastern Asia. Only record is from Shemya Island, Alaska, on 24 May 1985, supported by marginal photos. Note long wings. Adult male is striking, with blackish head and upper parts, white wing patch, short but broad downcurving white supercilium, and extensively orange underparts. Female is brownish above, burnt orange on throat and breast; has two thin pale wing bars. **First-year male** closer to female, but with partial, broad and downcurving supercilium. In Asian range feeds from mid- to upper canopy.

SPOTTED FLYCATCHER
Muscicapa striata | L 6" (15 cm)

Widespread breeder in the Palearctic, east to about Lake Baikal; winters in Africa, south of the Sahara. One photographed at Gambell, St. Lawrence Island, Alaska, on 14 Sept. 2002. Overall grayish brown above, with fine streaking on crown and forecrown, and indistinct whitish eye ring; lacks distinct malar and submoustachial markings; below indistinctly streaked, not spotted, across throat and breast. Juvenile is spotted with buff above and has dark mottling below.

fall adult

1st fall
inornatus

1st fall
blythi

1st year ♂

adult

adult ♀

SIBERIAN BLUE ROBIN
Luscinia cyane | L 5½" (14 cm)

Highly migratory Asian species. One certain North American specimen record from Attu Island, Alaska, 21 May 1985; an additional spring sighting of an adult male from the Yukon is disputed. Adult male is deep blue above, clear white below. **Adult female** brownish above with faint buffy eye ring; buffy wash across breast with faint stippling, and most have some bluish on tail (lacking on immature females). Immature male has some blue on scapulars, wings, and tail. Frequently vibrates tail.

ORANGE-BILLED NIGHTINGALE-THRUSH
Catharus aurantiirostris | L 6½" (17 cm)

Widespread in Neotropics. Two spring records from south Texas: one photographed in hand on 8 Apr. 1996 at Laguna Atascosa National Wildlife Refuge; a specimen from Edinburg on 28 May 2004. Orange-brown above, pale gray and whitish below with distinctive bright orange bill, legs, orbital ring. In flight, lacks pale underwing bar of northern breeding *Catharus* thrushes.

BLACK-HEADED NIGHTINGALE-THRUSH
Catharus mexicanus | L 6½" (17 cm)

Found from northeastern Mexico to western Panama. Only record was one at Pharr, in south Texas, from 28 May to 29 Oct. 2004. Distinctive, with blackish crown and face, whitish throat and belly; otherwise gray below. Bright orange orbital ring, bill, and legs.

♂

EURASIAN BLACKBIRD
Turdus merula | L 10½" (27 cm)

Palearctic species. Only accepted record was a **male** found dead on 16 Nov. 1994 at Bonavista, Newfoundland. Other records, mainly from northeastern Canada, are of uncertain origin. There are a dozen Greenland records. Male is all-black; orange-yellow orbital eye ring and bill (duller on immatures). Female is browner with pale throat and dark-streaked chest; soft parts duller.

winter
adult ♂

CITRINE WAGTAIL
Motacilla citreola | L 6½" (17 cm)

Palearctic species that winters farther north than either Western or Eastern Yellow Wagtail. Only North American record was from Starkville, Mississippi, from 31 Jan. to 1 Feb. 1992. All plumages have gray back and bold, well-defined white wing bars. Breeding adult male has bright yellow head and underparts and a dark nape. Adult female and **winter adult male** have yellow-centered, grayish brown auriculars completely surrounded by yellow. Immatures lack all yellow, but have similar, winter-adult face pattern. **Call** is loud buzzy *tsweep*, like Eastern Yellow Wagtail (page 368).

TREE PIPIT
Anthus trivialis | L 6" (15 cm)

Palearctic breeding species, wintering mainly in Africa south of the Sahara and in India. Three records for western Alaska: a specimen from Cape Prince of Wales on 23 June 1978, and two photographed at Gambell, St. Lawrence Island, Alaska (6 June 1995; 21, 27 Sept. 2002). Resembles Olive-backed Pipit (page 370), but browner above with distinct back streaks; face pattern more blended, no dark spot, and fine flank streaks; **call** similar.

GRAY SILKY-FLYCATCHER
Ptilogonys cinereus | L 7½" (20 cm)

Resident (some seasonal movement) from northern Mexico to Guatemala; largely montane. Two accepted Texas records: from 31 Oct. to 11 Nov. 1985 at Laguna Atascosa National Wildlife Refuge and from 12 Jan. to 5 Mar. 1995 in El Paso. Four well-documented southern California records have been questioned on origin. Structured like Phainopepla. **Adult males** are gray with orange-yellow flanks and bright yellow undertail coverts; note white eye ring and white base of tail. Females and juveniles are similar, but duller.

WORTHEN'S SPARROW
Spizella wortheni | L 5½" (14 cm)

Critically endangered. Only extant populations are in northeastern Mexico in Coahuila and Nuevo Léon. Only U.S. record at Silver City, New Mexico, on 16 June 1884, the type specimen; probably part of a small resident population, subsequently extirpated. Resembles *aranacea* Field Sparrow (page 410), but crown solidly rufous, rump grayish, legs and feet dark; vocalizations differ.

PINE BUNTING
Emberiza leucocephalos | L 6½" (17cm)

Primarily an eastern Palearctic species. Two fall records from Attu Island, Alaska, both males: one photographed 18 to 19 Nov. 1995 and a specimen (nominate *leucocephalos*) on 6 Oct. 1993. Breeding male is rusty overall, with chestnut head, and bold white eyebrow and cheek patch; in winter, duller head lacks white crown patch. Female is duller still with a weak malar, but has a rusty eyebrow and at least a small white spot at rear of cheek.

YELLOW-THROATED BUNTING
Emberiza elegans | L 6" (15 cm)

East Asian species. One record of a male photographed on Attu Island, Alaska, 25 May 1998. Note prominent crest. Yellow-and-black head pattern of **adult male** is striking. Female and immature male are similar, but duller, with brownish auriculars.

TAWNY-SHOULDERED BLACKBIRD
Agelaius humeralis | L 8" (20 cm)

Resident on Cuba and in Haiti. Only U.S. record was two secured (specimens) at the Key West Lighthouse on 27 Feb. 1936. Smaller and slimmer than Red-winged Blackbird and with a slim, pointed bill; lesser coverts tawny, not red, and rear border has a narrow blended edge. More arboreal than Red-winged (page 444) and buzzy, muffled **song** is more drawn out.

EURASIAN SISKIN
Spinus spinus | L 4¾" (12 cm)

Palearctic species. Two records of males from Attu Island: a specimen, 21 to 22 May 1993 and a sight record on 4 June 1978. About six records from northeast North America, but the origin of these has been questioned; a male photographed at Saint-Pierre and Miquelon on 23 June 1983 is perhaps the most compelling. Unrecorded from Greenland. **Male** is distinctive with black forecrown and chin, olive above, and extensively yellow below. Female is much duller, the yellow restricted to sides of breast, and a wash of yellow on face, eyebrow, and rump; juvenile duller still. Some Pine Siskins (page 460) are very similar, but wing coverts of Eurasian average darker.

adult ♂

fall adult ♂

adult ♂

adult ♂

fall adult ♂

AOU AND ABA CHECKLIST DIFFERENCES

As noted in the introduction of this book, the AOU and ABA checklist committees serve different purposes. The differences can be summarized as follows:

Light-mantled Albatross *Phoebetria palpebrata*. Origin questioned by ABA (page 466). **Azure Gallinule** *Porphyrula flavirostris*. A specimen from Suffolk County, New York, on 14 Dec. 1986 was initially accepted by both committees, but subsequent information revealed that it may have escaped from a local aviculturalist. Still accepted by AOU. **Caribbean Elaenia** *Elaenia martinica*. Not accepted by AOU (page 475). **White-chinned Petrel**, *Procellaria aequinoctialis*. One near Rollover Pass, Texas, 27 Apr. 1986 (photos), was finally accepted by the ABA, but continues to be rejected by the AOU (origin questioned).

GREENLAND

Greenland is the largest island in the World, most of it lying north of the Arctic Circle, and forms the most northeastern part of North America. Over 230 species have been recorded there, and nearly all are substantiated by specimens. The definitive ornithological reference is *An Annotated Checklist to the Birds of Greenland* (1994) by David Boertmann (*Bioscience* 38). Boertmann divides Greenland into four regions (north, west, northeast, and southeast) and details the bird distribution for each. Greenland's avifauna is a mix of both Palearctic and Nearctic species, many of which are strays respectively from Europe (including Iceland) and mainland North America. The breeding avifauna includes Pink-footed, White-fronted (the endemic breeding subspecies *flavirostris*), and Barnacle Geese, White-tailed Eagle, Fieldfare, Redwing, White Wagtail (nominate *alba*), and Meadow Pipit. Three subspecies of Rock Ptarmigan are found in Greenland, two of which (*saturata* and *capta*) are endemic. Two subspecies of Dunlins breed in Greenland, one of which (*arctica*) is an endemic breeder in the northeast; the other (*schinzii*) breeds in southern Greenland as well as northwestern Europe. The Black Scoters recorded are of the nominate *nigra* subspecies from northwest Europe. The Merlin specimens from Greenland are of *subaesalon*, an endemic breeder on Iceland; also recorded is the more widespread *aesolon*, breeding mainly in northern Europe. The Red Crossbills collected are of nominate *curvirostra* from the Palearctic.

The above species, along with the following list of species *not* recorded in our area of coverage, should alert observers to the potential visitors to northeast North America. For polytypic species, the trinomial subspecies name is given, if known:

Eurasian Spoonbill *Platalea leucorodia leucorodia*. One Oct. record (1909) for the western region of Greenland.

Ruddy Shelduck *Tadorna ferruginea*. Four collected in the summer of 1892, an invasion year for the species in northwest Europe. (See also page 96.)

Water Rail *Rallus aquaticus hibernans*. Four records. The subspecies *hibernans* is endemic to Iceland.

Spotted Crake *Porzana porzana*. Eleven records, nearly all in fall, from the western region of Greenland!

Oriental Plover *Charadrius veredus*. One May record (1948) from western region. A remarkable record, breeds on the steppes of eastern Asia; winters in Australia.

Rook *Corvus frugilegus frugilegus*. One Mar. record (1901) for the southeastern region of Greenland.

Carrion Crow *Corvus corone cornix*. Two spring records (1897, 1907) were of the "Hooded Crow."

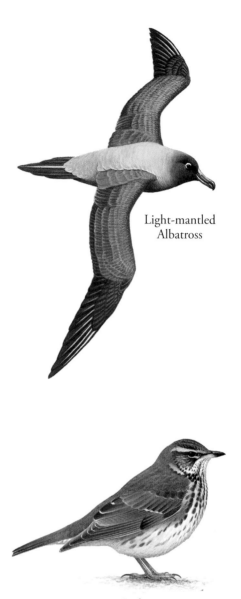

Light-mantled
Albatross

Redwing

White Wagtail
alba

Meadow Pipit *Anthus pratensis pratensis.* Scarce breeder in eastern Greenland.
White's Thrush *Zoothera aurea aurea.* One Oct. record (1954) for the northeastern region. It is here treated as a separate species from the Scaly Thrush, *Z.
dauma,* following other authorities.
Song Thrush *Turdus philomelos philomelos.* One June record (1982) for the northeastern region. (There is now an excepted record from Québec, 11-17 Nov. 2006.)
Blackcap *Sylvia atricapilla atricapilla.* One Nov. record (1916) for the southeastern region of Greenland.
Lesser Redpoll *Acanthis cabaret.* One Sept. record (1933) for the southeastern
region of Greenland.

Dunlin
schinzii

BERMUDA

Bermuda is a series of some 300 small islands, most of which are uninhabited. It is best
known as the only. breeding site of the Bermuda Petrel (also called the Cahow), which was
discovered early in the 17th century and then thought to be extinct until it was rediscovered
in 1951. The White-tailed Tropicbird reaches its northernmost breeding range here. The
only endemic breeding land bird is a nonmigratory subspecies of the White-eyed Vireo (*ber-
mudianus*). Established exotics not found in North America include European Goldfinch,
Common Waxbill, and Orange-cheeked Waxbill. Bermuda is well known for its migrants
and vagrants, which make up most of Bermuda's extensive species list—in excess of 360 species. Surprising northern species that have occurred include Northern Hawk Owl, Snowy
Owl, Bohemian Waxwing, White-winged Crossbill, and Pine Grosbeak. The single Snowy
Owl (1987) took to predating the endangered Bermuda Petrels and had to be collected—
sometimes conservationists have to make hard choices! Most of the vagrants come from
North America (including the West) or Europe, but a few (Large-billed Tern and Fork-
tailed Flycatcher) are from South America. The Red-necked Stint and, even more surpris-
ingly, the Dark-sided Flycatcher (a late Sept. specimen of nominate *sibirica*) are from Asia.
The most recent birding references for Bermuda are *A Guide to the Birds of Bermuda* (1991)
by Eric Amos and A *Birdwatching Guide to Bermuda* (2002) by Andrew Dobson.

Bermuda Petrel

The following species are recorded from the island of Bermuda but not from mainland
North America:

West Indian Whistling-Duck *Dendrocygna arborea.* One record (1907). This species is resident on some of the Bahamian islands. (A recent record from Virginia
is of uncertain origin.)
Ferruginous Duck *Aythya nyroca.* A winter sight record (1987).
Striated Heron *Butorides striata.* One record (1985) of a long-staying bird.
Booted Eagle *Hieraaetus pennatus.* A Sept. sight record (1989).
White Tern *Gygis alba.* A remarkable Dec. record (1972). Photographic evidence
indicates that it was not the expected nominate race from the south Atlantic but,
rather, one of the Pacific races!

White-tailed
Tropicbird

The main entry for each species is listed in **boldface** type and refers to the text page opposite the illustration.

A

Acanthis
    flammea 462
    hornemanni 462
Accentor, Siberian 368
Accipiter
    cooperii 126
    gentilis 126
    striatus 126
Accipitridae (family) 118–134
Acridotheres
    cristatellus 366
    tristis 366
Actitis
    hypoleucos 168
    macularius 168
Aechmophorus
    clarkii 76
    occidentalis 76
Aegithalidae (family) 338
Aegolius
    acadicus 264
    funereus 264
Aeronautes saxatalis 270
Aethia
    cristatella 238
    psittacula 238
    pusilla 238
    pygmaea 238
Agelaius
    humeralis 479
    phoeniceus 444
    tricolor 444
Aimophila
    aestivalis 408
    botterii 408
    carpalis 410
    cassinii 408
    quinquestriata 414
    ruficeps 410
Aix
    galericulata 50

    sponsa 28
Alauda arvensis 328
Alaudidae (family) 328
Albatross
    Black-browed 80
    Black-footed 78
    Laysan 78
    Light-mantled **466**
    Short-tailed 78
    Shy **80**
    Wandering 467
    Yellow-nosed 80
Alca torda 232
Alcedinidae (family) 280
Alcidae (family) 232–240
Alectoris chukar 56
Alle alle 232
Alopochen aegyptiacus 50
Amazilia
    beryllina 272
    rutila 474
    violiceps 272
    yucatanensis 272
Amazona
    amazonica 250
    finschi 250
    oratrix 250
    viridigenalis 250
Ammodramus
    bairdii 416
    caudacutus 418
    henslowii 416
    leconteii 418
    maritimus 418
    nelsoni 418
    savannarum 416
Amphispiza
    belli 414
    bilineata 414
Anas
    acuta 34
    americana 34
    bahamensis 34
    clypeata 36
    crecca 32
    cyanoptera 36
    discors 36

    falcata 32
    formosa 32
    fulvigula 30
    penelope 34
    platyrhynchos 30
    querquedula 36
    rubripes 30
    strepera 32
    zonorhyncha 30
Anatidae (family) 20–55
Anhinga 104
Anhinga anhinga 104
Anhingidae (family) 104
Ani
    Groove-billed 254
    Smooth-billed 254
Anous
    minutus 226
    stolidus 226
Anser
    albifrons 20
    anser 50, 466
    brachyrhyncus 20
    cygnoides 50
    erythropus 466
    fabalis 20
    indicus 50
    serrirostris 20
Anthracothorax prevostii 272
Anthus
    cervinus 370
    gustavi 370
    hodgsoni 370
    rubescens 370
    spragueii 370
    trivialis 478
Aphelocoma
    californica 322
    coerulescens 322
    insularis 322
    ultramarina 322
Aphriza virgata 176
Apodidae (family) 270
Apus
    apus 270
    pacificus 270
Aquila chrysaetos 124

Aramidae (family) 146
Aramus guarauna 146
Aratinga
    acuticaudata 248
    erythrogenys 248
    holochlora 248
    mitrata 248
    weddelli 248
Archilochus
    alexandri 276
    colubris 276
Ardea
    alba 112
    cinerea 468
    herodias 114
Ardeidae (family) 108–114
Ardeola bacchus 469
Arenaria
    interpres 176
    melanocephala 176
Arremonops rufivirgatus 404
Asio
    flammeus 256
    otus 256
    stygius 473
Athene cunicularia 264
Atthis heloisa 474
Auk
    Great 472
Auklet
    Cassin's 236
    Crested 238
    Least 238
    Parakeet 238
    Rhinoceros 240
    Whiskered 238
Auriparus flaviceps 338
Avocet, American 160
Aythya
    affinis 40
    americana 38
    collaris 40
    ferina 38
    fuligula 40
    marila 40
    valisineria 38

B

Baeolophus
    atricristatus 334
    bicolor 334
    inoratus 334
    ridgwayi 334
    wollweberi 334
Bananaquit 402
Bartramia longicauda 186
Basileuterus
    culicivorus 396
    rufifrons 396
Beardless-Tyrannulet, Northern 300
Becard, Rose-throated 310
Bishop, Orange 464
Bittern
    American 108
    Least 108
    Yellow 468
Black-Hawk, Common 128
Blackbird
    Brewer's 448
    Eurasian 478
    Red-winged 444
    Rusty 448
    Tawny-shouldered 479
    Tricolored 444
    Yellow-headed 444
Bluebird
    Eastern 354
    Mountain 354
    Western 354
Bluetail, Red-flanked 352
Bluethroat 352
Bobolink 442
Bobwhite, Northern 68
Bombycilla
    cedrorum 372
    garrulus 372
Bombycillidae (family) 372
Bonasa umbellus 60
Booby
    Blue-footed 100
    Brown 100
    Masked 100

    Nazca 467
    Red-footed 100
Botaurus lentiginosus 108
Brachyramphus
    brevirostris 234
    marmoratus 234
    perdix 234
Brambling 454
Brant 24
Branta
    bernicla 24
    canadensis 24
    hutchinsii 24
    leucopsis 22
Brotogeris
    chiriri 248
    versicolurus 248
Bubo
    scandiacus 258
    virginianus 256
Bubulcus ibis 112
Bucephala
    albeola 46
    clangula 46
    islandica 46
Budgerigar 250
Bufflehead 46, 55
Bulbul, Red-whiskered 366
Bullfinch, Eurasian 462
Bulweria bulwerii 90
Bunting
    Blue 440
    Gray 432
    Indigo 440
    Lark 420
    Lazuli 440
    Little 434
    McKay's 432
    Painted 440
    Pallas's 434
    Pine 479
    Reed 434
    Rustic 434
    Snow 432
    Varied 440
    Yellow-breasted 432

Yellow-throated 479

Burhinus bistriatus 470

Bushtit 338

Buteo

    albicaudatus 134

    albonotatus 128

    brachyurus 128

    jamaicensis 132

    lagopus 134

    lineatus 130

    magnirostris 128

    nitidus 130

    platypterus 130

    regalis 134

    swainsoni 132

Buteogallus anthracinus 128

Butorides virescens 110

C

Cairina moschata 28

Calamospiza melanocorys 420

Calcarius

    lapponicus 430

    mccownii 428

    ornatus 428

    pictus 430

Calidris

    acuminata 186

    alba 178

    alpina 178

    bairdii 184

    canutus 178

    ferruginea 178

    fuscicollis 184

    himantopus 190

    maritima 176

    mauri 180

    melanotos 186

    minuta 182

    minutilla 180

    ptilocnemis 176

    pusilla 180

    ruficollis 182

    subminuta 182

temminckii 182

tenuirostris 178

Callipepla

    californica 66

    gambelii 66

    squamata 68

Calliphlox evelynae 272

Calonectris

    diomedea 88

    edwardsii 88

    leucomelas 92

Calothorax lucifer 272

Calypte

    anna 276

    costae 276

Campephilus principalis 292

Camptorhynchus labradorius 466

Camptostoma imberbe 300

Campylorhyncus brunneicapillus 344

Canvasback 38, 54

Caprimulgidae (family) 266–268

Caprimulgus

    carolinensis 268

    indicus 473

    ridgwayi 268

    vociferus 268

Caracara, Crested 136, 145

Caracara cheriway 136

Cardellina rubrifrons 398

Cardinal, Northern 438

Cardinalidae (family) 436–440

Cardinalis

    acanthis 482

    cardinalis 438

    chloris 485

    spinus 496

    sinica 454

    sinuatus 438

Carpodacus

    cassinii 456

    erythrinus 456

    mexicanus 456

    purpureus 456

Catbird, Gray 362

Cathartes aura 118

Cathartidae (family) 118

Catharus

    auantiirostris 478

    bicknelli 356

    fuscescens 356

    guttatus 356

    mexicanus 478

    minimus 356

    ustulatus 356

Catherpes mexicanus 344

Centrocercus

    minimus 64

    urophasianus 64

Cepphus

    columba 234

    grylle 234

Cerorhinca monocerata 240

Certhia americana 340

Certhiidae (family) 340

Chachalaca, Plain 56

Chaetura

    pelagica 270

    vauxi 270

Chaffinch, Common 454

Chamaea fasciata 334

Charadriidae (family) 154–158

Charadrius

    alexandrinus 156

    collaris 470

    dubius 158

    hiaticula 156

    leschenaultii 470

    melodus 156

    mongolus 158

    montanus 158

    morinellus 158

    semipalmatus 156

    vociferus 158

    wilsonia 156

Chat, Yellow-breasted 396

Chen

    caerulescens 22

    canagica 22

    rossi 22

Chickadee

    Black-capped 336

    Boreal 338

Carolina 336
Chestnut-backed 338
Gray-headed 338
Mexican 336
Mountain 336
Chlidonias
 hybridus 472
 leucopterus 224
 niger 224
Chloris sinica 454
Chloroceryle americana 280
Chondestes grammacus 414
Chondrohierax uncinatus 122
Chordeiles
 acutipennis 266
 gundlachii 266
 minor 266
Chroicocephalus
 cirrocephalus 471
 philadelphia 200
 ridibundus 200
Chuck-will's-widow 268
Chukar 56
Ciccaba virgata 473
Ciconiidae (family) 114
Cinclidae (family) 344
Cinclus mexicanus 344
Circus cyaneus 122
Cistothorus
 palustris 344
 platensis 344
Clangula hyemalis 46
Coccothraustes
 coccothraustes 462
 vespertinus 462
Coccyzus
 americanus 252
 erythropthalmus 252
 minor 252
Coereba flaveola 402
Colaptes
 auratus 284
 chrysoides 284
Colibri thalassinus 272
Colinus virginianus 68
Columba livia 242

Columbidae (family) 242–246
Columbina
 inca 246
 passerina 246
 talpacoti 246
Condor, California 118
Contopus
 caribaeus 294
 cooperi 294
 pertinax 294
 sordidulus 294
 virens 294
Conuropsis carolinensis 472
Coot
 American 150
 Eurasian 150
Coragyps atratus 118
Cormorant
 Brandt's 106
 Double-crested 106
 Great 104
 Neotropic 104
 Pelagic 106
 Red-faced 106
Corvidae (family) 320–326
Corvus
 brachyrhynchos 326
 caurinus 326
 corax 326
 cryptoleucus 326
 imparatus 326
 monedula 326
 ossifragus 326
Coturnicops noveboracensis 148
Cowbird
 Bronzed 448
 Brown-headed 448
 Shiny 448
Cracidae (family) 56
Crake
 Corn 148
 Paint-billed 469
Crane
 Common 152
 Sandhill 152
 Whooping 152

Creeper, Brown 340
Crex crex 148
Crossbill
 Red 458
 White-winged 458
Crotophaga
 ani 254
 sulcirostris 254
Crow
 American 326
 Fish 326
 Northwestern 326
 Tamaulipas 326
Cuckoo
 Black-billed 252
 Common 254
 Mangrove 252
 Oriental 254
 Yellow-billed 252
Cuculidae (family) 252–254
Cuculus
 canorus 254
 optatus 254
Curlew
 Bristle-thighed 172, 194
 Eskimo 170, 194
 Eurasian 172, 194
 Far Eastern 172, 194
 Little 170, 194
 Long-billed 172, 194
 Slender-billed 471
Cyanocitta
 cristata 320
 stelleri 320
Cyanocompsa parellina 440
Cyanocorax
 morio 324
 yncas 324
Cygnus
 buccinator 26
 columbianus 26
 cygnus 26
 olor 26
Cynanthus latirostris 274
Cypseloides niger 270
Cyrtonyx montezumae 68

D

Delichon urbicum 330

Dendragapus

fuliginosus 60

obscurus 60

Dendrocopos major 288

Dendrocygna

autumnalis 28

bicolor 28

Dendroica

caerulescens 382

castanea 388

cerulea 382

chrysoparia 384

coronata 380

discolor 386

dominica 386

fusca 382

graciae 386

kirtlandii 386

magnolia 380

nigrescens 384

occidentalis 384

palmarum 388

pensylvanica 380

petechia 390

pinus 388

striata 388

tigrina 380

townsendi 384

virens 384

Dickcissel 438

Diomedea exulans 467

Diomedeidae (family) 78–80

Dipper, American 344

Dolichonyx oryzivorus 442

Dotterel, Eurasian 158, 196

Dove

African Collared- 244

Common Ground- 246

Eurasian Collared- 244

Inca 246

Key West Quail- 246

Mourning 244

Oriental Turtle- 244

Ruddy Ground- 246

Ruddy Quail- 246

Spotted 244

White-tipped 246

White-winged 244

Zenaida 244

Dovekie 232

Dowitcher

Long-billed 188

Short-billed 188, 195

Dryocopus pileatus 292

Duck

American Black 30, 52

Eastern Spot-billed 30

Falcated 32, 53

Harlequin 44, 55

Labrador 466

Long-tailed 46, 55

Mandarin 50

Masked 50, 53

Mottled 30

Muscovy 28

Ring-necked 40, 54

Ruddy 50, 52

Tufted 40, 54

Whistling *see*
Whistling-Duck

Wood 28, 52

Dumetella carolinensis 362

Dunlin 178, 197

E

Eagle

Bald 124, 145

Golden 124, 145

Steller's Sea- 124

White-tailed 124

Ectopistes migratorius 472

Egret

Cattle 112

Chinese 468

Great 112

Little 112

Reddish 110

Snowy 112

Egretta

caerulea 110

eulophotes 468

garzetta 112

gularis 468

rufescens 110

thula 112

tricolor 110

Eider

Common 42, 54

King 42, 55

Spectacled 42, 55

Steller's 42, 55

Elaenia

Caribbean 475

Greenish 474

Elaenia martinica 475

Elanoides forficatus 120

Elanus leucurus 120

Emberiza

aureola 432

elegans 479

leucocephalos 479

pallasi 434

pusilla 434

rustica 434

schoeniclus 434

variabilis 432

Emberizidae (family) 402–434

Empidonax

alnorum 296

difficilis 300

flaviventris 296

fulvifrons 300

hammondii 298

minimus 298

oberholseri 298

occidentalis 300

traillii 296

virescens 296

wrightii 298

Empidonomus varius 310

Eremophila alpestris 328

Estrildidae (family) 464

Eudocimus albus 116

Eugenes fulgens 274
Euphagus
   carolinus 448
   cyanocephalus 448
Euplectes franciscanus 464
Euptilotis neoxenus 278
Eurynorhynchus pygmeus 184
Euthlypis lachrymosa 396

F

Falcipennis canadensis 60
Falco
   columbarius 138
   femoralis 136
   mexicanus 140
   peregrinus 140
   rusticolus 140
   sparverius 138
   subbuteo 136
   tinnunculus 138
   vespertinus 469
Falcon
   Aplomado 136
   Collared Forest- 469
   Peregrine 140, 142
   Prairie 140, 142
   Red-footed 469
Falconidae (family) 136–140
Ficedula
   albicilla 350
   mugimaki 477
   narcissina 350
Fieldfare 358
Finch
   Black Rosy 454
   Brown-capped Rosy- 454
   Cassin's 456
   Gray-crowned Rosy- 454
   House 456
   Purple 456
   Rosy *see* Finch, Black Rosy-;
     Finch, Brown-capped Rosy-;
     Finch, Gray-crowned Rosy
   see also Chaffinch,

Common; Greenfinch, Oriental; Hawfinch;_Rosefinch, Common
Flamingo, American 114
Flicker
   Gilded 284
   Northern 284
Flycatcher
   Acadian 296
   Alder 296
   Ash-throated 304
   Asian Brown 350
   Brown-crested 304
   Buff-breasted 300
   Cordilleran 300
   Dark-sided 350
   Dusky 298
   Dusky-capped 304
   Fork-tailed 308
   Gray 298
   Gray-streaked 350
   Great Crested 304
   Hammond's 298
   La Sagra's 304
   Least 298
   Mugimaki 477
   Narcissus 350
   Nutting's 304
   Olive-sided 294
   Pacific-slope 300
   Piratic 310
   Scissor-tailed 310
   Silky *see* Silky-flycatcher, Gray
   Social 475
   Spotted 477
   Sulphur-bellied 310
   Taiga 350
   Tufted 302
   Variegated 310
   Vermilion 302
   Willow 296
   Yellow-bellied 296
Fratercula
   arctica 240
   cirrhata 240
   corniculata 240

Fregata
   ariel 468
   magnificens 98
   minor 467
Fregatidae (family) 98
Fregetta tropica 467
Frigatebird
   Great 467
   Lesser 468
   Magnificent 98
Fringilla
   coelebs 454
   montifringilla 454
Fringillidae (family) 454–462
Fulica
   americana 150
   atra 150
Fulmar, Northern 82
Fulmarus glacialis 82

G

Gadwall 32, 53
Gallinago
   delicata 190
   gallinago 190
   stenura 190
Gallinula chloropus 150
Gallinule, Purple 150
Gannet, Northern 102
Garganey 36, 53
Gavia
   adamsii 72
   arctica 70
   immer 72
   pacifica 70
   stellata 70
Gaviidae (family) 70–72
Gelochelidon nilotica 220
Geococcyx californianus 252
Geothlypis
   poliocephala 396
   trichas 396
Geotrygon
   chrysia 246

montana 246
Geranospiza caerulescens 469
Glareola maldivarum 471
Glaucidium
    brasilianum 262
    gnoma 262
Gnatcatcher
    Black-capped 346
    Black-tailed 346
    Blue-gray 346
    California 346
Godwit
    Bar-tailed 174, 194
    Black-tailed 174, 194
    Hudsonian 174, 194
    Marbled 174, 194
Goldeneye
    Barrow's 46, 55
    Common 46, 55
Goldfinch
    American 460
    Lawrence's 460
    Lesser 460
Goose
    Bar-headed 50
    Barnacle 22
    Cackling 24
    Canada 24
    Egyptian 50
    Emperor 22
    Graylag 50, 466
    Greater White-fronted 20
    Lesser White-fronted 466
    Pink-footed 20
    Ross's 22
    Snow 22
    Swan 50
    Taiga Bean- 20
    Tundra Bean- 20
Goshawk, Northern 126, 143
Grackle
    Boat-tailed 446
    Common 446
    Great-tailed 446
Gracula religiosa 366
Grassquit
    Black-faced 402
    Yellow-faced 402

Grebe
    Clark's 76
    Eared 74
    Horned 74
    Least 74
    Pied-billed 74
    Red-necked 76
    Western 76
Greenfinch, Oriental 454
Greenshank, Common
    164, 195
Grosbeak
    Black-headed 436
    Blue 438
    Crimson-collared 436
    Evening 462
    Pine 458
    Rose-breasted 436
    Yellow 436
Grouse
    Dusky 60
    Greater Sage- 64
    Gunnison Sage- 64
    Ruffed 60
    Sharp-tailed 64
    Sooty 60
    Spruce 60
Gruidae (family) 152
Grus
    americana 152
    canadensis 152
    grus 152
Guillemot
    Black 234
    Pigeon 234
Gull
    Belcher's 204
    Black-headed 200, 216
    Black-tailed 204
    Bonaparte's 200, 216
    California 204, 217
    Franklin's 198, 216
    Glaucous 208, 217
    Glaucous-winged 210, 217
    Gray-hooded 471
    Great Black-backed
        212, 217
    Heermann's 198

    Herring 206, 217
    Iceland 208, 217
    Ivory 214
    Kelp 204
    Laughing 198, 216
    Lesser Black-backed
        212, 217
    Little 200, 216
    Mew 202, 216
    Ring-billed 202, 216
    Ross's 200, 216
    Sabine's 214, 216
    Slaty-backed 212, 217
    Thayer's 208, 217
    Western 210, 217
    Yellow-footed 210, 217
    Yellow-legged 206
Gymnogyps californianus 118
Gymnorhinus cyanocephalus 322
Gyrfalcon 140, 142

H

Haematopodidae (family) 160
Haematopus
    bachmani 160
    ostralegus 470
    palliatus 160
Haliaeetus
    albicilla 124
    leucocephalus 124
    pelagicus 124
Harrier, Northern 122, 143
Hawfinch 462
Hawk
    Broad-winged 130, 143
    Common Black- 145
    Cooper's 126, 143
    Crane 469
    Ferruginous 134, 144
    Gray 130, 143
    Harris's 128, 144
    Red-shouldered 130, 143
    Red-tailed 132, 144
    Roadside 128
    Rough-legged 134, 144

Sharp-shinned **126**, 143
Short-tailed **128**
Swainson's **132**, 144
White-tailed **134**, 144
Zone-tailed **128**, 145
Heliomaster constantii **274**
Helmitheros vermivorus **394**
Heron
    Chinese Pond- **469**
    Gray **468**
    Great Blue **114**
    Green **110**
    Green-backed *see*
        Heron, Green
    Little Blue **110**
    Tricolored **110**
    Western Reef **468**
    see also Night-Heron
Himantopus
    himantopus **471**
    mexicanus **160**
Hirundapus caudacutus **270**
Hirundinidae (family) 330–332
Hirundo rustica **332**
Histrionicus histrionicus **44**
Hobby, Eurasian 136
Hoopoe, Eurasian 474
Hummingbird
    Allen's **278**
    Anna's **276**
    Berylline **272**
    Black-chinned **276**
    Blue-throated **274**
    Broad-billed **274**
    Broad-tailed **278**
    Buff-bellied **272**
    Bumblebee **474**
    Calliope **278**
    Cinnamon **474**
    Costa's **276**
    Lucifer **272**
    Magnificent **274**
    Ruby-throated **276**
    Rufous **278**
    Violet-crowned **272**
    White-eared **274**

Xantus's **274**
Hydrobates pelagicus **94**
Hydrobatidae (family) 94–96
Hydrocoloeus minutus **200**
Hydroprogne caspia **218**
Hylocharis
    leucotis **274**
    xantusii **274**
Hylocichla mustelina **356**

I

Ibis
    Glossy **116**
    White **116**
    White-faced **116**
Icteria virens **396**
Icteridae (family) 442–452
Icterus
    bullockii **450**
    cucullatus **450**
    galbula **450**
    graduacauda **452**
    gularis **452**
    parisorum **452**
    pectoralis **452**
    pustulatus **452**
    spurius **450**
    wagleri **452**
Ictinia mississippiensis **120**
Ixobrychus
    exilis **108**
    sinensis **468**
Ixoreus naevius **358**

J

Jabiru 114
Jabiru mycteria **114**
Jacana, Northern 160
Jacana spinosa **160**
Jacanidae (family) 160
Jackdaw, Eurasian 326
Jaeger
    Long-tailed **230**

Parasitic **230**
Pomarine **230**
Jay
    Blue **320**
    Brown **324**
    Florida Scrub- **322**
    Gray **320**
    Green **324**
    Island Scrub- **322**
    Mexican **322**
    Pinyon **322**
    Scrub *see* Jay, Florida Scrub-; Jay,
        Island Scrub-; Jay, Western
        Scrub-
    Steller's **320**
    Western Scrub- **322**
Junco
    Dark-eyed **426**
    Yellow-eyed **426**
Junco
    hyemalis **426**
    phaeonotus **426**
Jynx torquilla **474**

K

Kestrel
    American **138**, 142
    Eurasian **138**
Killdeer 158, 196
Kingbird
    Cassin's **306**
    Couch's **306**
    Eastern **308**
    Gray **308**
    Thick-billed **308**
    Tropical **306**
    Western **306**
Kingfisher
    Belted **280**
    Green **280**
    Ringed **280**
Kinglet
    Golden-crowned **346**
    Ruby-crowned **346**

Kiskadee, Great 310
Kite
  Hook-billed **122, 142**
  Mississippi **120, 142**
  Snail **122, 142**
  Swallow-tailed **120**
  White-tailed **120, 142**
Kittiwake
  Black-legged **214, 216**
  Red-legged **214, 216**
Knot
  Great **178**
  Red **178, 195**

L

Lagopus
  lagopus **62**
  leucura **62**
  muta **62**
Lampornis clemenciae **274**
Laniidae (family) 312
Lanius
  cristatus **312**
  excubitor **312**
  ludovicianus **312**
Lapwing, Northern 158
Laridae (family) 198–226
Lark
  Horned **328**
  Sky **328**
Larus
  argentatus **206**
  belcheri **204**
  californicus **204**
  canus **202**
  crassirostris **204**
  delawarensis **202**
  dominicanus **204**
  fuscus **212**
  glaucescens **210**
  glaucoides **208**
  heermanni **198**
  hyperboreus **208**
  livens **210**

  marinus **212**
  michahellis **206**
  occidentalis **210**
  schistisagus **212**
  thayeri **208**
Laterallus jamaicensis **148**
Legatus leucophaius **310**
Leptotila verreauxi **246**
Leucophaeus
  atricilla **198**
  pipixcan **198**
Leucosticte
  atrata **454**
  australis **454**
  tephrocotis **454**
Limicola falcinellus **184**
Limnodromus
  griseus **188**
  scolopaceus **188**
Limnothlypis swainsonii **394**
Limosa
  fedoa **174**
  haemastica **174**
  lapponica **174**
  limosa **174**
Limpkin 146
Locustella
  lanceolata **348**
  ochotensis **348**
Lonchura punctulata **464**
Longspur
  Chestnut-collared **428**
  Lapland **430**
  McCown's **428**
  Smith's **430**
Loon
  Arctic **70**
  Common **72**
  Pacific **70**
  Red-throated **70**
  Yellow-billed **72**
Lophodytes cucullatus **48**
Loxia
  curvirostra **458**
  leucoptera **458**
Luscinia

  calliope **352**
  cyane **478**
  svecica **352**
Lymnocryptes minimus **190**

M

Magpie
  Black-billed **324**
  Yellow-billed **324**
Mallard 30, 53
Mango, Green-breasted 272
Mannikin, Nutmeg 464
Martin
  Brown-chested **476**
  Common House- **330**
  Cuban **475**
  Gray-breasted **476**
  Purple **330**
  Southern **476**
Meadowlark
  Eastern **442**
  Western **442**
Megaceryle
  alcyon **280**
  torquatus **280**
Megascops
  asio **260**
  kennicottii **260**
  trichopsis **260**
Melanerpes
  aurifrons **284**
  carolinus **284**
  erythrocephalus **282**
  formicivorus **282**
  lewis **282**
  uropygialis **284**
Melanitta
  fusca **44**
  nigra **44**
  perspicillata **44**
Melanotis caerulescens **362**
Meleagris gallopavo **58**
Melopsittacus undulatus **250**
Melospiza
  georgiana **422**

lincolnii 422
melodia 422
Merganser
   Common 48, 54
   Hooded 48, 54
   Red-breasted 48, 54
Mergellus albellus 48
Mergus
   merganser 48
   serrator 48
Merlin 138, 142
Micrastur semitorquatus 469
Micrathene whitneyi 262
Mimidae (family) 362–364
Mimus
   gundlachii 362
   polyglottos 362
Mitrephanes phaeocercus 302
Mniotilta varia 382
Mockingbird
   Bahama 362
   Blue 362
   Northern 362
Molothrus
   aeneus 448
   ater 448
   bonariensis 448
Moorhen, Common 150
Morus bassanus 102
Motacilla
   alba 368
   cinerea 368
   citreola 478
   tschutschensis 368
Motacillidae (family) 368–370
Murre
   Common 232
   Thick-billed 232
Murrelet
   Ancient 236
   Craveri's 236
   Kittlitz's 234
   Long-billed 234
   Marbled 234
   Xantus's 236
Muscicapa

dauurica 350
griseisticta 350
sibirica 350
striata 477
Muscicapidae (family) 350
Myadestes townsendi 354
Mycteria americana 114
Myiarchus
   cinerascens 304
   crinitus 304
   nuttingi 304
   sagrae 304
   tuberculifer 304
   tyrannulus 304
Myioborus
   miniatus 398
   pictus 398
Myiodynastes luteiventris 310
Myiopagis viridicata 474
Myiopsitta monachus 248
Myiozetetes similis 475
Myna
   Common 366
   Crested 366
   Hill 366

N

Nandayus nenday 248
Needletail, White-throated 270
Neocrex erythrops 469
Night-Heron
   Black-crowned 108
   Yellow-crowned 108
Nighthawk
   Antillean 266
   Common 266
   Lesser 266
Nightjar
   Buff-collared 268
   Gray 473
Noddy
   Black 226
   Brown 226
Nomonyx dominicus 50

Nucifraga columbiana 320
Numenius
   americanus 172
   arquata 172
   borealis 170
   madagascariensis 172
   minutus 170
   phaeopus 170
   tahitiensis 172
   tenuirostris 471
Nutcracker, Clark's 320
Nuthatch
   Brown-headed 340
   Pygmy 340
   Red-breasted 340
   White-breasted 340
Nyctanassa violacea 108
Nycticorax nycticorax 108
Nyctidromus albicollis 266

O

Oceanites oceanicus 94
Oceanodroma
   castro 94
   furcata 96
   homochroa 96
   hornbyi 467
   leucorhoa 94
   melania 96
   microsoma 96
   tethys 96
Odontophoridae (family) 66–68
Oenanthe oenanthe 352
Onychoprion
   aleuticus 222
   anaethetus 226
   fuscatus 226
Oporornis
   agilis 390
   formosus 392
   philadelphia 390
   tolmiei 390
Oreortyx pictus 66
Oreoscoptes montanus 364

Oriole
  Altamira 452
  Audubon's 452
  Baltimore 450
  Black-vented 452
  Bullock's 450
  Hooded 450
  Nothern *see* Oriole, Baltimore;
    Oriole, Bullock's
  Orchard 450
  Scott's 452
  Spot-breasted 452
  Streak-backed 452
Ortalis vetula 56
Osprey 118, 145
Otus
  flammeolus 262
  sunia 473
Ovenbird 394
Owl
  Barn 256
  Barred 258
  Boreal 264
  Burrowing 264
  Eastern Screech- 260
  Elf 262
  Ferruginous Pygmy- 262
  Flammulated 262
  Great Gray 258
  Great Horned 256
  Long-eared 256
  Mottled 473
  Northern Hawk 264
  Northern Pygmy- 262
  Northern Saw-whet 264
  Oriental Scops- 473
  Short-eared 256
  Snowy 258
  Spotted 258
  Stygian 473
  Western Screech- 260
  Whiskered Screech- 260
Oxyura jamaicensis 50
Oystercatcher
  American 160
  Black 160
  Eurasian 470

P
Pachyramphus algaiae 310
Pagophila eburnea 214
Pandion haliaetus 118
Parabuteo unicinctus 128
Parakeet
  Black-hooded 248
  Blue-crowned 248
  Carolina 472
  Dusky-headed 248
  Green 248
  Mitred 248
  Monk 248
  Red-masked 248
  Rose-ringed 250
  White-winged 248
  Yellow-chevroned 248
Pardirallus maculatus 470
Paridae (family) 334–338
Parrot
  Lilac-crowned 250
  Orange-winged 250
  Red-crowned 250
  Thick-billed 250
  Yellow-headed 250
Partridge, Gray 56
Parula
  Northern 378
  Tropical 378
Parula
  americana 378
  pitiayumi 378
  superciliosa 378
Parulidae (family) 374–398
Passer
  domesticus 464
  montanus 464
Passerculus sandwichensis 420
Passerella iliaca 420
Passeridae (family) 464
Passerina
  amoena 440
  caerulea 438
  ciris 440
  cyanea 440
  versicolor 440

Patagioenas
  fasciata 242
  flavirostris 242
  leucocephala 242
  squamosa 472
Pauraque, Common 266
Pelagodroma marina 94
Pelecanidae (family) 102
Pelecanus
  erythrorhynchos 102
  occidentalis 102
Pelican
  American White 102
  Brown 102
Perdix perdix 56
Perisoreus canadensis 320
Petrel
  Bermuda 86
  Black-capped 86
  Bulwer's 90
  Cook's 84
  Fea's 86
  Great-winged 82
  Hawaiian 84
  Herald 86
  Mottled 84
  Murphy's 82
  Parkinson's 82
  Stejneger's 84
  Storm- *see* Storm-Petrel
Petrochelidon
  fulva 332
  pyrrhonota 332
Peucedramidae (family) 398
Peucedramus taeniatus 398
Pewee
  Cuban 294
  Eastern Wood- 294
  Greater 294
  Western Wood- 294
Phaethon
  aethereus 98
  lepturus 98
  rubricauda 98
Phaethontidae (family) 98
Phaetusa simplex 226

Phainopepla 372
Phainopepla nitens 372
Phalacrocoracidae (family) 104–106
Phalacrocorax
    auritus 106
    brasilianus 104
    carbo 104
    pelagicus 106
    penicillatus 106
    urile 106
Phalaenoptilus nuttallii 268
Phalarope
    Red 192, 197
    Red-necked 192, 197
    Wilson's 192, 195
Phalaropus
    fulicarius 192
    lobatus 192
    tricolor 192
Phasianidae (family) 56–64
Phasianus colchicus 58
Pheasant, Ring-necked 58
Pheucticus
    chrysopeplus 436
    ludovicianus 436
    melanocephalus 436
Philomachus pugnax 186
Phoebastria
    albatrus 78
    immutabilis 78
    nigripes 78
Phoebe
    Black 302
    Eastern 302
    Say's 302
Phoebetria palpebrata 466
Phoenicopteridae (family) 114
Phoenicopterus ruber 114
Phylloscopus
    borealis 348
    fuscatus 348
    inornatus 477
    sibilatrix 477
    trochilus 476
Pica
    hudsonia 324

nuttalli 324
Picidae (family) 282–292
Picoides
    albolarvatus 282
    arcticus 290
    arizonae 288
    borealis 288
    dorsalis 290
    nuttallii 288
    pubescens 290
    scalaris 288
    villosus 290
Pigeon
    Band-tailed 242
    Passenger 472
    Red-billed 242
    Rock 242
    Scaly-naped 472
    White-crowned 242
Pinguinus impennis 472
Pinicola enucleator 458
Pintail
    Northern 34, 52
    White-cheeked 34
Pipilo
    aberti 404
    chlorurus 404
    crissalis 404
    erythrophthalmus 406
    fuscus 404
    maculatus 406
Pipit
    American 370
    Olive-backed 370
    Pechora 370
    Red-throated 370
    Sprague's 370
    Tree 478
Piranga
    bidentata 400
    flava 400
    ludoviciana 400
    olivacea 400
    rubra 400
Pitangus sulphuratus 310
Platalea ajaja 116

Plectrophenax
    hyperboreus 432
    nivalis 432
Plegadis
    chihi 116
    falcinellus 116
Ploceidae (family) 464
Plover
    American Golden- 154, 196
    Black-bellied 154, 196
    Collared 470
    Common Ringed 156, 196
    European Golden- 154, 196
    Greater Sand- 470
    Lesser Sand- 158, 196
    Little Ringed 158, 196
    Mongolian 196
    Mountain 158, 196
    Pacific Golden- 154, 196
    Piping 156, 196
    Semipalmated 156, 196
    Snowy 156, 196
    Wilson's 156, 196
Pluvialis
    apricaria 154
    dominica 154
    fulva 154
    squatarola 154
Pochard, Common 38, 54
Podiceps
    auritus 74
    grisegena 76
    nigricollis 74
Podicipedidae (family) 74–76
Podilymbus podiceps 74
Poecile
    atricapillus 336
    carolinensis 336
    cinctus 338
    gambeli 336
    hudsonicus 338
    rufescens 338
    sclateri 336
Polioptila
    caerulea 346
    californica 346

melanura 346

nigriceps 346

Polysticta stelleri 42

Pooecetes gramineus 422

Poorwill, Common 268

Porphyrio

   martinica 150

   porphyrio 150

Porzana carolina 148

Prairie-Chicken

   Greater 64

   Lesser 64

Pratincole, Oriental 471

Procellaria parkinsoni 82

Procellariidae (family) 82–92

Progne

   chalybea 476

   crytoleuca 475

   elegans 476

   subis 330

   tapera 476

Protonotaria citrea 374

Prunella montanella 368

Prunellidae (family) 368

Psaltiparus minimus 338

Psittacidae (family) 248–250

Psittacula krameri 250

Ptarmigan

   Rock 62

   White-tailed 62

   Willow 62

Pterodroma

   arminjoniana 86

   cahow 86

   cookii 84

   feae 86

   hasitata 86

   inexpectata 84

   longirostris 84

   macroptera 82

   sandwichensis 84

   ultima 82

Ptilogonatidae (family) 372

Ptilogonys cinereus 479

Ptychoramphus aleuticus 236

Puffin

Atlantic 240

Horned 240

Tufted 240

Puffinus

   assimilis 88

   bulleri 92

   carneipes 90

   creatopus 92

   gravis 88

   griseus 90

   lherminieri 88

   opisthomelas 92

   pacificus 90

   puffinus 88

   tenuirostris 90

Pycnonotidae (family) 366

Pycnonotus jocosus 366

Pyrocephalus rubinus 302

Pyrrhula pyrrhula 462

Pyrrhuloxia 438

Q

Quail

   California 66

   Gambel's 66

   Montezuma 68

   Mountain 66

   Scaled 68

Quetzal, Eared 278

Quiscalus

   major 446

   mexicanus 446

   quiscala 446

R

Rail

   Black 148

   Clapper 146

   King 146

   Spotted 470

   Virginia 148

   Yellow 148

Rallidae (family) 146–150

Rallus

elegans 146

limicola 148

longirostris 146

Raven

   Chihuahuan 326

   Common 326

Razorbill 232

Recurvirostra americana 160

Recurvirostridae (family) 160

Redhead 38, 54

Redpoll

   Common 462

   Hoary 462

Redshank

   Common 164, 195

   Spotted 164, 195

Redstart

   American 398

   Painted 398

   Slate-throated 398

Redwing 358

Regulidae (family) 346

Regulus

   calendula 346

   satrapa 346

Remizidae (family) 338

Rhodostethia rosea 200

Rhodothraupis celaeno 436

Rhynchopsitta pachyrhyncha 250

Ridgwayia pinicola 360

Riparia riparia 332

Rissa

   brevirostris 214

   tridactyla 214

Roadrunner, Greater 252

Robin

   American 360

   Clay-colored 360

   Rufous-backed 360

   Siberian Blue 478

   White-throated 360

Rosefinch, Common 456

Rostrhamus sociabilis 122

Rubythroat, Siberian 352

Ruff 186, 197

Rynchops niger 226

S

Salpinctes obsoletus 344

Sanderling 178, 197

Sandpiper

Baird's 184, 197

Broad-billed 184

Buff-breasted 186, 195

Common 168

Curlew 178, 195

Green 166, 195

Least 180, 197

Marsh 164

Pectoral 186, 197

Purple 176, 197

Rock 176, 197

Semipalmated 180, 197

Sharp-tailed 186, 197

Solitary 166, 195

Spoon-billed 184

Spotted 168

Stilt 190, 195

Terek 166

Upland 186, 195

Western 180, 197

White-rumped 184, 197

Wood 166, 195

Sapsucker

Red-breasted 286

Red-naped 286

Williamson's 286

Yellow-bellied 286

Saxicola torquatus 352

Sayornis

nigricans 302

phoebe 302

saya 302

Scaup

Greater 40, 54

Lesser 40, 54

Scolopacidae (family) 162–192

Scolopax

minor 190

rusticola 471

Scoter

Black 44, 55

Surf 44, 54

White-winged 44, 54

Seedeater, White-collared 402

Seiurus

aurocapilla 394

motacilla 394

noveboracensis 394

Selasphorus

platycerus 278

rufus 278

sasin 278

Setophaga ruticilla 398

Shearwater

Audubon's 88

Black-vented 92

Buller's 92

Cape Verde 88

Cory's 88

Flesh-footed 90

Greater 88

Little 88

Manx 88

Pink-footed 92

Short-tailed 90

Sooty 90

Streaked 92

Wedge-tailed 90

Shelduck

Common 50

Ruddy 50

Shoveler, Northern 36, 53

Shrike

Brown 312

Loggerhead 312

Northern 312

Sialia

currucoides 354

mexicana 354

sialis 354

Silky-flycatcher, Gray 479

Siskin

Eurasian 479

Pine 460

Sitta

canadensis 340

carolinensis 340

pusilla 340

pygmaea 340

Sittidae (family) 340

Skimmer, Black 226

Skua

Great 228

South Polar 228

Skylark, Eurasian *see* Lark, Sky

Smew 48, 55

Snipe

Common 190

Jack 190

Pin-tailed 190

Wilson's 190

Snowcock, Himalayan 58

Solitaire, Townsend's 354

Somateria

fischeri 42

mollissima 42

spectabilis 42

Sora 148

Sparrow

American Tree 410

Bachman's 408

Baird's 416

Black-chinned 414

Black-throated 414

Botteri's 408

Brewer's 412

Cassin's 408

Chipping 412

Clay-colored 412

Eurasian Tree 464

Field 410

Five-striped 414

Fox 420

Golden-crowned 424

Grasshopper 416

Harris's 424

Henslow's 416

House 464

Lark 414

Le Conte's 418

Lincoln's 422

Nelson's 418

Olive 404

Rufous-crowned 410
Rufous-winged 410
Sage 414
Saltmarsh 418
Savannah 420
Seaside 418
Song 422
Swamp 422
Vesper 422
White-crowned 424
White-throated 424
Worthen's 479
Sphyrapicus
  nuchalis 286
  ruber 286
  thyroideus 286
  varius 286
Spindalis, Western 402
Spindalis zena 402
Spinus
  lawrencei 460
  pinus 460
  psaltria 460
  spinus 479
  tristris 460
Spiza americana 438
Spizella
  arborea 410
  atrogularis 414
  breweri 412
  pallida 412
  passerina 412
  pusilla 410
  wortheni 479
Spoonbill, Roseate 116
Sporophila torqueola 402
Starling, European 366
Starthroat, Plain-capped 274
Stelgidopteryx serripennis 332
Stellula calliope 278
Stercorariidae (family) 228–230
Stercorarius
  longicaudus 230
  maccormicki 228
  parasiticus 230
  pomarinus 230

skua 228
Sterna
  dougallii 220
  fosteri 220
  hirundo 222
  paradisaea 222
Sternula
  antillarum 224
Stilt
Black-necked 160
Black-winged 471
Stint
Little 182
Long-toed 182
Red-necked 182
Temminck's 182, 197
Stonechat 352
Stork, Wood 114
Storm-Petrel
Ashy 96
Band-rumped 94
Black 96
Black-bellied 467
European 94
Fork-tailed 96
Leach's 94
Least 96
Ringed 467
Wedge-rumped 96
White-faced 94
Wilson's 94
Streptopelia
  chinensis 244
  decaocto 244
  orientalis 244
  roseogrisea 244
Streptoprocne zonaris 270
Strigidae (family) 256–264
Strix
  nebulosa 258
  occidentalis 258
  varia 258
Sturnella
  magna 442
  neglecta 442
Sturnidae (family) 366
Sturnus vulgaris 366

Sula
  dactylatra 100
  granti 467
  leucogaster 100
  nebouxii 100
  sula 100
Sulidae (family) 100–102
Surfbird 176
Surnia ulula 264
Swallow
Bahama 330
Bank 332
Barn 332
Cave 332
Cliff 332
Mangrove 476
Northern Rough-winged 332
Tree 330
Violet-green 330
Swamphen, Purple 150
Swan
Mute 26
Trumpeter 26
Tundra 26
Whooper 26
Swift
Antillean Palm- 473
Black 270
Chimney 270
Common 270
Fork-tailed 270
Vaux's 270
White-collared 270
White-throated 270
Sylvia curruca 477
Sylviidae (family) 346–348
Synthliboramphus
  antiquus 236
  craveri 236
  hypoleucus 236

T

Tachornis phoenicobia 473
Tachybaptus dominicus 74
Tachycineta

albilinea 476

bicolor 330

cyaneoviridis 330

thalassina 330

Tadorna

  ferruginea 50

  tadorna 50

Tanager

  Flame-colored 400

  Hepatic 400

  Scarlet 400

  Summer 400

  Western 400

Tarsiger cyanurus 352

Tattler

  Gray-tailed 168

  Wandering 168

Teal

  Baikal **32**, 52

  Blue-winged **36**, 53

  Cinnamon **36**, 53

  Green-winged **32**, 52

Tern

  Aleutian 222

  Arctic 222

  Black 224

  Bridled 226

  Caspian 218

  Common 222

  Elegant 218

  Forster's 220

  Gull-billed 220

  Large-billed 226

  Least 224

  Roseate 220

  Royal 218

  Sandwich 218

  Sooty 226

  Whiskered 472

  White-winged 224

Tetraogallus himalayensis 58

Thalassarche

  cauta 80

  chlororhynchos 80

  melanophris 80

Thalasseus

elegans 218

maximus 218

sandvicensis 218

Thick-knee, Double-striped 470

Thrasher

  Bendire's 364

  Brown 362

  California 364

  Crissal 364

  Curve-billed 364

  Le Conte's 364

  Long-billed 362

  Sage 364

Thraupidae (family) 400–402

Threskiornithidae (family) 116

Thrush

  Aztec 360

  Bicknell's 356

  Black-headed Nightingale- 478

  Dusky 358

  Eyebrowed 358

  Gray-cheeked 356

  Hermit 356

  Orange-billed Nightingale- 478

  Swainson's 356

  Varied 358

  Wood 356

Thryomanes bewickii 342

Thryothorus ludovicianus 342

Tiaris

  bicolor 402

  olivaceus 402

Timaliidae (family) 334

Tit

  Siberian *see* Chickadee, Gray-
    headed

Titmouse

  Black-crested 334

  Bridled 334

  Juniper 334

  Oak 334

  Plain *see* Titmouse, Juniper; Tit-
    mouse, Oak

  Tufted 334

Tityra, Masked 475

Tityra semifasciata 475

Towhee

  Abert's 404

  California 404

  Canyon 404

  Eastern 406

  Green-tailed 404

  Spotted 406

Toxostoma

  bendirei 364

  crissale 364

  curvirostre 364

  lecontei 364

  longirostre 362

  redivivum 364

  rufum 362

Tringa

  brevipes 168

  erythropus 164

  flavipes 162

  glareola 166

  incana 168

  melanoleuca 162

  nebularia 164

  ochropus 166

  semipalmata 162

  solitaria 166

  stagnatillis 164

  totanus 164

Trochilidae (family) 272–278

Troglodytes

  aedon 342

  troglodytes 342

Troglodytidae (family) 342–344

Trogon, Elegant 278

Trogon elegans 278

Trogonidae (family) 278

Tropicbird

  Red-billed 98

  Red-tailed 98

  White-tailed 98

Tryngites subruficollis 186

Turdidae (family) 352–360

Turdus

  assimilis 360

  grayi 360

  iliacus 358

merula 478
migratorius 360
naumanni 358
obscurus 358
pilaris 358
rufopalliatus 360
Turkey, Wild 58
Turnstone
  Black 176
  Ruddy 176
Tympanuchus
  cupido 64
  pallidicinctus 64
  phasianellus 64
Tyrannidae (family) 294–310
Tyrannulet, Northern
 Beardless- 300
Tyrannus
  couchii 306
  crassirostris 308
  dominicensis 308
  forficatus 310
  melancholicus 306
  savana 308
  tyrannus 308
  verticalis 306
  vociferans 306
Tyto alba 256
Tytonidae (family) 256–264

U

Upupa epops 474
Uria
  aalge 232
  lomvia 232

V

Vanellus vanellus **158**
Veery 356
Verdin 338
Vermivora
  bachmanii 376
  celata 376

chrysoptera 374
crissalis 378
luciae 378
peregrina 376
pinus 374
ruficapilla 378
virginiae 378
Violetear, Green 272
Vireo
  Bell's 316
  Black-capped 314
  Black-whiskered 318
  Blue-headed 316
  Cassin's 316
  Gray 316
  Hutton's 316
  Philadelphia 318
  Plumbeous 316
  Red-eyed 318
  Solitary see Vireo, Blue-headed;
      Vireo, Cassin's; Vireo,
      Plumbeous
  Thick-billed 314
  Warbling 318
  White-eyed 314
  Yellow-green 318
  Yellow-throated 314
  Yucatan 475
Vireo
  altiloquus 318
  atricapilla 314
  bellii 316
  cassinii 316
  crassirostris 314
  flavifrons 314
  flavoviridis 318
  gilvus 318
  griseus 314
  huttoni 316
  magister 475
  olivaceus 318
  philadelphicus 318
  plumbeus 316
  solitarius 316
  vicinior 316
Vireonidae (family) 314–318
Vulture

Black **118**, 145
Turkey **118**, 145

W

Wagtail
  Citrine 478
  Eastern Yellow 368
  Gray 368
  White 368
Warbler
  Arctic 348
  Bachman's 376
  Bay-breasted 388
  Black-and-white 382
  Black-throated Blue 382
  Black-throated Gray 384
  Black-throated Green 384
  Blackburnian 382
  Blackpoll 388
  Blue-winged 374
  Canada 392
  Cape May 380
  Cerulean 382
  Chestnut-sided 380
  Colima 378
  Connecticut 390
  Crescent-chested 378
  Dusky 348
  Fan-tailed 396
  Golden-cheeked 384
  Golden-crowned 396
  Golden-winged 374
  Grace's 386
  Hermit 384
  Hooded 392
  Kentucky 392
  Kirtland's 386
  Lanceolated 348
  Lucy's 378
  MacGillivray's 390
  Magnolia 380
  Middendorf's Grasshopper- 348
  Mourning 390
  Nashville 378

Olive **398**
Orange-crowned **376**
Palm **388**
Pine **388**
Prairie **386**
Prothonotary **374**
Red-faced **398**
Rufous-capped **396**
Swainson's **394**
Tennessee **376**
Townsend's **384**
Virginia's **378**
Willow **476**
Wilson's **392**
Wood **477**
Worm-eating **394**
Yellow **390**
Yellow-browed **477**
Yellow-rumped **380**
Yellow-throated **386**
Waterthrush
Louisiana **394**
Northern **394**
Waxwing
Bohemian **372**
Cedar **372**
Wheatear, Northern 352
Whimbrel 170, 194
Whip-poor-will 268
Whistling-Duck
Black-bellied **28**
Fulvous **28**
Whitethroat, Lesser 477
Wigeon
American **34, 52**
Eurasian **34, 52**
Willet 162, 195
Wilsonia
canadensis **392**
citrina **392**
pusilla **392**
Woodcock
American **190**
Eurasian **471**
Woodpecker
Acorn **282**

American Three-toed **290**
Arizona **288**
Black-backed **290**
Downy **290**
Gila **284**
Golden-fronted **284**
Great Spotted **288**
Hairy **290**
Ivory-billed **292**
Ladder-backed **288**
Lewis's **282**
Nuttall's **288**
Pileated **292**
Red-bellied **284**
Red-cockaded **288**
Red-headed **282**
White-headed **282**
Woodstar, Bahama 272
Wren
Bewick's **342**
Cactus **344**
Canyon **344**
Carolina **342**
House **342**
Marsh **344**
Rock **344**
Sedge **344**
Winter **342**
Wrentit 334
Wryneck, Eurasian 474

X

Xanthocephalus xanthocephalus **444**
Xema sabini **214**
Xenus cinereus **166**

Y

Yellowlegs
Greater **162, 195**
Lesser **162, 195**
Yellowthroat
Common **396**
Gray-crowned **396**

Z

Zenaida
asiatica **244**
aurita **244**
macroura **244**
Zonotrichia
albicollis **424**
atricapilla **424**
leucophrys **424**
querula **424**

Jonathan Alderfer: Front cover; 11-Pacific Golden-Plover; 12-Black-bellied Plover and Short-tailed Albatross; 15-American Golden-Plover; 16-Short-billed Dowitcher; 18-Carolina Parakeet; 31-Spot-billed Duck; 45-White-winged Scoter *stejnegeri;* 47-Goldeneye hybrid; 49-Goosander; 75; 77-heads; 79; 81; 83-Parkinson's Petrel and Murphy's Petrel; 85; 89-Cory's Shearwater, Cape Verde Shearwater, and small comparison figures; 91-Wedge-tailed Shearwater, Bulwer's Petrel, left Short-tailed Shearwater, and heads; 93; 101; 107; 155; 175; 177-flying Black Turnstone; 189; 191; 233-flying winter Dovekie; 235-Long-billed Murrelet; 239; 241-flying Rhinoceros Auklet; 245-with N. John Schmitt; 307; 311-Rose-throated Becard; 347-female Blue-gray Gnatcatcher tail; 363-Blue Mockingbird; 367-Common Myna; 466-except Lesser White-fronted Goose; 467-except Great Frigatebird; 468-Gray Heron; 472-except Whiskered Tern and Scaly-naped Pigeon; 475-Social Flycatcher and Masked Tityra; 476-Mangrove Swallow; 479-Gray Silky-flycatcher; 480-except Redwing.

David Beadle: 9-Horned Lark; 11-Least Flycatcher; 13-Acadian Flycatcher; 14-Willow and Alder Flycatchers; 83-Great-winged Petrel; 295; 297; 299; 301; 303-Tufted Flycatcher; 309-Gray Kingbird and Thick-billed Kingbird; 315-Thick-billed Vireo; 319-Philadelphia Vireo and Warbling Vireo; 329; 343-western Winter Wren; 379-Crescent-chested Warbler; 389-fall Bay-breasted Warbler; 399-Red-faced Warbler; 415-Sage Sparrow *canescens*; 423-Vesper Sparrow; 427-Pink-sided Dark-eyed Junco; 469-Paint-billed Crake; 470-Spotted Rail and Double-striped Thick-knee; 472-Scaly-naped Pigeon; 473-Gray Nightjar and Antillean Palm-Swift; 474-Bumblebee Hummingbird and Greenish Elaenia; 475-except Social Flycatcher and Masked Tityra; 476-except Mangrove Swallow and Willow Warbler; 478-Orange-billed Nightingale-Thrush and Black-headed Nightingale Thrush; 479-Worthen's Sparrow and Tawny-shouldered Blackbird. **Peter Burke:** 6-Summer Tanager; 12-Flycatcher tails; 13-Scarlet Tanager; 17-Shiny Cowbird; 109; 117-Glossy Ibis (except flying) and White-faced Ibis; 305; 311-Piratic Flycatcher and Variegated Flycatcher; 317-Gray Vireo; 361-White-throated Robin; 397-except Common Yellowthroat; 401; 403-adult male White-collared Seedeater and Yellow-faced Grassquit; 405; 407; 437-Crimson-collared Grosbeak; 449-Shiny Cowbird;

451; 453. **Marc R. Hanson:** 83-Northern Fulmar; 89-Greater Shearwater, Manx Shearwater, Audubon's Shearwater, and Little Shearwater; 91-Flesh-footed Shearwater, Sooty Shearwater, and right Short-tailed Shearwater; 95-except European Storm-Petrel; 97; 147; 149; 151-except Purple Swamphen. **Cynthia J. House:** 10-Lesser and Greater Scaups; 12-American Black Duck; 21; 23; 25-except flying and Taverner's Cackling Goose and Lesser Canada Goose; 27-except juvenile Whooper Swan; 29-except flying Muscovy Duck; 31-except head of female American Black Duck and Spot-billed Duck; 33; 35; 37; 39; 41; 43; 45-except White-winged Scoter *stejnegeri*; 47-except Goldeneye hybrid; 49-except Goosander; 51-except Egyptian Goose; 52-53; 54-55. **H. Jon Janosik:** 77-except heads; 99; 103; 105; 161; 481-White-tailed Tropicbird. **Donald L. Malick:** 1; 8; 9-American Three-toed Woodpecker; 17-Snowy Owl; 19-Red-cockaded Woodpecker; 119; 121; 125-except Stellar's Sea-Eagle and 3rd year Bald Eagle; 127-except flying figures; 129-perched figures of juvenile Common Black-Hawk, Zone-tailed Hawk, and Short-tailed Hawk; 133; 135-except dark-morph Ferruginous Hawk and flying adult and dark juvenile White-tailed Hawks; 137-perched Aplomado Falcons and Crested Caracara; 139-all perched figures except Merlin *suckleyi* and upper flying American Kestrel; 141-except flying figures of adult male Prairie Falcon, Peregrine Falcon, and Gyrfalcon; 247; 257; 259; 261; 263; 265; 281; 283; 285; 287; 289-except Great Spotted Woodpecker; 291; 293. **Killian Mullarney:** 16-Ruff; 159; 187; 193; 196-Little Ringed Plover. **Michael O'Brien:** 87, 163-Willet; 335; 337; 339-except Verdin and Bushtit; 481-Bermuda Petrel. **John P. O'Neill:** 279-Elegant Trogan and Eared Quetzal; 339-Verdin and Bushtit. **Kent Pendleton:** 57-Gray Partridge and Chukars on ground; 59; 61; 63; 65; 67; 69; 123-except Northern Harrier; 141-flying figures of adult male Prairie Falcon, Peregrine Falcon, and Gyrfalcon; 142-except Hook-billed Kite; 143; 144; 145. **Diane Pierce:** 10-Chipping and Lark Sparrows; 16-Indigo Bunting; 19-Dusky Seaside Sparrow; 111; 113-except Little Egret; 115; 117-flying Glossy Ibis, White Ibis, Scarlet Ibis, and Roseate Spoonbill; 153; 409-except Bachman's Sparrow; 411; 413; 415-except Sage Sparrow *canescens*: 417-except Orange Bishop; 419-Seaside Sparrow; 421; 423-except Vesper Sparrow; 425; 427-except

"Pink-sided" Dark-eyed Junco and "Slate-colored" Dark-eyed Junco; 429; 431; 433-except Yellow-breasted Bunting; 435; 437-except Crimson-collared Grosbeak; 439; 441; 455-except Common Chaffinch; 457; 459; 461; 463; 468-Yellow Bittern. **John C. Pitcher**: 157; 163-except Willet and flying Yellowlegs; 165-standing Common Grenshank and standing Spotted Redshank; 167; 169; 177-except flying Black Turnstone; 181; 183; 185. **H. Douglas Pratt**: 7-Crested Myna; 9-Cerulean Warbler; 10-Northern Parula; 14-Painted Redstart; 16-European Starling; 243; 253; 255; 273-except immature Green Violet-ear, Green-breasted Mango, and Lucifer Hummingbird; 275-except Xantus's Hummingbird; 277-except wing figures; 279-except Elegant Trogan and Eared Quetzal; 303-except Tufted Flycatcher; 309-adult Fork-tailed Flycatcher and Eastern Kingbird; 311-except Piratic Flycatcher, Variegated Flycatcher, and Rose-throated Becard; 313-except Brown Shrike; 315-except Thick-billed Vireo; 317-except Gray Vireo; 319-except Philadelphia Vireo and Warbling Vireo; 321; 323-except Island Scrub-Jay and adult Mexican Jays; 325; 327; 331-except Common House-Martin; 333; 341; 343-except western Winter Wren; 345; 347-except female Blue-gray Gnatcatcher tail; 349-except Lanceolated Warbler; 355; 359-Varied Thrush; 361-except White-throated Robin; 363-except Blue Mockingbird; 367-except Common Myna; 369-except Siberian Accentor; 373; 375; 377; 379-except Virginia's Warbler and Crescent-chested Warbler; 381; 383-except Black-and-white Warbler; 385; 387; 389-except fall Bay-breasted Warbler; 393; 395; 397-Common Yellowthroat; 399-except Red-faced Warbler; 403-female and 1st winter male White-collared Seedeater, Bananaquit, and Black-faced Grassquit; 445; 447; 449-except Shiny Cowbird. **David Quinn**: 7-Asian Brown Flycatcher; 17-Redwing and Eurasian Hoopoe; 71; 73; 95-European Storm-Petrel; 113-Little Egret; 151-Purple Swamphen; 165-Marsh Sandpiper and Common Redshank; 195-Common Redshank; 289-Great Spotted Woodpecker; 313-Brown Shrike; 331-Common House-Martin; 349-Lanceolated Warbler; 351; 353; 359-except Varied Thrush; 369-Siberian Accentor; 371; 433-Yellow-breasted Bunting; 455-Common Chaffinch; 466-Lesser White-fronted Goose; 468-Chinese Egret and Western Reef-Heron; 469-Chinese Pond-Heron; 470-except Spotted Rail and Double-striped Thick-knee; 471; 472-Whiskered Tern; 474-except Bumblebee Hummingbird and Greenish Elaenia;

476-Willow Warbler; 477; 478-except Nightingale-Thrushes; 479-Pine Bunting, Yellow-throated Bunting and Eurasian Siskin; 480-Redwing. **Chuck Ripper**: 233-except flying winter Dovekie; 235-except Long-billed Murrelet; 237; 241-except flying Rhinoceros Auklet; 267; 269-except southwestern Whip-poor-will tail. **N. John Schmitt**: 6-Saltmarsh Sparrow; 7-Orange Bishop; 11-Common Black-Hawk; 13-House Sparrow; 25-flying and Taverner's Cackling Goose and Lesser Canada Goose; 27-juvenile Whooper Swan; 29-flying Muscovy Duck; 31-head of female American Black Duck; 51-Egyptian Goose; 57-Plain Chachalaca and flying Chukar; 123-Northern Harrier; 125-Stellar's Sea-Eagle and 3rd year Bald Eagle; 127-flying figures; 129-except perched figures of juvenile Common Black-Hawk, Zone-tailed Hawk, and Short-tailed Hawk; 131; 135-dark-morph Ferruginous Hawk and flying adult and dark juvenile White-tailed Hawks; 137-Eurasian Hobby and flying Aplomado Falcons; 139-Merlin *suckleyi* and all flying figures except upper American Kestrel; 142-Hook-billed Kite; 171-Little Curlew; 173-standing figures of Bristle-thighed Curlew and Eurasian Curlew; 194-Little Curlew; 245-with Jonathan Alderfer; 249; 251; 271; 309-juvenile Fork-tailed Flycatcher; 323-Island Scrub-Jay and adult Mexican Jays; 365; 417-Orange Bishop; 419-except Seaside Sparrow; 427-Slate-colored Dark-eyed Junco; 465. **Thomas R. Schultz**: 3; 11-Great Black-backed Gull wing, and Western and California Gulls; 12-Common Tern; 14-Herring Gull; 179; 199; 201; 203; 205; 207; 209; 211; 213; 215; 216-217; 219; 221; 223; 225; 227; 229; 231, 357; 379-Virginia's Warbler; 383-Black-and-white Warbler; 391; 403-Western Spindalis; 409; 443; 467-Great Frigatebird; 468-Lesser Frigatebird; 469-except Chinese Pond-Heron and Paint-billed Crake; 473-except Gray Nightjar and Antillean Palm-Swift; 481-Dunlin. **Daniel S. Smith**: 163-flying Yellowlegs; 165-flying Common Grenshank and flying Spotted Redshank; 171-except Little Curlew; 173-except standing figures of Bristle-thighed Curlew and Eurasian Curlew; 194-except Little Curlew; 195-except Common Redshank; 196-except Little Ringed Plover; 197. **Sophie Webb**: 269-southwestern Whip-poor-will tail; 273-immature Green Violetear, Green-breasted Mango, and Lucifer Hummingbird; 275-Xantus's Hummingbird; 277-wing figures.

THE EDITORS WISH TO THANK THE FOLLOWING INDIVIDUALS AND INSTITUTIONS FOR THEIR VALUABLE ASSISTANCE IN THE PREPARATION OF THE FIFTH EDITION:

M. Adams, Natural History Museum, Tring, UK; Jim Arterburn, Tulsa, Oklahoma; Ken Behrens, Pittsburg, Pennsylvania; Gavin Bieber, Tucson, Arizona; Chuck Carlson, Fort Peck, Montana; John Carlson, Helena, Montana; Robin Carter, Columbia, South Carolina; Allen Chartier, Inkster, Michigan; Ricky Davis, Rocky Mount, North Carolina; James Dinsmore, Iowa City, Iowa; Richard Erickson, Irvine, California; Doug Faulkner, Denver, Colorado; Dr. C. T. Fisher, World Museum Liverpool, UK; Robert Fisher, Independence, Missouri; Rick Fridell, Hurricane, Utah; Kimball L. Garrett, Los Angeles County Museum of Natural History, Los Angeles, California; Daniel D. Gibson, University of Alaska, Fairbanks, Alaska; Britt Griswold, Annapolis, Maryland; Robert Hamilton, Long Beach, California; Tom and Jo Heindel, Big Pine, California; Rich Hoyer, Tucson, Arizona; Dan Kassebaum, Belleville, Illinois; Tom Kent, Iowa City, Iowa; Rudolf Koes, Winnipeg, Manitoba; Dr M. Largen, World Museum Liverpool, UK; Tony Leukering, Brighton, Colorado; Mark Lockwood, Alpine, Texas; Derek Lovitch, Portland, Maine; Rich MacIntosh, Kodiak, Alaska; Ron Martin, Sawyer, North Dakota; Terry McEneaney, Yellowstone NP, Wyoming; Mick McHugh, Shawnee Mission, Kansas; Ian McLaren, Halifax, Nova Scotia; Steve Mlodinow, Everett, Washington; Glenn Murphy, Royal Ontario Museum, Ontario, Canada; Kenny Nichols, Pangburn, Arkansas; Jerry Oldenettel, Socorro, New Mexico; Mike Overton, Boone, Iowa; A. Parker, World Museum Liverpool, UK; John Parmeter, Albuquerque, New Mexico; Michael Patten, Bartlesville, Oklahoma; Mark Peck, Royal Ontario Museum, Ontario, Canada; Bill Pranty, Bayonet Point, Florida; David Quady, Berkeley, California; Mark Robbins, Lawrence, Kansas; Don Roberson, Pacific Grove, California; Bill Rowe, St Louis, Missouri; Larry Sansone, Hollywood, California; Larry Semo, Westminster, Colorado; John Sterling, Woodland, California; Mark Stevenson, Tucson, Arizona; Sherman Suter, Alexandria, Virginia; Peder Svingen, Duluth, Minnesota; Charles Trost, Pocatello, Idaho; David Willard, Field Museum of Natural History, Chicago, Illinois; Chris Wood, Ithaca, New York.

THE EDITORS ALSO WISH TO THANK THE FOLLOWING FOR THEIR CONTRIBUTIONS TO PREVIOUS EDITIONS OF THIS GUIDE:

Thomas A. Allen; David Agro; J. Phillip Angle; Stephen Bailey; Lawrence G. Balch; Dr. Richard C. Banks; John Barber; Jon Barlow; Jen and Des Bartlett; Giff Beaton; Louis Bevier; Eirik A.T. Blom; Daniel Boone; Jack Bowling; Edward S. Brinkley; the Department of Ornithology at the British Museum, Tring; Dawn Burke; Danny Bystrak; Richard Cannings; Steven W. Cardiff; Charles Carlson; Graham Chisholm; Carla Cicero; Charles T. Clark; William S. Clark; Rene Corado; Marian Cressman; Denver Museum of Natural History, Colorado; Bruce Deuel; Donna L. Dittman; Robert Dixon; Peter J. Dunn; Peter Dunne; Cameron Eckert; Victor Emanuel; Richard Erickson; Field Museum of Natural History, Chicago; Dr. Clemency Fisher; John W. Fitzpatrick; David Fix; Kimball L. Garrett; Freida Gentry; Daniel D. Gibson; Peter Grant; John A. Gregoire; Dr. James L. Gulledge; Jon S. Greenlaw; Dr. George A. Hall; J.B. Hallett, Jr.; Robert Hamilton; Jo and Tom Heindel; Matt Heindel; Steve Heinl; Paul M. Hill; Chris Hobbs; Phill Holder; Steve N.G. Howell; Rebecca Hyman; Frank Iwen; Greg Jackson; Alvaro Jaramillo; Joseph R. Jehl, Jr.; Ned K. Johnson; Roy Jones; Colin Jones; Lars Jonsson; Kenn Kaufman; Wayne Klockner; Marianne G. Koszorus; Lasse J. Laine; Los Angeles County Museum of Natural History; Daniel Lane; Dr. Malcolm Largen; Greg Lasley; Paul E. Lehman; Nick Lethaby; Tony Leukering; Rich Levad; Liverpool Museum, England; Mark Lockwood; Los Angeles County Museum of Natural History; Tim Loseby; Aileen Lotz; Rich MacIntosh; Bruce Mactavish; Laura Martin; Guy McCaskie; Terry McEneaney; Doug McRae; Dominic Mitchell; Steve Mlodinow; Joseph Morlan; Killian Mullarney; Museum of Natural History, Santa Barbara, California; Museum of Natural Science, Louisiana State University, Baton Rouge; Museum of Vertebrate Zoology, University of California, Berkeley; Glen Murphy; National Museum of Natural History, Smithsonian Institution, Washington, D.C.; Natural History Museum, San Diego, California; Harry Nehls; Michael O'Brien; Jerry Oldenettel; Gerald Oreel; Tony Parker; John Parmeter; Michael Patten; Brian Patteson; Dennis Paulson; Mark Peck; Paul Prior; Peter Pyle; Betsy Reeder; Dr. J.V. Remsen; Robert F. Ringler; Don Roberson; Mark Robins; Gary Rosenberg; Philip D. Round; John Rowlett; Rose Ann Rowlett; Royal Ontario Museum; Will Russell; San Diego Natural History Museum; Larry Sansone; Rick Saval; Robert T. Scholes; Brad Schram; Thomas Schulenberg; Scott Seltman; David Sibley; Ross Silcock; Mark Stackhouse; James Stasz; Rick Steenberg; Andrew Stepniewski; Doug Stotz; Sherman Suter; Thede Tobish; Dr. John Trochet; Charles Trost; Laurel Tucker; Nigel Tucker; Bill Tweit; Philip Unitt; U.S. Fish and Wildlife Service's Patuxent Wildlife Research Center in Laurel, Maryland; Arnoud van den Berg; T.R. Wahl; George Wallace; Western Foundation of Vertebrate Zoology; Mel White; Tony White; Hal Wierenga; Claudia P. Wilds; David W. Willard; Jeff Wilson; Alan Wormington; Louise Zemaitis; Barry Zimmer; Kevin Zimmer.

ILLUSTRATED BIRDS OF NORTH AMERICA
FOLIO EDITION

Edited by Jon L. Dunn and Jonathan Alderfer

Published by the National Geographic Society

John M. Fahey, Jr., President and Chief
    Executive Officer
Gilbert M. Grosvenor, Chairman of the Board
Tim T. Kelly, President, Global Media Group
John Q. Griffin, Executive Vice President;
    President, Publishing
Nina D. Hoffman, Executive Vice President;
    President, Book Publishing Group

Prepared by the Book Division

Barbara Brownell Grogan, Vice President
    and Editor in Chief
Marianne R. Koszorus, Director of Design
Carl Mehler, Director of Maps
R. Gary Colbert, Production Director
Jennifer A. Thornton, Managing Editor
Meredith C. Wilcox, Administrative Director,
    Illustrations

Staff for This Book

Barbara Levitt, Project Editor
Carol Farrar Norton, Art Director
Linda Johansson, Designer
Jennifer Conrad Seidel, Text Editor
Paul E. Lehman, Chief Map Researcher/Editor
Sven M. Dolling, Map Production
Richard S. Wain, Production Project Manager

Manufacturing and Quality Management

Christopher A. Liedel, Chief Financial Officer
Phillip L. Schlosser, Vice President
Chris Brown, Technical Director
Nicole Elliott, Manager
Rachel Faulise, Manager

Special thanks to National Geographic Digital Imaging for their careful production of the art enlargements for this book.

The National Geographic Society is one of the world's largest nonprofit scientific and educational organizations. Founded in 1888 to "increase and diffuse geographic knowledge," the Society works to inspire people to care about the planet. It reaches more than 325 million people worldwide each month through its official journal, *National Geographic,* and other magazines; National Geographic Channel; television documentaries; music; radio; films; books; DVDs; maps; exhibitions; school publishing programs; interactive media; and merchandise. National Geographic has funded more than 9,000 scientific research, conservation and exploration projects and supports an education program combating geographic illiteracy.

For more information, please call 1-800-NGS LINE (647-5463) or write to the following address:

National Geographic Society
1145 17th Street N.W.
Washington, D.C. 20036-4688 U.S.A.

Visit us online at www.nationalgeographic.com

For information about special discounts for bulk purchases, please contact National Geographic Books Special Sales: ngspecsales@ngs.org

For rights or permissions inquiries, please contact National Geographic Books Subsidiary Rights: ngbookrights@ngs.org

Library of Congress Cataloging-in-Publication Data
Dunn, Jon, 1954-
   Birds of North America / Jon L. Dunn and Jonathan Alderfer. -- Folio ed.
      p. cm.
   Based on: National Geographic field guide to the birds of North America / edited by Jon L. Dunn and Jonathan Alderfer. 2006
   Includes bibliographical references and index.
   ISBN 978-1-4262-0525-5 (trade hardcover : alk. paper) --
ISBN 978-1-4262-0577-4 (deluxe hardcover : alk. paper)
   1. Birds--North America--Identification. I. Alderfer, Jonathan K. II. Title.
   QL681.D874 2009
   598.097--dc22
                                    2009020381

ISBN: 978-1-4262-0525-5
ISBN: 978-1-4262-0577-4 (deluxe)

Printed in U.S.A.

09/RRDW/1